Scanning Probe Microscopy in Industrial Applications

Scanning Probe Microscopy in Industrial Applications

Nanomechanical Characterization

Edited by

Dalia G. Yablon

Published by John Wiley & Sons, Inc., Hoboken, New Jersey
Published simultaneously in Canada

For general information on our other products and services or for technical support, Please contact our Customer Care Department within the United States at (800) 762-2974, outside the United States at (371) 572-3993 or fax (317) 572-4002.

Wiley also publishes its books in variety of electronic formats. Some content that appears in print may not be available in electronic formats. For more information about Wiley products, visit our web site at www.wiley.com.

Library of Congress Cataloging-in-Publication Data:

Scanning probe microscopy for industrial applications : nanomechanical characterization / edited by Dalia G. Yablon.
pages cm
Includes bibliographical references and index.
ISBN 978-1-118-28823-8 (hardback)
1. Materials–Microscopy. 2. Scanning probe microscopy–Industrial applications. I. Yablon, Dalia G., 1975–
TA417.23.S33 2013
620.1′127–dc23

2013009638

Printed in the United States of America

10 9 8 7 6 5 4 3 2 1

Dedicated with love to:
Andrew
and
Ayelet, Daniela, Mati, and Eli

Contents

Contributors List

Sudharsan Balasubramaniam
Department of Mechanical Engineering
Purdue University
West Lafayette, IN

Steve J. Bull
School of Chemical Engineering and Advanced Materials
University of Newcastle
Newcastle upon Tyne, United Kingdom

Robert W. Carpick
Department of Mechanical Engineering and Applied Mechanics
University of Pennsylvania
Philadelphia, PA

Geraud Dubois
IBM Almaden Research Center
San Jose, CA

Johann Foucher
LETI–CEA
Grenoble, France

Jane Frommer
IBM Almaden Research Center
San Jose, CA

Anthony Galliano
L'Oreal Research and Innovation
Aulnay sous Bois, France

A. Patrick Gunning
Institute of Food Research
Norwich Research Park
Norwich, United Kingdom

Greg Haugstad
University of Minnesota
Minneapolis, MN

Donna C. Hurley
National Institute of Standards and Technology
Boulder, CO

Tevis D. B. Jacobs
Department of Materials Science and Engineering
University of Pennsylvania
Philadelphia, PA

Jason P. Killgore
National Institute of Standards and Technology
Boulder, CO

Robin King
IBM Almaden Research Center
San Jose, CA

Daniel Kiracofe
Department of Mechanical Engineering
Purdue University
West Lafayette, IN

Matthew S. Lamm
Merck Research Laboratories
Merck & Co.
Summit, NJ

Gustavo S. Luengo
L'Oreal Research and Innovation
Aulnay sous Bois, France

C. Mathew Mate
HGST, A Western Digital Company
San Jose, CA

Victor J. Morris
Institute of Food Research
Norwich Research Park
Norwich, United Kingdom

Arvind Raman
Department of Mechanical Engineering
Purdue University
West Lafayette, IN

Alan M. Schilowitz
ExxonMobil Research and Engineering
Annandale, NJ

Andy H. Tsou
ExxonMobil Research and Engineering
Annandale, NJ

Kevin T. Turner
Department of Mechanical Engineering
and Applied Mechanics
University of Pennsylvania
Philadelphia, PA

Kumar Virwani
IBM Almaden Research Center
San Jose, CA

Willi Volksen
IBM Almaden Research Center
San Jose, CA

Katharine Walz
IBM Almaden Research Center
San Jose, CA

Klaus Wormuth
Institute for Physical Chemistry
University of Cologne
Cologne
Germany

Dalia G. Yablon
SurfaceChar, LLC.
Sharon, MA

Preface

The idea for this book came a couple of years ago after perusing atomic force microscopy (AFM) book collections at many professional society conferences over the years and finding that there was constantly a key topic missing: how AFM was used practically in industrial R&D. This topic is close to my heart as I have pursued a career developing novel methods and applying scanning probe microscopy (SPM) to a wide variety of industrial R&D problems at ExxonMobil Research and Engineering's Corporate Strategic Research in New Jersey for over a decade. At ExxonMobil, I have had the opportunity to explore application of AFM in the three vast sectors in the petroleum industry: the downstream, chemicals, and upstream sector studying a wide variety of problems from corrosion to lubrication to geology to polymer materials. All the AFM-focused books I found were research reviews on focused topical areas or textbooks on operating principles and theory. So though these books offered excellent technical treatments of a variety of topics, the incredible utility, applicability, and flexibility of AFM to solve real-world everyday problems was lacking.

My background in graduate school was actually in scanning tunneling microscopy (STM), as I studied self-assembled monolayers at the liquid–solid interface in George Flynn's lab at Columbia University's chemistry department, not the most relevant "real-world, industrial" problem. However, being an STM lab in the 1990s, we also had AFMs in the lab (since they shared the same "brain," the controller), so every once in a while we would help out a colleague and study their samples. As I moved to ExxonMobil after graduate school in 2002, my intent was not necessarily to continue in scanning probe microscopy. In fact, it was serendipity that upon arriving at ExxonMobil's Corporate Strategic Research labs as a postdoctoral fellow, I found an unused Digital Instruments Nanoscope III controller with a fully accessorized Multimode. The rest, as they say, is history.

Over my time at ExxonMobil, I ran into many like-minded colleagues in other industrial facilities using AFM, and we would hold fascinating discussions of how the technology was used in one's particular field. On the other side, I could sense great interest from my academic colleagues that industrial researchers were using SPM to such a great extent, as well as recent graduates trying to connect their hard-earned knowledge and expertise in the field as students to practical, useful areas. A book covering this material for the wider audience did indeed seem missing.

And so I began to survey the industrial R&D community—both members I had known for years and new ones—to gauge interest in putting such a volume together. I was inspired by the enthusiasm I received from my colleagues and decided to go forward. Then came the question of what kind of AFM? As surveyed in Chapter 1,

AFM is a tremendously broad field that characterizes a wide variety of material and surface properties. However, at its heart, AFM is a high-resolution nanomechanical probe of the surface. That probe can be coated with an electronically conducting layer or magnetic material, or even replaced by an optical fiber tip in order to probe a variety of properties. However, a strong majority of the industrial R&D both in my own direct personal experience and those that I was familiar with was conducted with mechanical probing of properties. Of course, there are some key counterexamples that stand out, such as electrostatic force microscopy to probe electrical properties of conducting and semiconducting materials and magnetic force microscopy to probe magnetic storage devices. Perhaps those applications will be covered in a future volume! But the overwhelming use of AFM in industrial R&D was nanomechanical characterization, and thus it was quickly chosen as the focus for this book.

It was important that this book be self-contained. While the book's focus is on the applications, it is intended to be accessible to a broad audience both from a professional and educational background so that everyone from an advanced undergraduate to a seasoned professional will benefit. Various professional backgrounds should benefit as well from those in an academic or academic-like environment who wish to learn how AFM is used in an industrial environment to those in an industrial environment eager to learn what other applications are possible. While a freshman chemistry and physics background is assumed, no prior experience with AFM is assumed. There is certainly advanced information in each chapter so that the content should appeal to a variety of backgrounds in the AFM community from novice to expert.

With this aim of appealing to a broad audience, I include a tutorial on nanomechanical methods in the first half of the book so that applications could be understood and appreciated with context without having to refer to other books or references. I chose the topics that I felt were most relevant to studying nanomechanical characterization in the practical industrial environment: contact mechanics, force curves, phase imaging, dynamic contact methods, and nanoindentation. Note that this is not an exhaustive list of all the nanomechanical characterization methods available with AFM.

I worked closely with my co-authors to make these overview chapters as *practical* as possible and not overwhelming in theoretical detail; several chapters include worked examples of useful calculations (e.g., using Hertz mechanics with and without adhesion to model a contact in Chapter 2 or data analysis of contact resonance measurements in Chapter 5). These chapters are meant to be overviews to introduce the terminology and key concepts. They are not meant to be an exhaustive step-by-step guide to the operation of various techniques; those treatments are done elsewhere and are referenced in the book. Chapter 6 is a somewhat unique chapter on "Best Practices in AFM Imaging" that I co-wrote with my colleague Greg Haugstad. This chapter outlines some of the common pitfalls encountered by AFM users, again with a practical approach to what the most common and most important problems are and how to avoid them.

Chapters 7–15 then dive into how AFM has been used in a variety of industrial R&D areas to explore (a) phenomena and processes such as new formulation development in pharmaceuticals, the effect of humidity and temperature on biomaterials, and nanostructure formation in food processing, (b) characterization of various materials including polymer blends and composites in the chemicals sector, skin and hair in personal care products, coatings and thin dielectric films in the semiconductor industry, (c) AFM's function as a key quality control tool in semiconductor metrology, and (d) a slightly different application of AFM microcantilevers as a physical and chemical sensors in the petroleum industry.

My hope is that this book will show what an incredibly useful tool AFM has become to industrial R&D in its short 25 years. It is currently an indispensable tool in any microscopy laboratory, despite its commercial youth. So while the AFM field continues to grow and develop in terms of its capability and understanding, spearheaded by many notable groups in academia and elsewhere, it has already penetrated significantly into a wide variety of industrial labs. This trend will only continue and grow. As the penetration into industrial labs continues, it is my hope that the reader understands not only how the AFM capability increasingly benefits industry but also that industrial labs are quickly becoming a partner in driving AFM's innovation and research as its commercial and practical utility and importance continue to thrive.

Dalia G. Yablon
Sharon, MA

Acknowledgments

I have been blessed to have wonderful colleagues to share ideas with and gain inspiration from during this process. First and foremost, I must thank my co-authors on this book. Working with each and every author has been an absolute pleasure. I had the distinct pleasure of working closely with Andy Tsou and Alan Schilowitz for many years at ExxonMobil. Some of the other co-authors I have been fortunate to count as longstanding collaborators including Rob Carpick, Greg Haugstad, Jason Killgore, and Donna Hurley; I thank them for their support and enthusiastic participation for this particular project. From all I learned so much technically, and made many new friends in the process. This book would not have been possible without everyone's collective hard work to share their insights and wisdom with us.

I would like to thank my colleagues at ExxonMobil with whom I have had the pleasure of developing our AFM capabilities in our AFM lab. Specifically, Jean Grabowski deserves a special callout and has been a tremendous asset to our efforts as well as a genuine friend along the way. I also would like to thank Rebecca Locker, Ishita Chakraborty, and Daniel Kiracofe. I also would like to thank Steve Minne and Roger Proksch for help and guidance in the incipient stages of this project, as well as Emily Rapalino and Susan Klein for advice and proofreading!

Most of all, I have to thank my husband Andrew for his absolute unwavering support and cheerleading throughout the past year and a half. His confidence in me was a source of real motivation and inspiration to carry forward despite obstacles. And logistically, of course, I have to thank him for providing childcare to our children in the evenings and weekends (covering for me on many, many bath nights and storyreading—such as on this night) so that I could complete this book, one which hopefully one day they might be interested in enough to read and understand.

Chapter 1

Overview of Atomic Force Microscopy

Dalia G. Yablon

SurfaceChar LLC, Sharon, MA

Atomic force microscopy (AFM) is a family of nanoscale characterization techniques that has exploded onto the overall characterization and nanotechnology field as its versatility and high resolution continue to feed a dizzying variety of disciplines from biology to physics and chemistry to engineering. AFM entered the scientific arena in 1981 with the now famous invention of its older sibling in the scanning probe microscopy family, the scanning tunneling microscope (STM) in the IBM Zurich labs of Gerd Binnig and Heinrich Rohrer, for which they received the Nobel Prize in Physics in 1986. The AFM was then invented in 1986 by Gerd Binnig, Cal Quate, and Christoph Gerber [1]. Together, STM and AFM formed the scanning probe microscopy (SPM) family, which includes other methods such as near-field scanning optical microscopy (NSOM).

The STM spawned the next 25 years of the continuously developing field of scanning probe microscopy and specifically atomic force microscopy, which now includes dozens of different methods under its name to probe various properties—including mechanical, electrical, magnetic, chemical, and optical—of materials and surfaces. Atomic force microscopy is a powerful tool in various research enterprises. It is found in practically any university characterization facility alongside optical and electron microscopes, and most undergraduates in science or engineering fields at this point are at least familiar with it, if not have performed a laboratory experiment in their undergraduate curriculum.

The focus of this book is to understand and appreciate the role this young technique has played in *industrial* research and development (R&D). AFM has penetrated into a variety of industrial research sectors as witnessed by the diverse set of applications described in this book. Alongside electron and optical microscopy, which have been

Scanning Probe Microscopy in Industrial Applications: Nanomechanical Characterization, First Edition. Edited by Dalia G. Yablon.

around for decades and have reached an impressive level of commercial maturity and ease of use, AFM has become a vital characterization method despite its youth and continued technical evolution. So, though AFM is still an active area of academic research as its capabilities continue to develop and be better understood, it has proven to be a useful microscopy to address industrial and commercial needs from quality control and assurance to product formulation and process monitoring.

The goal of this introductory chapter is to provide an overview of AFM to nonspecialists and introduce the various topics that are the subject of subsequent individual chapters. As such, this chapter will provide a brief review of the beginnings of AFM and special features that make it particularly suited for industrial research. Then a brief overview of AFM operation will be presented including the hardware, software, calibrations involved, and finally the different nanomechanical methods that will be described in detail both in theory and application. Entire books are written on AFM operation, and this chapter is not meant to be an exhaustive introduction to its operation, merely serving to provide enough information for the rest of the book to be followed intelligently. Readers interested in more detail about AFM operation can consult a number of excellent books on the topic [2–5].

1.1 A WORD ON NOMENCLATURE

Before the rest of this chapter continues, some definitions are in order. Similar to many other surface science techniques, AFM has succumbed to a somewhat unwieldy mess of abbreviations and jargon that has become a hard-to-navigate alphabet soup. SPM refers to an umbrella of a variety of methods. Methods that fall under SPM include perhaps its most famous member, atomic force microscopy (AFM), in addition to others such as scanning tunneling microscopy (STM), near-field scanning optical microscopy (NSOM or SNOM, depending on the continent), and other lithographic methods. And then within AFM there are dozens of methods that rely on the AFM probe–sample interactions to provide a variety of material properties including electrical, optical, magnetic, and mechanical properties. To date, there are dozens of SPM/AFM-based methods. It is beyond the scope of this book to list and/or define all the related methods. These methods characterize a huge variety of material properties. Some of the more common methods have been included in Figure 1.1 with the category of properties that they measure. The wide diversity of properties of materials that can be measured with AFM is clear. The focus of this book is on nanomechanical characterization, due to its broad appeal in industrial R&D. Again, Figure 1.1 is offered to demonstrate the variety and flavor of properties that can be probed with SPM.

1.2 ATOMIC FORCE MICROSCOPY—THE APPEAL TO INDUSTRIAL R&D

This book specifically focuses on AFM, which is the method that has most penetrated the general characterization and industrial research fields. AFM has key features that make it especially attractive to industrial research and development. First, its resolution,

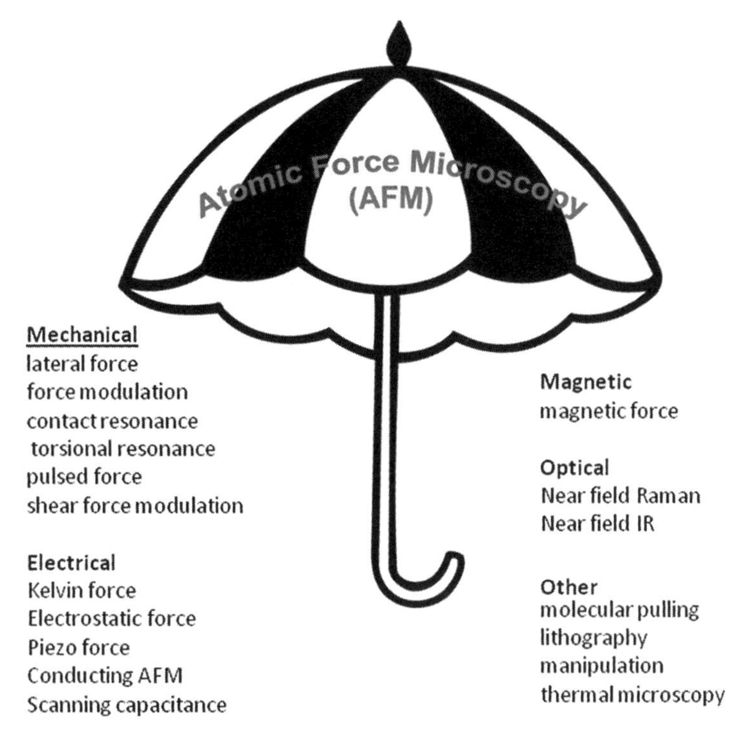

Figure 1.1 Schematic of AFM as an umbrella technique that includes a wide variety of methods to probe various properties of materials. Some of these methods are shown here.

the heart of its utility as a characterization tool, is indeed impressive with 5- to 10-nm lateral resolution and angstrom vertical resolution achieved routinely on commercial instruments with commercial cantilevers. The limits of lateral resolution continues to be pushed with specialized techniques and cantilevers, and it has reached true "atomic" resolution where point defects of certain materials can now be imaged under certain operating parameters and imaging modes, especially under liquid [6–8]. True atomic resolution is still currently mostly achieved through STM, which operates through an entirely different tip–sample interaction mechanism based on quantum mechanical tunneling. Many of these STM studies are conducted in ultrahigh vacuum (UHV), though some are performed in ambient and liquid conditions. It is fair to say that AFM is catching up, however, with recent groundbreaking work imaging cyclic aromatic molecules on Cu[111] [10, 11].

A second critical feature of AFM that makes it particularly amenable to industrial research is the flexibility of the environment in which it can operate. Despite being a high-resolution microscopy that rivals the resolution of electron microscopy, AFM can operate in an ambient or even liquid environment with minimal compromise to its resolution. The ability to work in "real-world" environments makes it a critical tool for many industrial R&D applications, where the research emphasis is consistently to understand mechanisms and materials in real-world situations as opposed to idealized materials and conditions that typically exist in academic research endeavors. In many cases, the researcher may not want the sample to be forced into a pristine vacuum environment where perhaps key materials or components will be evacuated and thus missed in the characterization effort. In addition, the flexibility of an environment permits in situ characterization of processes. For example, processes such as corrosion, lubrication, catalyst dissolution, crystal growth (for active pharmaceutical ingredient, API) can all be studied in situ with AFM as a unique attribute of this microscopy. The power of in situ measurements for biological applications is simple as many biological processes and materials simple cannot survive ex situ.

A corollary of the utility of the AFM's flexible environment to industrial R&D is the flexibility of type and shape of surface or material that AFM can image. There are virtually no restrictions on the size of a material that can be probed with AFM. Many commercial "large-sample" AFMs currently exist that can image any design or shape of a surface from engine parts to thin films. Indeed, one of the AFMs built in the early 1990s (Topometrix Explorer) was advertised as being able to be placed on top of jet wings to image fractures. In addition, there are practically no mechanical or electrical restrictions on the type of sample an AFM can image. Unlike electron microscopy or STM that require conducting or semiconducting surfaces to image or avoid artifacts [or in the case of insulating surface in scanning electron microscopy (SEM) that requires deposition of a thin conducting layer such as Cr], AFM can image insulating surfaces in addition to conducting or semiconducting surfaces with minimal sample prep. Typically, the most significant requirement on the sample for an AFM is that it be smooth. Commercial instruments will have a maximum z (vertical) range that can be imaged, dictated by the design and desired resolution of the particular instrument. The maximum z range is typically from 1 μm (for high-resolution studies) to several microns (for lower resolution studies), meaning that the sample cannot have features, a tilt, or overall roughness that exceed that limit in order for the AFM to image effectively. Smooth surfaces are often prepared with the use of an ultramicrotome. An ultramicrotome is simply a very controlled and precise way of cutting of thin (down to 100 nm) sections, at controlled speeds and in a controlled fashion to result in ultrasmooth surfaces of nanometer or less than nanometer roughenss. The ultramicrotome typically employs the use of a glass knife for a coarse cut and then a diamond knife for a fine cut. Cryomicrotomy is simply ultramicrotomy at cold temperatures. Cryomicrotomy is typically used to prepare polymer surfaces to cut below the glass transition temperature of the material; depending on the instrument, cryomicrotomes can cool down to –180°C.

Finally, AFM is fundamentally a *mechanical* probe of the surface through actual contact between the tip and sample. So, in addition to the ability to provide high resolution on surface features, the goal of any form of microscopy is the unique tip–sample interaction mechanism enabling the probing of many properties, most obvious of which is mechanical properties. Mechanical properties covers a wide range of materials and properties. Entire textbooks and courses are devoted to the studies of mechanics, for example, by Callister and Rethwisch [12]. We very briefly review some of the main concepts below to provide enough of a background for the rest of the chapter.

This edition focuses specifically on nanomechanical characterization. Probing other material properties such as optical, electrical, or magnetic properties is not covered in this book. The reason is that though these other properties have been successfully probed in various industrial R&D applications, using AFM to probe mechanical properties of surfaces is more ubiquitous in industrial R&D, and so was a logical focus for a first book on SPM in industrial applications.

1.3 MECHANICAL PROPERTIES

When one considers mechanical properties, a range of descriptors come to mind, including "stiffness," "toughness," "hardness," "brittleness," and "ductility" to name a few. Each of these properties has their own mathematical expressions with associated experimental methods of measuring them. An interaction between two materials can be typically categorized into either elastic (or recoverable) or dissipative or plastic (permanent or nonrecoverable) interaction. On the bulk scale, most mechanical analysis of materials relies on measuring their strain (material displacement) as a function of stress (pressure exerted on the material) or vica versa. The measurement and analysis of stress versus strain behavior of various materials is a rich and active field, and there are many excellent textbooks that describe it [12]. The description provided below is simplified to provide a context for these measurements ultimately described on the nanoscale.

Elastic deformation of a material is typically nonpermanent or reversible and exists in a regime where the strain (material displacement) is proportional to the stress (pressure exerted on the material) by a proportionality factor known as a modulus (though in reality, no material is truly only elastic). Elastic materials thus typically follow Hooke's law. Figure 1.2 shows a representative stress versus strain curve for materials highlighting the elastic region (blue). Additionally, in elastic

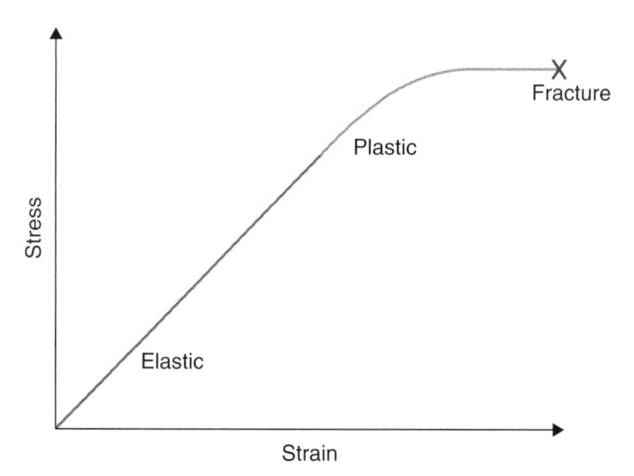

Figure 1.2 Basic stress vs. strain curve showing three mechanical regimes of elastic deformation, plastic deformation, and fracture. For color details, please see color plate section.

materials, the loading curve (as force is applied) is identical to the unloading curve (as the force is removed.) Modulus is a key parameter used to describe elastic behavior. Depending on the direction in which a material is being stressed, the material will have different moduli (e.g., a tensile modulus, a shear modulus, etc.).

Plastic deformation is, on the other end, permanent and nonrecoverable. Plastic deformation does not follow Hooke's law and so stress is not proportional to strain. In Figure 1.2, the region of plastic deformation is shown in green. In contrast to elastic deformation, the loading curve is not identical to the unloading curve, and, in fact, the unloading curve returns to a nonzero strain at a zero stress level. There are many common properties associated with plastic deformation including yielding (onset of plastic deformation), ductility (plastic deformation at a fracture), and hardness (resistance to localized plastic deformation).

Finally, there is another very important class of materials called *visco*elastic materials, where the elastic properties are frequency dependent. For example, when a stress such as pulling is exerted on the material, the material's response is rate dependent. Perhaps the most familiar viscoelastic material to most readers from childhood is Silly Putty, where its response to pulling depends on how fast it is pulled. If Silly Putty is pulled very slowly, the material stretches out very nicely. If Silly Putty is pulled very fast, it will quickly break. Polymers and rubbers, two materials that play a significant role in consumer products, and which are characterized quite heavily with AFM, are also viscoelastic materials.

1.4 OVERVIEW OF AFM OPERATION

1.4.1 AFM Hardware

A schematic of AFM hardware is shown in Figure 1.3. The main components of the AFM are (1) the cantilever, (2) optical detection system, (3) x-y-z scanner, and (4) feedback loop.

1.4.2 Cantilevers and Probes

The heart of the AFM lies in its cantilever/probe assembly that interacts or probes materials to provide the information of interest. The technology to fabricate AFM cantilevers takes advantage of technology developed for the semiconductor industry to make similar scale devices and features out of silicon. Thus, today AFM cantilevers are typically made by micromachining silicon (single-crystal Si) or silicon nitride (Si_3N_4). The dimensions of a cantilever vary and dictate the stiffness, or spring constant, of the cantilever. Depending on the material that needs to be probed and the mode of operation, one might pick a lever with a stiffer (higher) or softer (lower) spring constant. Cantilevers are generally either rectangular in geometry (Si or Si_3N_4) or triangular (Si_3N_4), with now hybrid cantilevers being manufactured (e.g., Si tips on SiN cantilevers). Generally, the Si_3N_4 probes have lower force constants and might be used for imaging in liquid, among other applications, whereas the Si probes can be brittle and tend to break or chip in contact when scanning a surface. Typically, the

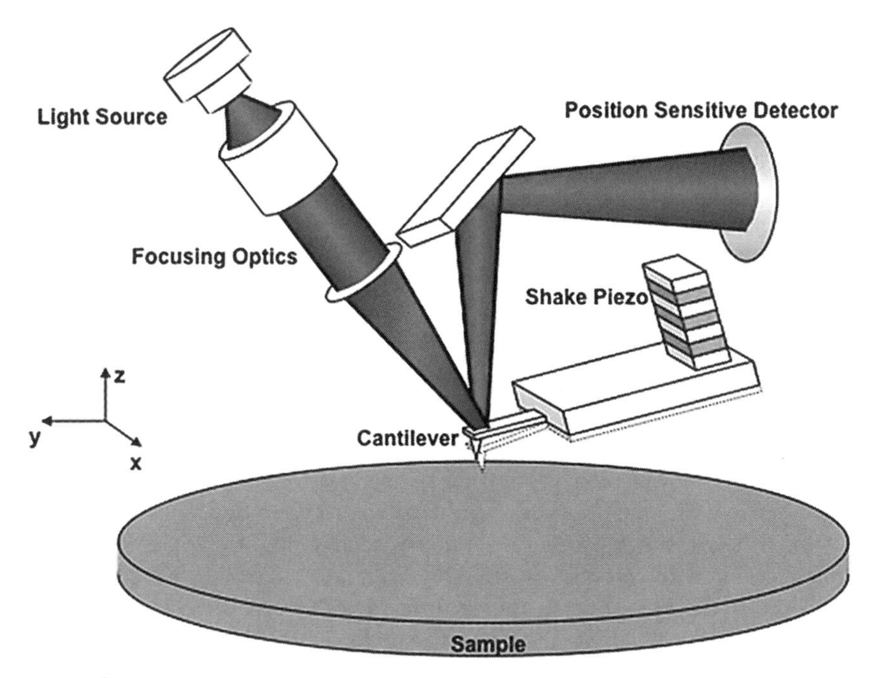

Figure 1.3 Schematic of basic operating principles of AFM. A laser is focused and directed off the back end of a cantilever and directed toward a position-sensitive director (PSD) to monitor its vertical and lateral motion as it interacts with the surface. Depending on the mode used to image the sample, the cantilever can be vibrated with a shake piezo for dynamic imaging modes that involve the resonance of the cantilever.

cantilevers are hundreds of microns in length, tens of microns in width, and a few microns in thickness; for reference, the diameter of a human hair is approximately 100 μm. Cantilevers can be coated with gold or aluminum to provide high reflectivity. Commercial cantilevers are also available with other thin metal coatings for magnetic or electrical imaging. Cantilevers and probes can also be functionalized chemically or biologically for specific interactions.

Probes are grown at the end of the lever and vary in shape and sharpness. A sharper probe can often improve the lateral resolution of the AFM. Probe design is a continuously evolving field as manufacturers strive for the most robust, reproducible, and sharp tips. Cantilever/probe assemblies are mounted into a cantilever holder that typically includes a piezoceramic element (described below) that can vibrate the cantilever at a given frequency—either above, at, or below its resonance—depending on the desired tip–sample interaction or AFM mode.

1.4.3 Optical Detection System

In order to track the motion of the cantilever/probe assembly as it scans a surface, an AFM typically includes an optical detection system that consists of a laser reflected off the back side of the cantilever and directed toward a position sensitive detector

(PSD) as seen in Figure 1.3. Optical detection systems are the most common way to track cantilever motion in commercial AFM systems. However, other methods exist to bypass optical detection systems (in, e.g., a dark medium where the laser light would be absorbed) with self-actuated cantilevers where their motion is read by piezos integrated into the levers. The laser is typically a visible photodiode, though some commercial instruments have implemented superluminescent diode (SLD) laser. The laser is detected by a four-segment position-sensitive detector (PSD), which can track the vertical and lateral motion of the cantilever accurately, which can be converted to units of nanometers or displacement through careful calibration described below.

1.4.4 *x-y-z* Scanner

An AFM can operate either by scanning the sample relative to a stationary tip (sample scanning) or scanning the tip relative to a stationary sample (tip scanning.) Each has its own advantages in terms of size of sample it can accommodate (tip scanning more flexible), ability to accessorize (tip scanning again more flexible), and stability and signal to noise (sample scanning easier to build with better specifications). Each shares the requirement to move the tip relative to the sample in a highly accurate way with minimal noise.

Tip–sample motion is often accomplished in commercial instruments by piezoelectric materials, which are materials that respond either by expansion or contraction in response to an applied voltage. These kinds of materials provide the ability for very fine motion (nanometers to microns). Piezoelectric materials can be made in several shapes including tubes, disks, and bimorphs, depending on the range and geometry of motion required. Although piezoelectric materials are very effective at moving tip or sample, they are plagued by nonlinear behavior such as hysteresis and creep that has serious consequences for accurate AFM imaging and interpretation of data. Artifacts induced by piezoelectric materials are reviewed in detail in Chapter 6.

Piezoelectric scanners can be programmed to move with either open-loop or closed-loop feedback circuits. In open-loop scanners, the piezo is moved by programs to ramp x and y bias voltages to move the prescribed amounts. So in open-loop scanners, if the piezo has to move a large distance or move around on the sample, there can be considerable hysteresis of creep with the best recourse of patience or moving/zooming in and out in small increments since hysteresis/creep is proportional to the displacement or motion the scanner is trying to execute.

One solution to this problem is to use closed-loop scanners that use additional x-y-z sensors to independently measure and then correct for scanner movement. In closed-loop operation, feedback circuits are continually adjusted to the applied scanner biases to correct for the x-y motion and to achieve the desired motion. It is important to note that correction of x-y motion in an AFM is slightly different than correction of z motion. This is due to the fact that x-y motion can be predicted, whereas z motion cannot as it depends on the surface topography of the sample being scanned. Therefore, typically there are active feedback circuits to control x and y

motion as corrected by the motion detected by the external sensors. Closed-loop z operation monitors z motion with a passive feedback circuit to obtain an image of the z sensor signal, in addition to the z piezo signal. The z sensor signal will be a more accurate representation of the surface topography, though it also is usually noisier as well, which could affect topographic measurements on the subnanometer length scale. Finally, it is important to note that the piezo/other electromechanical method used to move the AFM tip or sample in x, y, and z is independent from the piezo used to actuate or vibrate the cantilever. The piezo used to vibrate the cantilever is often referred to as a shake or dither piezo and is integrated into the cantilever holder and is specific to oscillating the cantilever.

1.4.5 AFM Software

Atomic force microscopy software is primarily used for controlling the AFM stage and then doing subsequent data/image analysis to extract the information of interest. The software interface with the electronics controller is critical and responsible for setting motion of the x-y stage, controlling the probe approach to the surface, setting feedback parameters to optimize the image quality.

Image processing is an integral part of effective AFM characterization and analysis. Many image processing tools are integrated into commercial AFM software, though several third-party software packages specific to AFM also exist including SPIP, Gwyddion, and WSxM among others.

1.4.6 Calibrations

Often the AFM measures the force exerted onto the material with a cantilever/tip assembly. The challenge is that what is actually being measured is the cantilever deflection in volts by the PSD. So this system requires several different calibrations to accurately know the actual force between the tip and sample. First, one has to start with a calibration of the cantilever stiffness or spring constant (N/m). Then, one has to measure the optical lever sensitivity or conversion of volts to nanometers on the photodetector (though the order of these two measurements can be reversed depending on the method being used). This calibration factor is specific to cantilever and alignment. Only then can the conversion be done of measured volts on the PSD to units of force, or Newtons, that the cantilever exerts on the surface. We now briefly go through each of these calibrations.

1.4.7 Cantilever Spring Constant

For accurate mechanical measurements of tip–sample interactions, precise calibration of the cantilever spring constant is required in order to measure the force exerted on your sample. Cantilevers have spring constants associated with all degrees of motion, for example, normal and lateral. Typically, it is the normal spring constant that requires calibrating for most force measurements; lateral or torsional spring constants are required for lateral force or frictional force microscopy. We discuss here calibration of normal spring constants.

The normal spring constant of a cantilever can be measured in several ways. For a perfectly rectangular cantilever, the spring constant, k_L (L for lever), is given by

$$k_L = \frac{E\omega t^3}{4l^3}$$

where E is the Young's modulus of the material, ω is the width, l is the length, and t is the thickness, all of the cantilever. While width and length of the cantilevers can be fairly easily measured by SEM, the accurate measurement of thickness (t) is more difficult. With the value of k_L depending on the cube of the thickness, a small inaccuracy in the value of t will lead to great uncertainty in the value of k_L. Therefore, for more accurate measurements of k_L, other methods exist. A commonly used method is the Sader method [13], which requires only measurements of length, width, resonance frequency, and quality factor of the cantilever. Based on the viscous damping of the cantilever, the spring constant can then be evaluated [14]. Another method involves pressing the lever against a reference cantilever with known spring constant [15]. A common method implemented on several commercial instruments is known as the thermal noise method where the thermal noise spectrum is measured [16, 17]. Other methods for normal spring constant calibration include the somewhat tedious but nondestructive method of the added mass method [18], where the change in frequency is measured as known added masses are added to the cantilever. With all these methods present, practically it is reasonable to measure a spring constant to within 20% error. Getting below that level of error is experimentally challenging and has to include among other things accounting of the exact geometry of the tip, how far from the end of the tip the spot is from the lever, and the tilt of the cantilever. For a detailed discussion see chapter in Haugstad [3].

In addition to the spring constant of the lever, which is in units of newtons/meter and ultimately gives the newtons, or force, exerted by the cantilever, the optical lever sensitivity of the photodetector needs to be calibrated to convert photodetector voltage to meters. This is typically accomplished by conducting a static force curve to measure the deflection of the cantilever against a very stiff surface. With such a measurement, it is assumed that the deflection of the cantilever follows a 1:1 relationship with the piezo pushing down on the surface, so that by measuring the slope of the repulsive wall of the force curve, an accurate nanometer/Volt calibration of the photodetector is obtained. With the calibration of the cantilever spring constant (N/m) and optical lever sensitivity (m/V), a calibration of newtons of force per voltage on the photodetector is obtained.

Note that calibration of the fundamental or first eigenmode cantilever spring constants is discussed here. As multifrequency methods are developed further, calibration protocols for higher order eigenmode spring constants are needed and are currently an active area of research. For more detail and definition on cantilever eigenmodes, see Chapter 3.

1.4.8 Tip Shape Calibration

The AFM tip shape is a critical parameter for many measurements. The tip shape (and diameter) plays a significant role in defining the resolution. The tip shape also is a critical parameter that needs to be well known in order to extract any quantitative information from the tip–sample interaction such as material properties (e.g., modulus) of the surface. Knowing the shape of the AFM probe is a moving target, as the shape is most likely changing during the course of imaging either through contamination or wear. This problem is on top of the fact that AFM probes are realistically not necessarily manufactured in the cone geometry that is commonly idealized, and manufacturing is not wholly reproducible so that there is variety between batches. Direct imaging of the AFM tips by high-resolution techniques such as SEM or transmission electron microscopy (TEM) [19, 20] do not provide true three-dimensional (3D) information of the tip to fully characterize the entire probe. Also, such techniques tend to damage the tips when trying to obtain the highest resolution images necessary.

Other common ways to calibrate tip shapes are by reverse imaging of the tip [21] or blind reconstruction [22].

1.5 NANOMECHANICAL METHODS SURVEYED IN BOOK

As the reader will observe, many different nanomechanical AFM modes are implemented into industrial R&D. These modes range from the "simple" modes such as force curves, nanoindentation, and phase imaging to more sophisticated modes such as contact resonance, Harmonix™, dynamic indentation, and bimodal imaging as well as many of these methods conducted at environmental conditions such as temperature and humidity taking advantage of the flexible environment in which the AFM can operate. Below, the different nanomechanical methods implemented in the industrial R&D applications described in this book are listed. Note that this is *not* an exhaustive list of all AFM-based nanomechanical methods, but rather the nanomechanical methods that are somewhat well established and have entered the industrial R&D market surveyed here. Most methods have a dedicated chapter (Chapters 3–7) that describes the underlying theory and experimental practice of each technique. Note that Chapter 2 on contact mechanics provides the background theory that underpins all these methods.

Force Curves (Chapter 3) Single-point measurements where the AFM tip is directed in and out of a surface while the cantilever deflection is measured as a function of tip–sample separation. Force curves can be conducted in static mode or in dynamic mode. In the latter, the tip is oscillated at its resonant frequency as it approaches and retracts from the surface. Mapping of force curves exists in various forms such as force volume imaging, pulsed force mode, and peak force QNM.

Contact Mode In this mode, the tip is in constant contact with the surface at a load (force) set by the user. The tip is then raster scanned while in contact the whole time. This mode is useful for imaging topography but can be harsh or abrasive on the surface, depending on the material. If the tip is scanned in a direction perpendicular to the cantilever axis, the side-to-side force exerted on the cantilever can be measured in a mode called lateral force microscopy, which can provide information on friction of the surface [23].

Tapping Mode or Intermittent Contact Mode or Amplitude Modulated AFM (Chapter 4) One of the most common modes, the tip is oscillated at or near a resonant frequency as it raster scans along the surface. Either the amplitude is held constant by the feedback loop (amplitude modulated) or the frequency is held constant (frequency modulation.) This mode provides a more gentle interaction with the surface than contact mode and thus is preferred for soft materials such as polymers or biological materials. Topography is imaged as well as phase imaging, described below. Recently, multifrequency/higher mode tapping mode has become an object of much research where the tip is oscillated at a higher eigenmode than the tapping mode that occurs at the conventional first eigenmode [24–26].

Phase Imaging (Chapter 4) In tapping mode, the cantilever is oscillated or driven at a particular frequency and amplitude. When the cantilever interacts with the surface, the oscillation amplitude is reduced and a phase shift occurs between the drive and response. This phase shift can be plotted as the tip is raster scanned over the surface. Meaning and interpretation of phase shift remains a very active area of research, but it is sensitive to mechanical properties of the material.

Contact Resonance and Force Modulation (Chapter 5) In this dynamic contact method, a tip is in contact with the surface, and the tip–surface system is oscillated either below resonance (force modulation) or at a resonant frequency, that is typically much higher than the resonant frequency of the free cantilever (contact resonance). The advantage of this method is that the tip–sample system is now a linear system (as opposed to a nonlinear system such as in tapping mode where the tip is "striking" the surface as it taps) and so is easier to model mathematically to extract mechanical properties.

Best Practices (Chapter 6) This chapter outlines experimental best practices to maximize on the information obtained from AFM images. It is intended for an AFM user and outlines some of the common pitfalls and "things to watch out for" in the quest to attain repeatable, accurate, and quantitative measurements.

Nanoindentation (Chapter 7) Indentation is a classical method that has been around for over a century, starting with a test that used a 10-mm diameter steel ball to indent and was [27, 28] invented by J.A. Brinell. In the early 1920s, the now popular macroindentation test, the Vickers test, was invented using a pyramidal tip instead of a spherical tip, allowing hardness measurements independent of indenter size [29]. Approximately 20 years later, indenting with diamond pyramidal shapes was developed [30]. Thus the indentation world had quickly moved from macro- to microindentation and nanoindentation was not far behind. In nanoindentation, Young's modulus and hardness are measured in the few to 100s or 1000s of micronewtons with very low displacement of several to hundreds of nanometers.

The nanoindentor's fundamental measurement is a force versus displacement curve as a diamond tip indents into a material resulting in elastic and/or plastic deformation, not unlike the AFM force curve. The size of residual impressions or deformation is not typically measured. Tip–sample contact areas are instead calculated from the depth measurement together with a knowledge of the actual shape of the indentor (which is calibrated before). Nanoindentation can be performed with a diamond AFM tip (AFM-based nanoindentor) or an instrumented nanoindentor. In an instrumented nanoindentor, the vertical motion of the diamond tip is restricted in the z direction and thus is more accurate than an AFM-based nanoindentor where all the cantilever's degrees of freedom can convolute the measurement. Complexities and best practices for nanoindentation are covered thoroughly in Chapter 7.

Dynamic Nanoindentation (Chapter 7) Dynamic nanoindentation has an oscillation imposed onto the tip as it penetrates into the surface. This enables viscoelastic properties to be probed such as storage modulus (E′), loss modulus (E″), and loss tangent. It is also referred to as "continuous stiffness measurement" by some vendors.

1.6 INDUSTRIES REPRESENTED

Semiconductors AFM as a key metrology tool for semiconductor devices and to probe mechanical properties of key industrial materials (Chapters 8 and 14).

Chemicals Learn how AFM is used to characterize polymer materials, blends, and composites (Chapter 9).

Food Science Use of AFM to investigate the molecular nanostructures formed through food processing or those that are naturally present in foods (Chapter 10).

Petroleum (downstream) Physical and chemical sensing in hydrocarbon environments (Chapter 11).

Personal Care Products Tribology of hair and skin (Chapter 12).

Pharmaceuticals Study the effect of single-particle properties on bulk processing behavior of pharmaceutical powders and the role in developing new types of formulations (Chapter 13).

Biomaterials Use AFM to study morphology, temperature, and humidity effects of drug eluting biodegradable coatings (Chapter 15).

ACKNOWLEDGMENTS

I would like to thank Daniel Kiracofe and Jason Killgore for a careful review of this chapter.

REFERENCES

1. Binnig, G., Quate, C. F., and Gerber, C. *Physical Review Letters* **56** (1986): 930–933.
2. Eaton, P., and West, P. *Atomic Force Microscopy*, Oxford University Press: New York, 2011.
3. Haugstad, G. *Atomic Force Microscopy: Understanding Basic Modes and Advanced Applications*, Wiley: Hoboken, NJ, 2012.
4. Magonov, S. N., and Whangbo, M.-H. *Surface Analysis with STM and AFM: Experimental and Theoretical Aspects of Image Analysis*, Wiley-VCH: New York, 1996.
5. Meyer, E., Hug, H. J., and Bennewitz, R. *Scanning Probe Microscopy: The Lab on a Tip*, Springer: New York, 2004.
6. Bonnell, D., Ed. *Scanning Probe Microscopy and Spectroscopy*, 2nd ed., Wiley-VCH: New York, 2001.
7. Fukuma, T., Kimura, M., Kobayashi, K., Matsushige, K., and Yamada, H. *Review of Scientific Instruments* **76** (2005): 053704.
8. Fukuma, T. *Japanese Journal of Applied Physics* **48** (2009): 08JA01/1.
9. Voitchovsky, K., Kuna, J. J., Contera, S., Tosatti, E., and Stellacci, F. *Nature Nanotechnology* **5** (2010): 401.
10. Gross, L., Mohn, F., Moll, N., Liljeroth, P., and Meyer, G. *Science* **325** (2009).
11. Gross, L., Mohn, F., Moll, N., Schuler, B., Criado, A., Gutian, E., Pena, D., Gourdon, A., and Meyer, G. *Science* **337** (2012): 1326.
12. Callister Jr., W. D. C., and Rethwisch, D. G. *Materials Science and Engineering: An Introduction*, 8th ed., Wiley: Hoboken, NJ, 2010.
13. Sader, J. E., Chon, J. W. M., and Mulvaney, P. *Review of Scientific Instruments* **70** (1999): 3967.
14. Sader, J. E. http://www.ampc.ms.unimelb.edu.au/afm/calibration.html.
15. Torii, A., *Measurement Science and Technology* **7** (1996): 179–184.
16. Hutter, J. L., and Bechhoefer, J. *Review of Scientific Instruments* **64** (1993): 1868–1873.
17. Butt, J. H., and Jaschke, M. *Nanotechnology* **6** (1995): 1–7.
18. Cleveland, J. P. *Review of Scientific Instruments* **64** (1993).
19. Chung, K.-H., Lee, Y.-H., and Kim, D.-E. *Ultramicroscopy* **102** (2005): 161–171.
20. Kim, K.-H., Moldovan, N., Ke, C., Espinosa, H. D., Xiao, X., Carlisle, J. A., and Auciello, O. *Small* **1** (2005): 866–874.
21. Bloo, M. L., Haitjema, H., and Pril, W. O. *Measurement* **25** (1999): 203–211.
22. Bhushan, B., and Kwak, K. J. *Nanotechnology* **18** (2007): 345505.
23. Carpick, R. W. *Chemical Reviews* **97** (1997): 1163.
24. Lozano, J. R., and Garcia, R. *Physical Review Letters* **100** (2008): 076102.
25. Martinez, N. F., Lozano, J. R., Herruzo, E. T., Garcia, F., Richter, C., Sulzbach, T., Garcia, R. *Nanotechnology* **19** (2008): 384001.
26. Proksch, R. *Applied Physics Letters* **89** (2006): 11312/1.
27. Tabor, D. *The Hardness of Metals*, Oxford University Press: London, 1951.
28. E28 Committee, *Test Method for Brinnell Hradness of Metallic Materials*, ASTM International, 2001.
29. Smith, R. L., and Sandly, G. E. *Proceedings of the Institution of Mechanical Engineerings* **102** (1922): 623.
30. Knoop, F., Peters, C. G., and Emerson, W. B. *A Sensitive Pyramidal Diamond Tool for Indentation Measurements*, National Bureau of Standards: Washington, DC, 1939.

Chapter 2

Understanding the Tip–Sample Contact: An Overview of Contact Mechanics from the Macro- to the Nanoscale

Tevis D. B. Jacobs,[1] C. Mathew Mate,[2] Kevin T. Turner,[3] and Robert W. Carpick[3]

[1]Department of Materials Science and Engineering, University of Pennsylvania, Philadelphia, PA
[2]HGST, A Western Digital Company, San Jose, CA
[3]Department of Mechanical Engineering and Applied Mechanics, University of Pennsylvania, Philadelphia, PA

2.1 HERTZ EQUATIONS FOR ELASTIC CONTACT

2.1.1 Introduction

Just as we expect a solid body to have certain mechanical properties that are characteristic to its material and geometry, a mechanical contact between two bodies also has measurable properties that depend on analogous factors. So, just as we can perform a tension or compression experiment on a solid body, we can also perform similar types of experiments on contacts.

The field of contact mechanics was pioneered in 1880 by Heinrich Hertz who examined the problem of elastic deformation for two spheres being pressed into contact [1]. As it forms the basis of many other contact situations including nanoscale contacts, the Hertz model is treated here in some detail at the outset.

Scanning Probe Microscopy in Industrial Applications: Nanomechanical Characterization, First Edition. Edited by Dalia G. Yablon.

The Hertz model neglects any adhesion and friction forces between the two bodies in contact—a significant assumption given that these forces are always present to some degree and become more significant at smaller length scales. Yet in many situations, the assumption of negligible adhesion and friction forces works quite well, including for a wide variety of nanoscale contacts. For example, adhesion is rarely relevant for *nanoindentation* [see Chapter 7], where the applied loads are typically much larger than the adhesive loads, and for *contacts immersed in liquid*, because in many cases the liquid "screens" the adhesive force between the solid surfaces. In Section 2.2, we discuss contact models to be used when adhesion cannot be neglected.

2.1.2 Hertz Equations

Based on geometric considerations and the theory of elasticity [2], the Hertz analysis shows that when two spheres of radii R_1 and R_2 are pressed into contact with a loading force $F_{applied}$ (Fig. 2.1), the region around the contact deforms, with the center displacing by an amount δ in the normal direction and forming a circular contact area A having radius a, as given by [3]

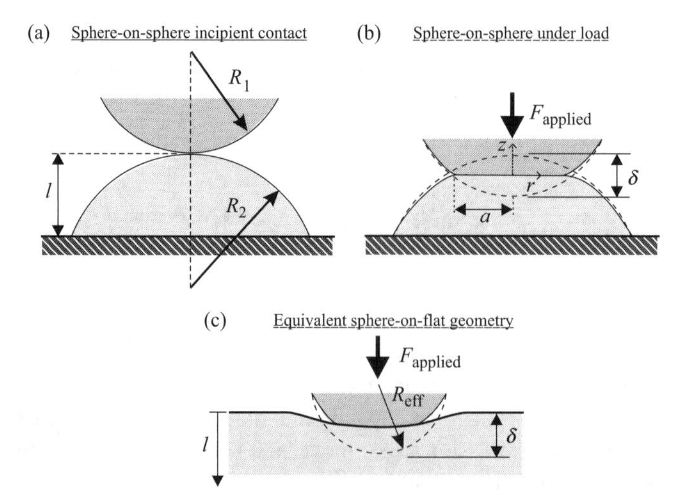

Figure 2.1 (a) Side view of two spherical protrusions that are just touching, with the lower protrusion mounted on a rigid flat surface. The lower body extends a distance l beneath the contact point. (b) The two protrusions are pressed together with a vertical loading force $F_{applied}$ to generate a circular contact area with radius a and area $A = \pi a^2$ due to elastic deformation. The total elastic deformation normal to the contact area is δ. The axes r, z (used in subsequent equations) are indicated, with an origin at the center of the contact. The dashed lines show the undeformed profiles of the protrusions. Note that the extent of compression has been significantly exaggerated in this figure for clarity; see Section 2.1 for limits of applicability of the Hertz model. (c) A sphere-on-flat geometry, in which the sphere has radius R_{eff} [defined in Eq. 2.3], with identical stress and displacement profiles to the case shown in (b).

$$a = \left(\frac{3F_{\text{applied}} R_{\text{eff}}}{4E^*} \right)^{1/3} \tag{2.1}$$

$$\delta = \frac{a^2}{R_{\text{eff}}} = \left(\frac{9F_{\text{applied}}^2}{16R_{\text{eff}} E^{*2}} \right)^{1/3} \tag{2.2}$$

The effective radius R_{eff} and the effective modulus E^* are defined as

$$R_{\text{eff}} = \left(\frac{1}{R_1} + \frac{1}{R_2} \right)^{-1} \tag{2.3}$$

$$E^* = \left(\frac{1-\nu_1^2}{E_1} + \frac{1-\nu_2^2}{E_2} \right)^{-1} \tag{2.4}$$

where E is Young's modulus, ν is Poisson's ratio, and subscripts 1, 2 designate the two spheres. The *normal stiffness* or contact stiffness k_{normal} (defined as the change in the load per unit change in deformation) for a Hertzian contact is given by

$$k_{\text{normal}} = \frac{\partial F_{\text{applied}}}{\partial \delta} = 2E^* \sqrt{R_{\text{eff}} \delta} = 2aE^* \tag{2.5}$$

The increase in normal stiffness with increasing deformation occurs because the stiffness is a function of the contact area, which increases with the applied load as can be seen in the expression for contact radius a, in Eq. (2.1). This dependence of stiffness on load is in stark contrast to the constant stiffness for a solid body undergoing small uniform deformations under axial loading, where Hooke's law applies and the deformation is linear with the applied load. The definition of contact stiffness in Eq. (2.5) is important and will recur in Chapter 3 on force curves and in Chapter 5 in dynamic contact methods, where derivation of mechanical properties directly from various atomic force microscopy (AFM) measurements is discussed. [Editor's note: contact stiffness is referred to as k_{ts} in Chapter 3 and k in Chapter 7.]

The pressure distributions at the surface and the stress distributions inside the two bodies can also be obtained from the Hertz model. We will follow the notation convention used by Johnson [3]. Thus, we use the symbol p to denote the contact pressure applied by one body to another at the contact interface; a compressive pressure is positive, and tensile contact pressures, that is, due to adhesion, are negative. In all other cases, stress components are denoted by σ and compressive stresses are negative (tensile stresses are positive), as is typical in an engineering/solid mechanics context. The compressive contact pressure $p(r)$ at the interface of the two bodies has a maximum value p_0 at the center of the contact, which is 50% higher than the mean pressure (F_{applied}/A):

$$p_0 = \frac{3}{2} p_{\text{mean}} = \left(\frac{6 F_{\text{applied}} E^{*2}}{\pi^3 R_{\text{eff}}^2} \right)^{1/3} \tag{2.6}$$

and decreases with the radial coordinate r [indicated in Fig. 2.1(b)] according to

$$p(r) = p_0 \left(1 - \frac{r^2}{a^2} \right)^{1/2} \tag{2.7}$$

The values of all three stress components at the interface ($z=0$) are shown in Figure 2.2(a); for consistency between all three components, we plot the stress component σ_z instead of the contact pressure p; the relation between the two is given by $p(r)=-\sigma_z(r, z=0)$. Equations for radial stress at the surface $\sigma_r(r, z=0)$, and circumferential stress at the surface $\sigma_\theta(r, z=0)$ can be found in Ref. [3] and are purely functions of r due to azimuthal symmetry.

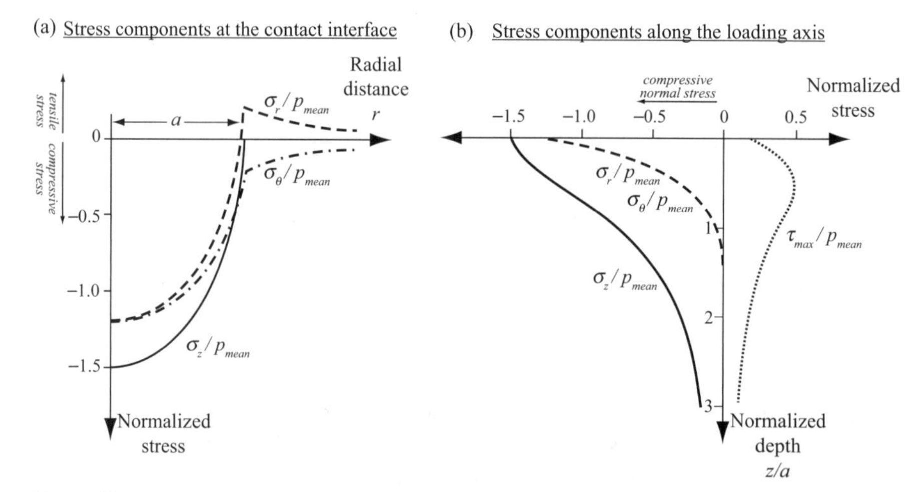

Figure 2.2 (a) Normal, radial, and circumferential stress components (σ_z, σ_r, σ_θ, respectively) on the contact interface (i.e., $z=0$). Stresses (vertical axis) are shown as a function of the radial distance r (horizontal axis). (The spatial directions r and z are defined in Fig. 2.1(b).) The contact stresses all have the highest magnitude at the center of the contact and decay with distance away from the center of contact. The normal pressure exerted by each body on the other is denoted in the text as $p=-\sigma_z(z=0)$. All values shown are normalized by the mean contact pressure p_{mean} as defined in Eq. (2.6). (b) The same three stress components (σ_z, σ_r, σ_θ, now on the horizontal axis) are shown as a function of depth z into the material (vertical axis). These stress components are highest at the surface and decay with depth into the material. These components are the principal stresses, and so the absolute maximum shear stress at each point (τ_{max}) reaches its largest value at a depth of roughly half the contact radius; the exact location depends on Poisson's ratio ν; for the calculated stresses in these figures, $\nu = 0.3$. Figures adapted with permission from Ref. [3].

The full equations for the axial stress σ_z, radial stress σ_r, and circumferential stress σ_θ at all points *inside the bodies* can be found elsewhere [4]. However, often it is sufficient just to know the stresses along the axis of loading z [indicated in Fig. 2.1(b)], which are given by

$$\sigma_z(z)\big|_{r=0} = -p_0\left(1+\frac{z^2}{a^2}\right)^{-1} \tag{2.8}$$

$$\sigma_r(z)\big|_{r=0} = \sigma_\theta(z)\big|_{r=0} = -p_0\left\{(1+\nu)\left[1-\left(\frac{z}{a}\right)\tan^{-1}\left(\frac{a}{z}\right)\right]-\frac{1}{2}\left(1+\frac{z^2}{a^2}\right)^{-1}\right\} \tag{2.9}$$

These are graphed in Figure 2.2(b). In the r, θ, z coordinate system, the shear stress components are all zero along the z axis; thus, the normal stress components in Eqs. (2.8) and (2.9) are principal stresses. These principal stress components are largest (i.e., most compressive) at the contact interface and decay with depth into the material, all falling to less than 10% of their maximum (surface) value at a depth of approximately three times the contact radius.

In many materials, including most metals, yielding and the onset of plastic deformation are determined by the largest value of the absolute maximum *shear stress* τ for each point ($\tau_{max} = \frac{1}{2}$|largest principal stress difference|) rather than by the normal stresses in the body. In a Hertzian contact, the τ_{max} is not highest at the surface but rather at a depth of approximately half the contact radius a (depending on Poisson's ratio) as shown in Figure 2.2(b); at this point, the largest value of τ_{max} ($=\frac{1}{2}|\sigma_z - \sigma_r|$) is approximately $\frac{1}{3}\,p_0$. This implies that permanent deformation will often begin beneath the surface, and this has been confirmed in experiments. If the von Mises criteria is used for yielding rather than the maximum shear stress, permanent deformation is expected to begin beneath the surface when

$$p_0 \cong 1.6\sigma_{\text{yield}} \tag{2.10}$$

where σ_{yield} is the normal yield strength of the softer material. Note that the onset of yielding depends on the value of Poisson's ratio; Eq. (2.10) assumes $\nu = 0.3$. For brittle materials, in contrast, tensile normal stresses are most critical. As shown in Figure 2.2(a), the radial stress σ_r at the surface becomes tensile, and its largest tensile value occurs at the outer ring of contact where $r=a$. Experiments indeed show that cracks are seen to initiate at this location in brittle contacts [5].

2.1.3 Assumptions of Hertz Model

The Hertz analysis makes the following assumptions:

1. The two bodies are homogeneous, that is, each body is a monolithic material with no surface coatings or films.

2. The materials are isotropic and exhibit linear elastic constitutive behavior.
3. The surfaces are frictionless, that is, there are no surface or traction forces acting parallel to the plane of the contact.
4. The mechanical response of the bodies in contact can be approximated as that of flat, infinite bodies under equivalent conditions, that is, the bodies are much larger than the size of their contact: R_1, R_2, $l >> a$, where l is the depth perpendicular to the contact surface to which the sphere or flat plane continues (Fig. 2.1). Note that, as a typical practical limit, $>>$ means "at least 10 times larger."
5. Deformations are small, that is, the bodies are much larger than the size of their deformations $\delta << R_1$, R_2, and l.
6. The surfaces are perfectly smooth.

The Hertz model is widely useful even though some assumptions are never strictly valid (e.g., some degree of friction will always occur between the two bodies, violating assumption 3). The Hertz model can be used as a first and often nearly exact approximation of the conditions in a contact. For instance, in many practical cases, it can be used to calculate the magnitude of the stresses that develop in a body; these can then be compared to the yield strength of the material to determine whether plastic deformation is expected. Later sections of this chapter will discuss how to address problems where one (or more) assumption is explicitly violated, as well as special considerations that become relevant for nanoscale contacts.

2.1.4 Worked Examples: Hertz Mechanics of Diamond Tips on Stiff and Compliant Substrates

Here we use the Hertz analysis to calculate the contact parameters for two practical examples: diamond AFM tips on stiff and compliant samples. As an example of a stiff sample, we first consider the case of the diamond AFM tip contacting a diamond-like carbon (DLC) film. The term DLC refers to a class of amorphous carbon films that exhibit some of the beneficial properties of diamond, such as high hardness, density, and thermal stability. These coatings are used in a wide range of applications to provide scratch and wear resistance and to protect the underlying materials from chemical attack and corrosion. In this example, we assume that the DLC film has a typical Young's modulus $E = 150\,\text{GPa}$ and Poisson's ratio $\nu = 0.2$; for the diamond tip, we assume $E = 1000\,\text{GPa}$ and $\nu = 0.1$: thus, $E^* = 135\,\text{GPa}$. We also assume that the AFM tip is paraboloidal in shape (i.e., a three-dimensional revolution of a parabola) with a radius of curvature of $R_{\text{tip}} = 50\,\text{nm}$. Finally, we assume that the tip and sample are immersed in a fluid medium that screens the interfacial forces so that the contact is nonadhesive and frictionless, making the Hertz model fully appropriate.

The values for contact radii, deformation, and maximum compressive stress calculated from the Hertz analysis (Eqs. (2.1), (2.2), and (2.6) are summarized in

Table 2.1 Calculated Values of Contact Radius, Contact Deformation, and Maximum Pressure (compressive stress) for a 50-nm Radius Diamond Tip on Diamond-like Carbon and Polyethylene Surfaces

Load $F_{applied}$ (nN)	Contact Radius a (nm)	Contact Deformation δ (nm)	Maximum Pressure p_0 (GPa)
Example 1	**Diamond Tip on Diamond-like Carbon Surface**		
0	0	0	0
0.1	0.30	0.002	0.52
1.0	0.65	0.01	1.1
10	1.4	0.04	2.4
100	3.0	0.18	5.2
Example 2	**Diamond Tip on Polyethylene Surface**		
0	0	0	0
0.1	2.5	0.12	0.01
1.0	5.3*	0.56	0.02
10	11*	2.6	0.04*
100	25*	12*	0.08*

* For contact with polyethylene, the assumptions of the Hertz model are violated at 10 nN and above, as indicated by an asterisk. For 1 nN, the contact radius exceeds 10% of the tip radius, but only slightly - the values will likely be approximately correct.

Table 2.1 for loads ranging from 0 to 100 nN. For DLC, the contact radii and deformations are extremely small ($a<3$ nm, $\delta<0.2$ nm) and close to atomic dimensions, while the peak contact stresses or maximum pressure are rather high (up to 5 GPa). Even with this high contact pressure, the DLC is expected to behave elastically up to the highest loads shown because the typical yield strength of DLC films is in the range of 5–20 GPa.

As an example of a compliant polymer sample, we consider the same tip ($R_{tip}=50$ nm) making contact with a sample of low-density polyethylene for which $E=0.2$ GPa and $\nu=0.45$, thus $E^*=250$ MPa. As before, the contact is immersed in a fluid such that there is negligible adhesion and friction. The significantly lower Young's modulus (by a factor of 750) of polyethylene leads to contact radii and deformations 10 times larger and contact pressures more than 10 times lower as compared to DLC (Table 2.1). In compliant materials such as polyethylene, the assumptions for Hertzian behavior are often violated. For the higher loads, the contact radius and deformation exceed 10% of the tip radius. Assuming a typical value of $\sigma_{yield}=25$ MPa, Eq. (2.10) predicts that yielding for the polyethylene sample will start when $p_0=40$ MPa, which is reached when the load is 10 nN. When the applied force is high enough for plastic deformation to occur, the true contact radius and deformation will be even higher than the values predicted by the Hertz model due to plastic flow around the penetrating tip.

2.2 ADHESIVE CONTACTS

2.2.1 Introduction to Adhesion

All materials experience some level of adhesion when in contact with other materials due to the interatomic forces, electrostatic attraction, and capillary forces acting across contacting interfaces. As these adhesive forces scale with surface area, adhesion can usually be neglected when relatively large objects contact since the objects' weights and the applied loads usually dominate over the surface adhesion forces. However, as the size of objects becomes smaller and the surface-to-volume ratios increase, adhesive forces become more significant. At the micrometer and nanometer scale, adhesion can often dominate over externally applied forces (and gravitational forces will be negligible). For example, adhesive forces are utilized effectively by flies and other small insects that adhere themselves on walls or ceilings by using surface structures on their feet to generate small but sufficient adhesion.

2.2.2 Basic Physics and Mathematics of Surface Interactions

By its design, AFM explores the surface forces between a sharp probe tip and the surface under examination, therefore some discussion is needed about the specifics of these interactions.

One interaction that exists between the atoms of every material—whether charged or neutral, solid or liquid or gas—is the subcategory of van der Waals forces known as "dispersion forces" or "London dispersion forces." Dispersion forces are quantum mechanical in nature but can be thought of as the process in which a random, instantaneous asymmetry in the electron cloud of one atom produces a momentary dipole in that electron cloud, which in turn induces, via polarization, momentary dipoles in the electron clouds of neighboring atoms. These induced dipoles attract one another, causing the atoms to attract with a magnitude that depends on the polarizability of the interacting atoms. The energy of these interactions depends on separation distance r between the atoms according to $E_{\text{interaction}} \propto r^{-6}$; therefore, the interaction force $F_{\text{interaction}} = -dE/dr \propto r^{-7}$. The reader should note that the symbol r in this section (Section 2.2.2) refers to the interatomic spacing; it should not be confused with the same symbol r used elsewhere in the chapter to describe the radial coordinate in contact mechanics.

Once the atoms are close enough that electron clouds start to overlap, the atoms begin to repel each other. These repulsive interactions are very short range and rise steeply as r decreases. The total interaction between two atoms as a function of separation distance is called the "interatomic pair potential" and is calculated as the sum of the repulsive and attractive interactions. The most common mathematical approximation for this interaction is the Lennard-Jones 6–12 potential [6], where the attractive part of the potential is assumed to be the van der Waals potential with the characteristic $-1/r^6$ dependence, and the repulsive part of the potential is modeled as having a $+1/r^{12}$ dependence, leading to an interaction force:

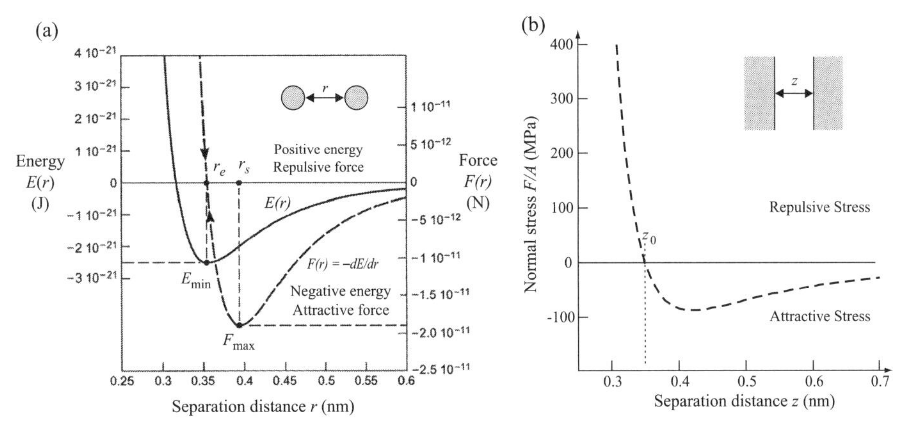

Figure 2.3 (a) Lennard-Jones interaction between two atoms (as sketched in the inset) yields the interaction energy $E(r)$ (solid curve, left axis) and the interaction force $F(r)$ [Eq. (2.11), dashed curve, right axis; attraction is represented as a negative force]. The equilibrium separation (r_e) and distance of maximum interaction force (r_s) are indicated. Values of $C = 2.5 \times 10^{-21}$ J and $D = 0.316$ nm have been assumed. Figure adapted with permission from Ref. [6]. (b) Lennard-Jones surface potential yields the force per unit area between two parallel solid surfaces [Eq. (2.12)], calculated with $W_{adh} = 0.060$ J/m^2 and $z_0 = 0.35$ nm.

$$F_{\text{interaction}}(r) = -\frac{dE_{\text{interaction}}(r)}{dr} = -24C\left(\frac{D^6}{r^7} - \frac{2D^{12}}{r^{13}}\right) \tag{2.11}$$

where C and D are empirical parameters used to scale the strength (C) and length scale (D) of the interaction. The plots of the interaction energy $E_{\text{interaction}}(r)$ and the force of interaction $F_{\text{interaction}}(r)$ are shown in Figure 2.3(a), assuming values of $C = 2.5 \times 10^{-21}$ J and $D = 0.316$ nm. Note that while the first term ($\propto r^{-7}$) is physically derived based on van der Waals interactions, the second term ($\propto r^{-13}$) arose from choosing a convenient mathematical expression for the repulsive energy. The atoms are at equilibrium when $F = 0$, or when $r = 2^{1/6}D$.

The net Lennard-Jones energy or force between two bodies can then be calculated by integrating Eq. (2.11) over the volumes of the two bodies. For two parallel surfaces separated by a distance z, the normal pressure acting between the two surfaces from the Lennard-Jones (L-J) intermolecular forces is [7]

$$p(z) = \frac{F_{\text{surfaces}}}{A_{\text{surfaces}}} = -\frac{8W_{\text{adh}}}{3z_0}\left[\left(\frac{z_0}{z}\right)^3 - \left(\frac{z_0}{z}\right)^9\right] \tag{2.12}$$

where z_0 is the equilibrium separation distance when the two surfaces are in contact and W_{adh} is the work of adhesion. Equation (2.12) is plotted in Figure 2.3(b) where we have assumed values of $W_{adh} = 0.060$ J/m^2 and $z_0 = 0.35$ nm. Note that, as in Section 2.1, a positive value of p represents compression; a negative value represents

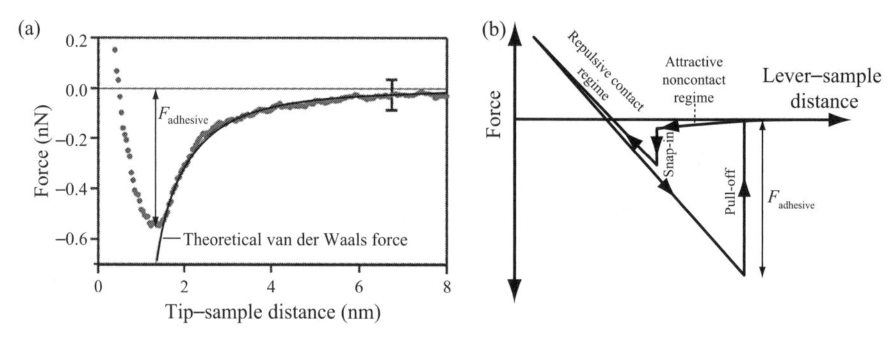

Figure 2.4 Examples of force curves measured by AFM. (a) The interaction force is shown as a function of tip–sample distance for a carboxyl-terminated AFM tip approaching a hydroxyl-terminated SAM in deionized water. The curve shows the fit to the theory for an attractive van der Waals force for a sphere approaching a flat surface. Reproduced with permission from Ref. [12]. Note that the tip–sample distance (horizontal axis) was calculated by subtracting the cantilever's displacement from the lever–sample separation distance (which is varied by ramping the *z* piezo). (b) More typically, the interaction force is plotted as a function of the lever–sample separation distance. In this case, as the AFM tip approaches and retracts, at a certain point the lever becomes unstable and enters and exits contact by sudden snap-in and pull-off. Figure adapted with permission from Ref. [64].

attraction. The work of adhesion is the energy per unit area required to separate two surfaces from intimate contact to a large separation distance. For two identical bodies (denoted as material 1) with surface energy γ_1, $W_{adh} = 2\gamma_1$; for two bodies of different materials with surface energies γ_1 and γ_2 and interfacial energy γ_{12}, the work of adhesion is defined as [6]

$$W_{adh} = \gamma_1 + \gamma_2 - \gamma_{12} \tag{2.13}$$

and is often approximated [8] as $W_{adh} \sim 2(\gamma_1\gamma_2)^{1/2}$. For a more in-depth explanation of surface energy and work of adhesion, see Appendix 2A.

The work of adhesion is often extracted by measuring force curves, which are discussed extensively in Chapter 3. The full force–separation curve [corresponding to integrating Eq. (2.12) over the tip–sample geometry of an AFM] can be extracted using non-contact AFM methods [9] or directly measured using force–feedback methods [10, 11]. An example of the latter approach is shown for an AFM tip approaching a sample in deionized water in Figure 2.4(a); both the tip and sample are coated with gold layers and functionalized with a hydroxyl-terminated self-assembled monolayer [12]. Here, the adhesion is due to hydrogen bonding and van der Waals interactions between the tip and sample. Dynamic or force–feedback methods are needed to plot the entire force–separation curve because with static measurements the tip can snap into contact with the sample when the gradient of the tip–sample interaction force exceeds the spring constant of the cantilever. Thus, more commonly, plots such as the schematic in Figure 2.4(b) are obtained. Here, the horizontal axis corresponds to the *z* motion of the piezo that moves either the sample or the cantilever chip, which is what is directly controlled and reported by the AFM. In addition to L-J forces, sometimes other forces exist between two surfaces that can be significantly stronger, such as electrostatic interactions (between charged surfaces or surfaces with permanent dipoles), chemical bond interactions (when covalent or metallic bonds form across interfaces), and capillary forces (from the menisci that form around contact points). A full description of the specific forces and interactions between materials is beyond the scope of this text (see Refs. [6, 13] for more discussion). Regardless of which specific forces are acting between the AFM tip and the substrate, however, a tensile force will be required to overcome the adhesion and to separate the AFM contact [as indicated by $F_{adhesive}$ in the schematic of an AFM force curve, where force is measured as a function of lever–sample distance in Figure 2.4(b). For a full treatment of AFM force curves, see Chapter 3]. To understand this adhesion with regard to AFM, we turn to adhesive contact mechanics. For the purposes of adhesive contact mechanics, knowledge of the exact nature of the adhesion is often not necessary; the models simply assume a value for the work of adhesion and then calculate the consequences of this additional term.

2.2.3 Derjaguin–Müller–Toporov and Johnson–Kendall–Roberts Models of Adhesion

The simplest extension of the Hertz model to account for adhesion is the Derjaguin–Müller–Toporov, or DMT, model [14, 15], which assumes that the adhesive forces are sufficiently weak or long range, or that the material is sufficiently stiff, that material around the contacting interface does not significantly deviate from the shape of a Hertzian contact due to the added stresses from adhesion. Rather, the primary effect of the adhesion is to increase the applied load by an amount equal to the adhesive force $F_{adhesive}$. Consequently, the Hertzian results can still be used with the simple replacement of the applied force $F_{applied}$ by the quantity $F_{total} = F_{adhesive} + F_{applied}$, where $F_{adhesive}$ is the adhesive force pulling the two bodies together. For a specific AFM contact, the adhesive force between the tip and sample can be determined by measuring the force acting on the tip as it is withdrawn from the surface: $F_{adhesive}$ is the maximum negative force achieved before the AFM tip snaps or pulls off the sample, the "pull-off" force shown in Figure 2.4(b).

In the DMT model, the adhesive force can be estimated using the Derjaguin approximation, which relates the work of adhesion to the adhesive force for a sphere-on-flat or sphere-on-sphere geometry [16]:

$$F_{adhesive,\ DMT} = 2\pi R_{eff} W_{adh} \tag{2.14}$$

In the DMT model, all of the equations for radius, deformation, and stress in the contact are the same as the Hertzian results [Eqs. (2.1)–(2.10)], with the simple replacement of $F_{applied}$ with $F_{total} = F_{adhesive} + F_{applied}$.

Around the same time that the DMT model was proposed (1971), Johnson, Kendall, and Roberts (JKR) introduced an adhesive contact model for situations where the adhesive stresses are strong enough to induce significant localized deformation around the contact zone, pulling more area into contact than expected for the Hertz theory without adhesion, or even for the DMT theory with adhesion. In the JKR model, the separation of the two bodies is analyzed using principles from fracture mechanics; details of this can be found in Refs. [3, 15, 17]. One of the principal results of the JKR theory is that the force required to separate the contact is given by

$$F_{\text{adhesive, JKR}} = \frac{3}{2}\pi R_{\text{eff}} W_{\text{adh}} \tag{2.15}$$

In the JKR model, the contact radius a_{JKR}, contact deformation δ_{JKR}, and position-resolved contact pressure $p_{\text{JKR}}(r)$ [where the symbol r in this and subsequent chapters refers to the radial coordinate as defined in Section 2.1 (see Fig. 2.1)] are given by the following equations [3]:

$$a_{\text{JKR}} = \left\{ \frac{3R_{\text{eff}}}{4E^*} \left[F_{\text{applied}} + 3W_{\text{adh}}\pi R_{\text{eff}} + \sqrt{6W_{\text{adh}}\pi R_{\text{eff}} F_{\text{applied}} + (3W_{\text{adh}}\pi R_{\text{eff}})^2} \right] \right\}^{1/3} \tag{2.16}$$

$$\delta_{\text{JKR}} = \frac{a_{\text{JKR}}^2}{R_{\text{eff}}} - \left(\frac{2\pi W_{\text{adh}} a_{\text{JKR}}}{E^*} \right)^{1/2} \tag{2.17}$$

$$p(r) = \left(\frac{2a_{\text{JKR}} E^*}{\pi R_{\text{eff}}} \right) \left(1 - \frac{r^2}{a^2} \right)^{1/2} - \left(\frac{4W_{\text{adh}} E^*}{\pi a_{\text{JKR}}} \right)^{1/2} \left(1 - \frac{r^2}{a^2} \right)^{-1/2} \tag{2.18}$$

2.2.4 More Realistic Picture of Adhesion

The DMT and JKR pictures of adhesive contacts predict different values for the contact radius, the contact stresses, and adhesive force for otherwise identical contacts. They were initially considered to be competing and incompatible models, but Tabor proposed [18] and Maugis later proved [15] that these two models are both correct but represent different ends of a spectrum defined by the ratio of the range of the adhesion forces relative to the elastic deformation those forces cause: The DMT limit applies to stiffer, smaller asperities in the presence of weaker, longer-range adhesion; while the JKR limit applies to more compliant, larger asperities under the action of stronger and shorter-range adhesion.

Contacts that fall between these extremes can be characterized by the Tabor parameter μ_T, defined as

$$\mu_T = \left(\frac{RW_{\text{adh}}^2}{E^{*2} z_0^3} \right)^{1/3} \tag{2.19}$$

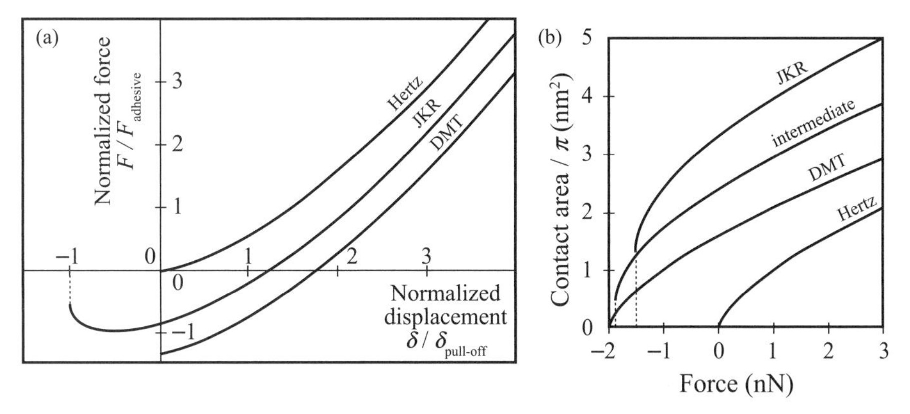

Figure 2.5 (a) Normalized load is shown as a function of displacement for the nonadhesive (Hertz) case, as well as both limits of the adhesive case (JKR and DMT). All loads are normalized by the JKR pull-off force given by Eq. (2.15); displacements are normalized by the displacement at the pull-off point in the JKR model. Reproduced with permission from Ref. [82]. (b) The contact area is shown as a function of load for all three models, as well as for an example contact from the intermediate region between JKR and DMT. It is assumed that $R_{eff} = 1$ nm, $E = 0.75$ GPa, and $W_{adh} = 0.318$ J/m^2. Reproduced with permission from Ref. [19].

where z_0 is the equilibrium spacing of the surfaces. Since z_0 is not directly measurable, for practical purposes physically reasonable upper and lower bounds for z_0 can be chosen and the Tabor parameter recalculated for each to give a range of possible values. A typical lower limit of z_0 is the interatomic spacing between the material used in the body (e.g., 0.154 nm for carbon atoms in diamond). A typical upper limit of z_0 is 0.5 nm based on the assumption that adhesion results from atomic bonding; larger values are possible if long-range electrostatic or van der Waals forces are substantial. The DMT limit is appropriate for cases where $\mu_T \leq 0.1$, whereas the JKR limit applies when $\mu_T \geq 5$. Values of μ_T between 0.1 and 5 correspond to contacts between the JKR and DMT limits, where the contact radius, deformation, and pull-off force are intermediate between the predictions of the two models. While equations for contacts in the transition region are beyond the scope of this text (see Ref. [15] or [19]), Figures 2.5(a) and 2.5(b), respectively, show how the elastic displacement and the contact area versus load vary for contacts described by different values of the Tabor parameter, including the JKR [from Eq. (2.16)] and DMT cases, as well as for the Hertz case [from Eq. (2.1)].

While the Tabor parameter can be easily calculated, three general rules of thumb are readily apparent from the form of Eq. (2.19):

1. For a given geometry, more compliant materials (rubber, polymers, biological materials) tend toward the JKR limit, while stiffer materials (silicon, diamond) tend toward the DMT limit;
2. For a given geometry, more adhesive materials (adhesive tape, gecko feet) tend toward JKR, while less adhesive materials (diamond-like carbon, lotus leaf) tend toward DMT;

3. While radius of curvature is a secondary effect, all else being equal, larger sphere radii move the contact toward JKR, while smaller tip radii push it toward DMT.

Finally, if the Maugis parameter is truly unknown, then both the DMT and JKR predictions can be calculated and used as upper and lower bounds for desired parameters.

2.2.5 Continuing the Worked Examples: Adding Adhesion to Diamond Tips on Stiff and Compliant Substrates

The examples described in Section 2.1.4 involved nonadhesive contact between a diamond AFM tip and a diamond-like carbon (DLC) or a polyethylene (PE) sample. Now these examples are extended to situations where adhesion becomes relevant.

For the first example, an AFM force curve between the 50-nm diamond tip on the DLC is used to measure a pull-off force of 3 nN. First, we calculate both the upper and lower bounds of the Tabor parameter by using upper and lower bounds of z_0 (0.15 and 1 nm, respectively) and for W_{adh} (using the DMT model with $W_{adh} = F_{adhesive}/(2\pi R_{eff})$, and the JKR model with $W_{adh} = F_{adhesive}/(1.5\pi R_{eff})$, respectively). In this way, the upper bound of μ_T (using the lower limit value of $z_0 = 0.15$ nm, and the larger value of the work of adhesion from the JKR model, $W_{adh} = 0.013$ J/m^2) is calculated to be $\mu_T = 0.051$. Similarly, the lower bound of μ_T is calculated as 0.006. Both of these lie firmly in the DMT limit; this is not surprising as these materials are relatively stiff and relatively low adhesion being probed by a tip with a small radius of curvature. Thus, the correct value of W_{adh} is obtained from Eq. (2.14), and so $W_{adh} = 0.010$ J/m^2.

We can then use Eqs. (2.1) and (2.2) with $F_{applied}$ replaced by $F_{total} = F_{adhesive} + F_{applied}$ (see Section 4.2.3) to find the contact deformation, radius, and stress at any value of applied load. At $F_{applied} = 0$ and continuing with the example where $F_{adhesive} = 3$ nN, the calculated values are $\delta = 0.018$ nm, and $a = 0.94$ nm [corresponding to a contact area of 2.8 nm^2, or approximately 50 atoms on a diamond (111) surface]. In fact, the contact area and displacement are so small that one may question whether continuum mechanics is appropriate; this issue is discussed in Section 2.6. The maximum compressive stress, which occurs at the center of the contact, is 1.62 GPa. This stress is quite high and demonstrates the significant effect that adhesion can have in nanoscale contacts. Although adhesion increases the contact radius, we still find that $a << R$, thus satisfying one of the conditions for the Hertz and DMT models to apply.

For the second example, the pull-off force between a 50-nm diamond tip and a PE substrate is measured to be 150 nN. As was done in the previous example, we first calculate the upper and lower bounds of the Tabor parameter for this situation. The upper limit value of $z_0 = 1.0$ nm is used, along with the smaller W_{adh} calculated using the DMT model [Eq. (2.14)], to calculate a lower bound Tabor parameter $\mu_T = 5.7$. The lower limit value of $z_0 = 0.15$ nm is used, along with the larger W_{adh} determined from the JKR model [Eq. (2.15)], to calculate an upper bound of $\mu_T = 46$. Both of these values are greater than 5, therefore the JKR model can be confidently used. Using the JKR model, we determine that $W_{adh} = 0.64$ J/m^2 [Eq. (2.15)], the contact deformation at $F_{applied}=0\,nN$ is $\delta=13.4\,nm$ [Eq. (2.17)], and the corresponding contact radius is $a=45\,nm$ [Eq. (2.16)]. Note the significantly larger contact radius on the compliant polymer sample due to the substantial adhesion and lower modulus. Also note that the Hertzian assumption that $a<<R$ is already violated.

2.3 FURTHER EXTENSIONS OF CONTINUUM CONTACT MECHANICS MODELS

The models described above have been widely applied to many contact situations, but it is important to understand all of the assumptions contained therein. The Hertz, JKR, and DMT models all assume that the materials are:

- *Homogeneous*—material properties uniform throughout space
- *Isotropic*—material properties uniform in all directions
- *Linear*—stress directly proportional to strain
- *Elastic*—all deformations are reversible (no plasticity or viscoelasticity)

The contact is also assumed to be between a sphere and a plane (or, equivalently, two spheres) where the contact radius is much smaller than the sphere radius (allowing it to be described mathematically as a paraboloid). Further, the loading is assumed to be purely in the normal direction—no tangential forces.

In applications, deviations from these assumptions can be significant, but, fortunately, solutions have been worked out for many such cases. Next, we discuss a few cases that we believe most likely to be of practical interest.

2.3.1 Tip Shape Differs from a Paraboloid

While in many cases approximating the tip shape as paraboloidal suffices to provide good estimates as argued by Greenwood [20], often the tip starts out nonparaboloidal or becomes so due to either excessive deformation at high loads or from wear from scanning over surfaces.

For the situation where a high load is combined with relatively low stiffness of the contacting materials, the effective radius of curvature can elastically distort away from a paraboloidal shape, leading to the contact radius being no longer small compared with the tip radius R. This situation frequently occurs with compliant polymers. These situations have been previously analyzed for nonadhesive contacts as summarized by Schwarzer [21]. The situation is more complex when adhesion is involved and has been discussed by Maugis [22] for the JKR case and by Lin and Chen [23] who took finite elasticity (required for large strains) into account.

Caution is required in the limit where the radius of the sphere is no longer much greater than z_0. Since z_0 should typically be less than 0.5–1 nm, this concern only arises for specialized AFM tips smaller than 5–10 nm. However, in this limit, one should question whether the corrections derivable from continuum mechanics approaches will be overwhelmed by atomistic effects due to the breakdown of continuum mechanics. This is discussed further in Section 2.6.

2.3.2 Flattened Tip Shapes

Even though AFM tips often start off very sharp with radii of curvature of a few nanometers or tens of nanometers, this sharpness, as discussed previously, results in high contact stresses, and these high stresses lead to wear of the tip surface. This wear reduces the tip sharpness until the contact stresses are below the threshold for plastic deformation and adhesive wear.

In cases of extreme wear, the tip develops a flat end and is better modeled as a flat punch pushed against the sample surface. Analytical solutions for the deformation and the stress field as a function of load for a flat-ended rigid circular punch of radius a indenting a flat elastic substrate date back to 1946 with the work of Sneddon [24]. The contact area is, by definition, constant at πa^2. The load F_{applied} and deformation δ are related by

$$\delta = (1-\nu_{\text{substrate}}^2)\frac{F_{\text{applied}}}{2aE_{\text{substrate}}} \tag{2.20}$$

Thus, the deformation increases linearly with F, instead of $F^{2/3}$ as it does in the Hertz case. The normal stress at the contact interface is given by

$$p(r) = p_{\min}\left(1-\frac{r^2}{a^2}\right)^{-1/2} \tag{2.21}$$

where the minimum contact pressure at $r=0$ is

$$p_{\min} = \frac{F}{2\pi a^2} = \frac{1}{2}p_{\text{mean}}$$

the pressure diverges at the contact edge ($r=a$) due to the sharp edge of the punch. A comparison between this stress profile and the Hertzian case is shown in Figure 2.6. This illustrates a general point regarding tips with increasingly flatter profiles: The contact stress profile evolves from being largest in the center to being largest at the edge.

The solutions for flat punches with adhesion have also been derived analytically [25–27]. The key practical point is that a flat punch profile creates a stress concentration at the sharp edge of the punch, which can then be a site for the initiation of plasticity, fracture, or wear.

2.3.3 Axisymmetric Power Law Tip Shapes

Various imaging techniques have been used to demonstrate that typical AFM tips start out well-approximated by a paraboloid and, after significant wear, can sometimes be described as a flat punch. As the tip is being blunted, it will lie somewhere between these two extremes and may be well described by an axisymmetric power-law-shaped tip given by [28]

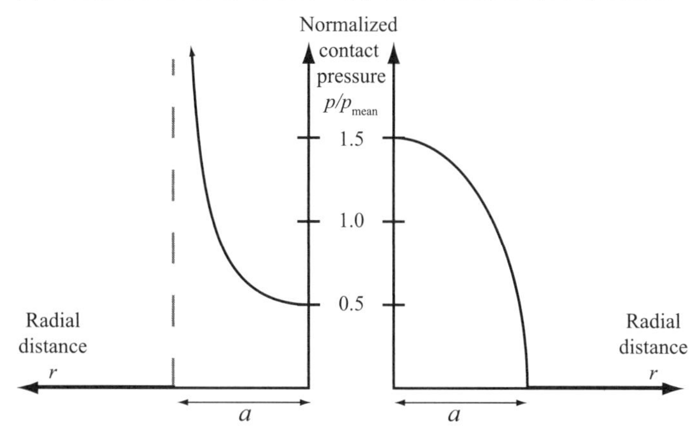

Figure 2.6 Compressive stress at the surface of the contact $p(r)$ normalized to the mean stress, for Sneddon's solution for a flat circular punch (a) and for Hertz's solution for a paraboloid (b).

$$z(r) = \frac{r^n}{nQ} \tag{2.22}$$

where the geometry is defined by two constants: n and Q. Contacting bodies with power law shapes have been examined using a JKR-like model [29, 30] and more recently using a Maugis–Dugdale type of model [31]. Zheng and Yu [31] provide a model (referred to as the M-D-n model) that addresses the full range of adhesive behavior from DMT-like to JKR-like adhesion behavior for power-law-shaped tips. This model can be used to examine contact stresses as well as the relationship between pull-off force and the adhesion properties at the interface. Tips with various power law shapes are shown in Figure 2.7(a). While power-law-shaped tips can be analyzed in a similar framework to the paraboloidal tips discussed earlier, the behavior is more complex. Notably, when $n > 2$, the pull-off force is a strong function of the adhesion range. This point is illustrated through the pull-off force for a rigid power-law-shaped tip that is given as [28]

$$F_{\text{adhesive, rigid}} = \frac{W_{\text{adh}}}{h} \pi (nhQ)^{\frac{2}{n}} \tag{2.23}$$

where a square potential (called a Dugdale potential, as described in Ref. [32] and elsewhere) is assumed at the interface and h is the range over which adhesion acts. The analysis uses a simple square potential, thus the length scale of the adhesion is described in terms of h rather than z_0. If the peak adhesive stress and the work of adhesion are matched between the square and Lennard-Jones potentials, then $z_0 = h/0.974$. When $n = 2$, Eq. (2.23) recovers the DMT pull-off force from Eq. (2.14); however, when $n > 2$, the pull-off force depends not only on the work of adhesion but also on the adhesion range.

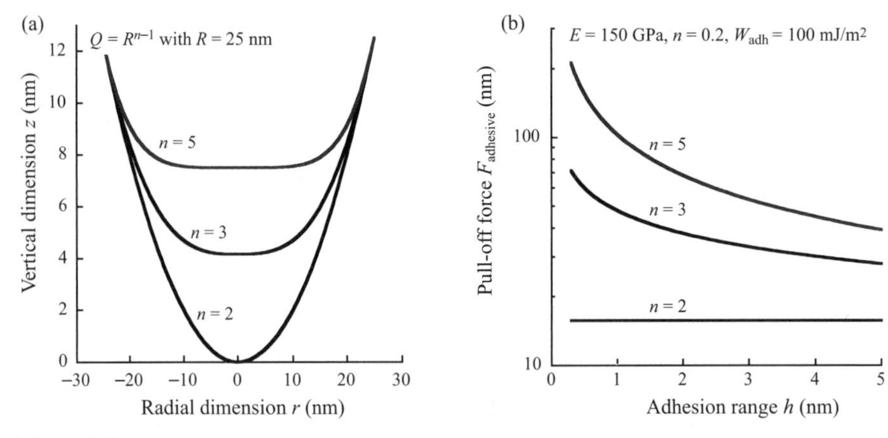

Figure 2.7 For tips with various power-law shapes (a), pull-off forces (b) are shown as function of adhesion range. The $n=2$ tip has a slight dependence on adhesion range (not visible because of plot scale) because of the transition from DMT to JKR behavior, while $n=3$ and $n=5$ tips have a strong dependence on adhesion range. Figure adapted with permission from Ref. [28].

This fact makes it hard to extract work of adhesion values from pull-off force measurements for power-law-shaped tips since the adhesion range is usually not known. Figure 2.7(b) shows the pull-off force versus adhesion range h for tips with various power indices. Varying the adhesion range is similar to changing the Tabor parameter, with small values of h leading to JKR-type behavior of power-law-shaped contacts and large values of h resulting in DMT-type behavior. The analysis in Zheng and Yu [31] also permits calculation of stresses in the contact.

2.3.4 Anisotropic Elasticity, Viscoelastic, and Plastic Effects

Beyond the tip shape itself, the contact behavior will be strongly affected by the mechanical properties of the materials in contact. All models discussed above assume that the materials are homogenous, isotropic, and linear elastic. This may not be the case in a range of practical applications.

The analyses discussed to this point have focused on cases in which the tip and substrate have isotropic elastic properties. Many single crystalline materials, for example, silicon, are used in AFM experiments as tips and substrates and have *anisotropic* elastic properties, that is, moduli that vary with direction. For example, for silicon, the effective Young's modulus varies from 130 to 169 GPa, and the effective Poisson's ratio varies from 0.06 to 0.28 across the different directions in the (100) plane [33]. Anisotropic elastic properties can be accounted for in the above contact and adhesion mechanics analyses. The contact of anisotropic materials can be quite complex but has been analyzed previously for generally anisotropic [34] and orthotropic [35] solids. Fortunately, most systems of interest are not fully anisotropic but rather have some degree of symmetry. A useful case, which has a rather straightforward solution, is that of transversely isotropic materials (five independent elastic constants) [36]. Many thin-film systems are well represented by transversely isotropic elasticity models, and the solution for systems in which there is cubic symmetry (three independent elastic constants, e.g., silicon) can be obtained from the transversely isotropic case. The detailed equations of the transversely isotropic case are beyond the scope of this chapter, but the analysis in Ref. [36] can be used to examine many practical cases.

Viscoelasticity refers to the time dependence of the mechanical response of materials [37]. Viscoelastic materials exhibit a time lag between the applied stress and the resulting strain. Polymers, biological materials, and even soft metals can exhibit appreciable viscoelasticity. There are several discussions of viscoelastic effects in contacts [38–43], which have been reviewed recently by Shull [44]. As one example, Wahl et al. observed consequences of viscoelasticity in an AFM of poly(vinylethylene) (PVE) [38]. The contact stiffness varies as a function of time as the load was increased and then decreased. The maximum contact stiffness [and, by implication, through Eq. (2.5), the contact area] occurred *after* the load had reached is maximum value. This allowed the relaxation time of the polymer to be determined. Modeling of this phenomena by Greenwood and Johnson indicated that viscoelastic effects can alter the apparent adhesion and stiffness of contacts [45].

Since time-dependent effects can arise from instrument artifacts, care must be taken to decouple these from the actual time dependence of the samples' response. This was highlighted by Tranchida et al. [46], who, after accounting for the instrumentation in effects, were able to measure the frequency-dependent Young's modulus of rubbery polymers, with quantitative agreement found between measurements conducted on two different AFM instruments.

The variation in the viscous response of materials is both of interest for measurement, but also of use for providing contrast to identify different phases of heterogeneous materials. Recent examples include the quantification and mapping of local storage and loss moduli of polymer blends using a high-frequency resonance technique [47, 48]. The Hertz model was used to extract quantitative values for the moduli since the applied forces were in most cases much larger than the adhesive forces. The moduli extracted were consistent with values expected by applying time–temperature superposition to dynamic mechanical analysis measurements of the same materials obtained at lower frequencies, with the exception of one phase of one of the samples, which had high adhesion [48]. The discrepancy could be explained by the lack of including adhesion in the analysis.

Finally, materials may exhibit *plastic deformation* if the stresses in the contact are sufficiently large to cause the material to yield. In a Hertzian contact, the stress field is triaxial [i.e., the axial stress, radial stress, and circumferential stress defined in Eqs. (2.8) and (2.9) are all nonzero in most locations], and thus it is important to not simply compare the maximum contact pressure to the yield strength of the material. Rather, an equivalent von Mises stress or maximum shear stress [3] that accounts for all stress components should be compared to the yield strength to assess plastic deformation as noted previously in this chapter. It is important to realize though that macroscopic values of yield strength are not necessarily meaningful at the nanoscale, as it is well known that strength values are size-dependent. Analysis of the contact after plastic deformation occurs is challenging as the contact radii typically become large relative to the size of the tip, and the exact geometry of the tip is often not known. Plastic deformation should occur first in the softer material in the contact pair. For example, a polymer film will typically yield before a harder silicon-based tip. If plastic deformation is limited to the sample, the problem is essentially an indentation problem and the large body of analysis developed for understanding micro- and nanoindentation measurements can be applied (e.g., Refs. [49, 50]) in many cases.

2.4 THIN FILMS

Many samples of interest for AFM studies are thin films. To ensure that the stress field largely involves the film and not the substrate, a rule of thumb is often used in nanoindentation hardness experiments [51] stating that the substrate can be ignored if the film thickness is more than 10 times the contact deformation δ, though this rule has limits when extracting hardness values [51]. (See Chapter 7.)

Consider the examples in Section 2.1.4, which assume the material is semi-infinite. For the diamond AFM tip contacting a DLC film, δ was only ~0.18 nm at a load of 100 nN. Thus, a DLC film needs to be only at least 1.8 nm thick to be treated as semi-infinite for experiments conducted at loads up to 100 nN. For contacting a polymer at 100 nN, δ was 12 nm (although we were exceeding the bounds for the Hertz model), so the rule of thumb would indicate that we need a film thickness greater than 120 nm to avoid substrate effects for experiments conducted at loads up to 100 nN. At a 1-nN load (within the Hertz model's bounds), the corresponding minimum thickness for the polymer film is 5.6 nm.

Since these calculations for δ assume contact with a semi-infinite medium, they will overestimate δ if the substrate is stiffer than the film. Including adhesion, as discussed in Section 2.2, will make δ larger. Furthermore, since the size of the elastic stress field scales with the contact radius and not δ, some have suggested that the rule of thumb should instead state that the *contact radius, a*, should be less than 10% of the film thickness to treat the film as a bulk entity [52]. This would lead to minimum film thicknesses of 3.0 and 250 nm, respectively, for the DLC and polyethylene films when contacted by the AFM tip at 100-nN load (ignoring adhesion) instead of the 1.8- and 120-nm film thicknesses, respectively, for DLC and PE film measurements calculated previously via the δ rule of thumb.

The large difference between DLC and PE samples illustrates the significant effect that the elastic properties of films have on the thickness required to ignore the substrate and show that, for compliant materials or highly adhesive materials, one should not treat materials with molecular-scale thicknesses as bulk entities. Rather, for films of sub-100-nm thickness, it is advisable to use a contact mechanics model that takes the finite film thickness into account. There are several treatments available: We focus on a few where the contact radius and elastic indentation depth are considered as a function of applied load.

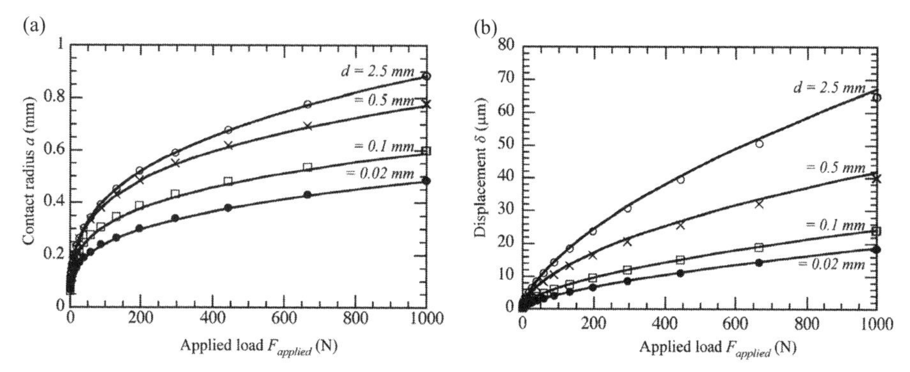

Figure 2.8 Hsueh and Miranda's [53] model (solid line) compared with finite-element simulations (symbols). Contact radius a [shown in (a)] and indenter displacement δ [shown in (b)] as a function of load F_{applied}, assuming $R_{\text{tip}} = 10\,\text{mm}$, $\nu_{\text{coating}} = \nu_{\text{substrate}} = 0.25$, $E_{\text{coating}} = 10\,\text{GPa}$, $E_{\text{substrate}} = 100\,\text{GPa}$, for four different coating thicknesses d. As the coating gets thin compared to a, the substrate's influence is clearly seen. Figures reproduced with permission from Ref. [53].

Hsueh and Miranda's [53] combined empirical-analytical method leads to an analytical equation that relates the actual contact radius to the Hertzian prediction as a function of the relative Young's moduli of the coating and substrate, and also as a function of the ratio of the coating thickness d to the contact radius (Eq. 13 in Ref. [53]). Thus, the equation requires iteration to solve for the contact radius; carrying out the iteration, good agreement is found with finite-element analysis, demonstrating the reliability of the technique. The contact radius a and indenter elastic displacement δ obtained from Hsueh and Miranda's method are typically plotted versus load for various ratios of coating to substrate modulus and values of d. Specific examples are shown in Figure 2.8. The solution assumes that the coating and substrate have identical Poisson's ratios of 0.25, that the coating and the substrate remain rigidly bonded during loading (no decohesion or slip at the film–substrate interface), and all other assumptions of the Hertz model are used.

The test values chosen for R_{tip}, F_{applied}, and d in Figure 2.8 correspond to macroscopic parameters; applying these results to any AFM experiments would require performing the iterative analysis numerically. A more convenient approach, referred to as "thin coating contact mechanics" (TCCM) has been worked out by Reedy [54, 55]. That work considers both nonadhesive (Hertz-like) and adhesive (JKR-like, DMT-like, and intermediate) cases, but with the limitation that the tip and sample are both composed of *rigid* substrates with elastic coatings on them. Using simplifying assumptions, Reedy derives a relation between contact radius and load for a nonadhesive contact, assuming a coating on both the tip and substrate (from Eqs. 4 and 5 in Ref. [54]):

$$a = \left(\frac{4R_{\text{eff}} h_{\text{tot}} F_{\text{applied}}}{\pi E_u} \right)^{1/4} \tag{2.24}$$

where a, R_{eff}, and $F_{applied}$ are all defined as in Section 2.1. This is related to the Hertz analysis (Section 2.1); however, the effective modulus E^* is replaced by

$$E_u = h_{tot}\left[\frac{h_1}{E_{u1}} + \frac{h_2}{E_{u2}}\right]^{-1}$$

where E_{u1} is the uniaxial strain modulus for the tip (material 1), defined as

$$E_{u1} = \frac{(1-\nu_1)E_1}{(1+\nu_1)(1-2\nu_1)}$$

and E_{u2} is defined equivalently for the substrate (material 2). The thickness of each coating is h_1 and h_2, giving a total thickness of $h_{tot} = h_1 + h_2$. All of the other assumptions of the Hertz model are used (frictionless contact, $a << R_{eff}$, etc.), and as well it is assumed that $h_{tot} << R_{eff}$. If either the tip or substrate lacks a coating, the respective value of h is simply set to zero. The equation assumes that strain is uniaxial and uniform throughout the coatings, which is reasonable if Poisson's ratio is $\leq \sim 0.4$. From Eq. (2.24), one sees that the contact radius varies as the load to the $\frac{1}{4}$ power, not the $\frac{1}{3}$ power as in the Hertz model.

When h_{tot}/R_{eff} and a/R_{eff} are not infinitesimal (but still relatively small), then an extended form of Eq. (2.24) is used (Eq. (2.6) in Ref. [54], presented here in an alternate form):

$$a^2 = \frac{2cR_{eff}h_{tot}}{\sqrt{\pi}}\left(\frac{F_{applied}}{E_u R_{eff} h_{tot}}\right)^d \tag{2.25}$$

where c and d are nondimensional constants that depend on ν and h_{tot}/R_{eff} (this exponent d used in Ref. [54] should not to be confused with the film thickness d used in Fig. 2.8 and in Ref. [53]). Values for c and d are determined by fitting Eq. (2.25) to results from finite element analysis, and several values are presented in Table I in Ref. [54].

Reedy also considers adhesion, and simple analytical equations are provided for JKR and DMT limits [54] as well as for the full JKR–DMT transition [55]. Interestingly, in the JKR limit, the pull-off force is found to be equal to $2\pi R_{eff} W_{adh}$, instead of $1.5\pi R_{eff} W_{adh}$ from the JKR solution for uniform, uncoated, materials.

An example of the variation of contact area with load is shown in Figure 2.9. Here, two rigid spheres are in contact with $R_1 = 200$ nm, $R_2 = 1000$ nm, $W_{adh} = 0.2$ J/m^2, and where both spheres have a 2-nm-thick coating with $E = 4$ GPa, $\nu = 0.4$ (possibly resembling a self-assembled monolayer). Reedy's model uses a "critical separation distance" δ_c which is nearly equivalent to the adhesion range z_0. In this example, it is assumed that $\delta_c = 1$ nm. The nondimensional contact area $\bar{A}$ has been normalized by $R_{eff}h_{tot}$ and the nondimensional load $\bar{P}$ has been normalized by $E_u R_{eff} h_{tot}$. Figure 2.9 shows the TCCM result (solid line), as well as the corresponding JKR-like and DMT-like results (corresponding to alternate cases of small and large values of δ_c, respectively). Finite-element analysis (triangles) results match the TCCM result very

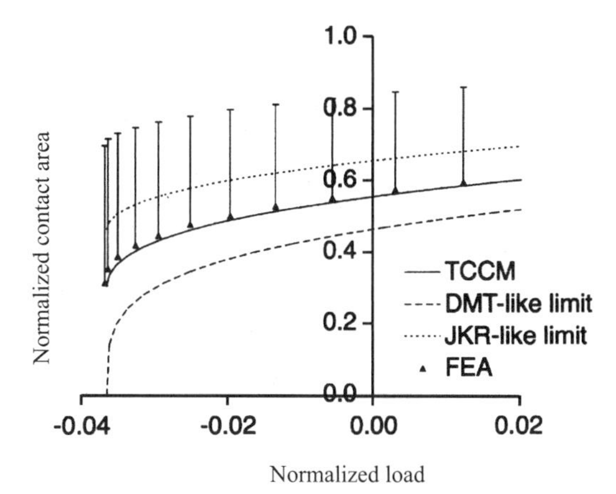

Figure 2.9 Comparison of TCCM theory that includes the transition between DMT-like and JKR-like response with finite-element results (symbols denote the contact radius while the bars indicate the length of the region where adhesion forces act across an open gap). Figure reproduced with permission from Ref. [55].

closely (the vertical bars represent the extent of the region where adhesion forces act, that is, where the gap is less than δ_c but greater than the equilibrium separation). The lack of dependence of the pull-off force on the JKR–DMT transition is readily seen. As with the regular JKR and DMT models, the contact area is larger in the JKR limit. Recently, a molecular dynamics simulation of thin self-assembled monolayer was shown to agree closely with the predictions of TCCM, in contrast to JKR or DMT predictions (i.e., models which ignore the thin-film nature of the sample) [56].

Contact between a rigid sphere and an adhesive elastic foundation has also been modeled, where analytical equations are derived from energy principles [57]. An elastic foundation is equivalent to a layer of laterally uncoupled elastic elements, like a bed of springs. The independence of the surface elements means no long-range strains assist in pulling the surfaces apart during unloading. Correspondingly, the pull-off force is found to be $7\pi RW_{adh}$, substantially larger than the JKR or DMT cases. This approach is appropriate for very thin layers (relative to the radius of the contact) of soft materials (polymers, gels, cells, and self-assembled monolayers) at high applied strains. The authors note how the independent elastic elements are perhaps analogous to the setae on the feet of geckos, a system that exhibits strongly enhanced adhesion.

In summary, recent work has provided analytical equations for both nonadhesive and adhesive coatings on rigid substrates, and other modeling work provides non-adhesive results for coatings on elastic substrates.

2.5 TANGENTIAL FORCES

Most of the above discussion has involved only loads acting perpendicular to the plane of contact. Often there will be additional loads that act parallel to the plane of the contact. In AFM specifically, the tip is often sliding along the surface such that a frictional force opposes the motion. Even if the tip is not sliding, there can be an additional component of in-plane or "tangential" loading (perhaps due to the tilt of the cantilever) in addition to the perpendicular loading force.

2.5.1 Three Possible Cases for a Tangentially Loaded Contact

Fully Slipped (i.e., Sliding) Contact

One common treatment of this problem assumes that Amontons' law of friction applies locally at all points in a sliding contact. More specifically, if we assume the sliding axis x is perpendicular to the loading axis z, then we can divide Amontons' law ($F_x = \mu F_z$, where μ is the experimentally measured coefficient of friction) by the contact area to yield the mean shear traction q_{mean} on the surface as a function of mean surface pressure p_{mean}:

$$q_{\text{mean}} = \mu p_{\text{mean}} \tag{2.26}$$

and it is assumed that this arises due to an averaging of the *local* stresses at all points within the contact. Thus, it has been presumed that

$$q(r) = \mu p(r) \tag{2.27}$$

everywhere in the contact. Therefore, the shear stress varies with position in proportion to the surface normal stress [cf. $p(r)$ in Eq. (2.7)].

Fully Unslipped (i.e., Static) Contact—Theoretical Picture

Since the mathematics of contact mechanics apply identically for normal stresses/deformations and for shear stresses/deformations, we do not need to derive additional equations to describe a tangentially loaded static contact. The boundary condition in this case is that all points within the radius of contact are displaced by the same amount in the lateral direction due to the tangential load. This is mathematically equivalent to the problem of a rigid flat punch normally indenting a surface (discussed in Section 2.3.2). Thus, the same solution (the Sneddon solution) is applied, and the shear traction q (analogous to contact pressure p [cf. $p(r)$ in Eq. (2.21)]) is described by [58]

$$q(r) = q_{\text{min}} \left(1 - \frac{r^2}{a^2}\right)^{-1/2} \tag{2.28}$$

where $q_{\text{min}} = F_x / (2\pi a^2)$. A schematic is shown in Figure 2.10(a) of two contacting spheres with normal force F_z and lateral force F_x applied with the total lateral displacement of the spheres δ. Note that this schematic also shows the effect of partial

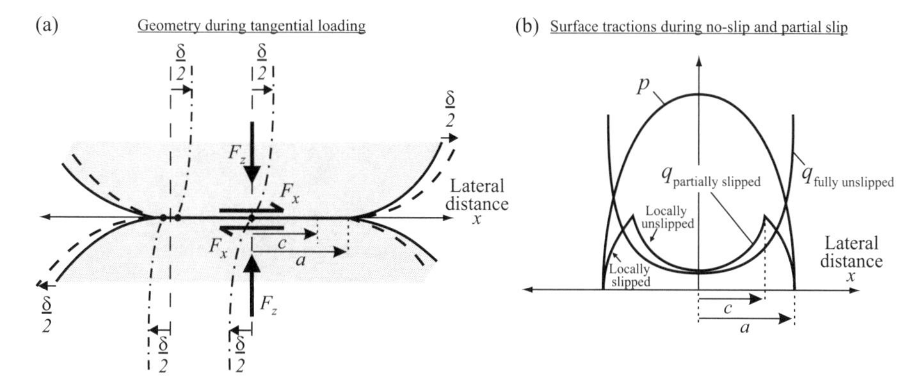

Figure 2.10 (a) Two contacting spheres are shown with a normal force F_z and lateral force F_x applied in a partially slipped condition. In gray, the position of the spheres is shown prior to the application of the lateral force; the dashed lines represent the position after. The lateral displacement of each sphere far from the contact is $\delta/2$ (total displacement δ is shared evenly for identical spheres). For points within the inner circle ($r < c$), where slip has not occurred, the relative displacements of the spheres at the surface is zero; for points in the slipped region ($c < r < a$), surface points that started out coincident (black dots) have now shifted relative to one another. Figure reproduced with permission from Ref. [3]. (b) Lateral stresses q on the contacting surface are shown for the fully unslipped case (which diverges as $r \to a$) and partially slipped case (which has a cusp at $r = c$ and goes to zero as $r \to a$), along with the normal stress p.

slip, which is discussed below. The shear stress diverges at the outermost edges of the contact, as shown in the plot of lateral stresses on the contact surface in Figure 2.10(b). In reality, this cannot occur; above some threshold, the contact cannot sustain the stress and the surfaces will locally slip, starting in the outer annulus of the contact zone, while the interior remains unslipped. The result is a *partially slipped contact.*

Partially Slipped Contact

The partially slipped contact is described by a combination of the above solutions, where an inner circle of contact is unslipped and an outer annulus has slipped. The lateral stress in the contact is given by [59]

$$q(r) = \mu p_0 \left[\left(1 - \frac{r^2}{a^2}\right)^{1/2} - \frac{c}{a}\left(1 - \frac{r^2}{c^2}\right)^{1/2} \right] \quad 0 < r < c \tag{2.29}$$

and

$$q(r) = \mu p_0 \left[\left(1 - \frac{r^2}{a^2}\right)^{1/2} \right] \quad c < r < a \tag{2.30}$$

The value c (the radius of the unslipped circle) is given by

$$\frac{c}{a}=\left(1-\frac{F_x}{\mu F_z}\right)^{1/3} \tag{2.31}$$

where F_x is the tangential force and F_z is the total normal force. A partially slipped contact is shown in Figure 2.10(a). For $0<r<c$, the dot-dashed contour, representing the lateral displacement in the materials, is continuous across the interface. However, for $c<r<a$, there is a discontinuity in the lateral displacement at the interface that is equal to the slip that has occurred, as indicated within the left half of the contact.

An alternate approach proposed by Johnson [60] assumes that the shear stress in the slipping region is a constant τ_0, rather than being linearly proportional to the normal stress as expressed by Eq. (2.27). This approach results in a modified form of Eq. (2.29) involving τ_0; full slip occurs when $F_x = \tau_0 A$. This is in agreement with AFM results discussed below. The effect of adhesion, from the JKR to the DMT limit, is also included. This work, which is justified in great detail in a complex analytical mechanics treatment [61], represents the most comprehensive continuum mechanics analysis of the problem of the contact mechanics of a sliding adhesive asperity.

2.5.2 Active Debate over the Behavior of the Shear Stress

While the above picture applies to macroscale contacts, how well it should apply at the micro- and nanoscales remains an open debate. In particular, the proper dependence of shear stress on normal stress during sliding has not been resolved.

One common approach to making the shear stress expression more general is to take the linear combination of the two approaches discussed above [62], that is, combine Eq. (2.27) with a constant:

$$q(r)=q_0+\alpha p(r) \tag{2.32}$$

where q_0 and α are constants [α effectively corresponds to μ in Eq. (2.27)]. Note that in most texts the nanoscale shear stress is referred to as τ; we chose to use q here to be consistent with Ref. [3] and with the previous sections of this chapter. Equation (2.32) can be integrated to yield the average behavior of the entire nanoscale contact:

$$q_{\text{mean}}=q_0'+\alpha' p_{\text{mean}} \tag{2.33}$$

where q_0' and α' are also constants. Note that this differs from the assumptions of continuum mechanics [i.e., Eqs. (2.26) and (2.27)] in two ways: if q_0' is nonzero, the lateral stress (and thus lateral force F_x) can be nonzero even at zero normal stress ($F_z = F_{\text{adhesive}} + F_{\text{applied}} = 0$); and the coefficient of proportionality may have a very different relation to the macroscopic, empirical friction coefficient μ than what is implied in Eq. (2.27).

There have been several experimental and simulation studies investigating the dependence of lateral force (or stress) on normal force (or stress). For example, Briscoe and Evans [63] used a surface force apparatus to measure friction and contact area between mica sheets coated with organic monolayers, and the results were consistent with Eq. (2.33), with both q_0' and α' being larger than zero. In AFM experiments, the contact area is not directly measured and only the total forces on the contact are measurable: F_x and F_z; thus, q_{mean} and p_{mean} can be estimated, but the local behavior [$q(r)$ and $p(r)$] must be inferred. Several AFM results, reviewed elsewhere [64, 65], have observed friction forces varying with load in a way that is consistent with Eq. (2.33) only if $\alpha' = 0$. This implies that the local shear stress is constant [$q(r) = q_0$] everywhere during sliding and is independent of the local normal stress $p(r)$. The notion of a pressure-independent shear stress is corroborated by molecular dynamics simulations of a single nanoscale asperity sliding on a flat surface in which the friction force is also demonstrated to be directly proportional to area [66, 67]. While the results in Ref. [60] provide a useful guide, further experiments and simulations are required to truly understand the local behavior within a contact, particularly since further effects in the limit of nanoscale contacts may occur, as discussed in Section 2.6.

2.5.3 Lateral Stiffness

As described in Section 2.1.2, the normal contact stiffness is defined as the derivative of force with respect to displacement. The lateral contact stiffness is also a useful property of the contact that can be measured in the AFM. According to continuum mechanics [59], the lateral stiffness is given by

$$k_{\text{lateral}} = 8G^* a \tag{2.34}$$

where

$$G^* = \left(\frac{2-\nu_1}{G_1} + \frac{2-\nu_2}{G_2} \right)^{-1} \tag{2.35}$$

where G_1 and G_2 are the tip and sample shear moduli, respectively. Thus, k_{lateral} is directly proportional to the contact radius. Equation (2.34) is a powerful relation because it is extremely general. It is valid regardless of the tip–sample interaction forces, unlike the analogous equation for normal stiffness [Eq. (2.5)], which must be modified for adhesive contacts [68]. If the value of G^* is reliably known, then a measurement of the lateral stiffness enables a to be measured [69, 70]. This in turn means that the relationship between friction force and contact area can be determined [69–74].

2.6 APPLICATION OF CONTINUUM MECHANICS TO NANOSCALE CONTACTS

2.6.1 Unique Considerations of Nanoscale Contacts

The previous sections have discussed continuum mechanics and its predictions in simple and more complex situations. The application of these predictions at the nanoscale is an active area of study in the AFM, nanomechanics, and nanotribology research communities. As mentioned, all of the models rely on a large number of assumptions, the most fundamental of which is that the solids in contact are comprised of a continuous material with smooth surfaces and uniform properties. Of course, materials are actually comprised of discrete atoms with characteristic (and often complex) interactions between them. For millimeter-sized objects, the length scale is at least seven orders of magnitude larger than the length scale of the atoms, and thus atomic-scale variations are smeared out (especially for noncrystalline solids). However, as the size of the bodies reduces to the nanometer scale, the atomic detail—both on the surface of the body and in its interior—begins to matter. Thus, there is good reason to think critically about applying continuum models to nanoscale contacts.

There are good reasons, however, to be assured that the concepts described in the previous sections of this chapter can be usefully applied at the nanoscale.

2.6.2 Evidence of Applicability of Continuum Contact Mechanics at the Nanoscale

While many of the continuum assumptions behind analysis of contact mechanics are likely to break down at the nanoscale, the deviations away from the expected continuum behavior are usually not so severe, and the models developed for the macroscale continue to produce physically reasonable results. While many studies have demonstrated the utility of adapting continuum mechanics to the nanoscale, we will demonstrate the effectiveness of these approaches by discussing four examples from the literature.

Measurement of Mechanical Properties of Metal Oxides Using AFM

Researchers used AFM-based nanoindentation on three different single-crystal metals to investigate the elastic modulus on a local level and as a function of depth of indentation [75]. Hertz mechanics was used to analyze results and extract relevant parameters. Specifically, they were able to show: an ultrasoft topmost region corresponding to hydrocarbon surface contamination; a much stiffer region slightly deeper corresponding to the native oxide layer; and finally an intermediate modulus at a larger depth, corresponding to the Young's modulus of the metal. The physical reasonable values extracted demonstrate that the Hertz model can be effectively applied to analyze load and displacement data down to the nanometer scale.

Friction of a Platinum Tip on Mica Demonstrated to Obey JKR Predictions

In an AFM investigation sliding an initially paraboloidal platinum-coated AFM tip on a mica substrate in vacuum showed a sublinear dependence of friction force on normal load [29]. The dependence was analyzed by assuming that the friction force under those conditions was directly proportional to contact area. Initially, the JKR model was used to calculate the predicted contact area at every tested value of normal load, and it was shown that friction force obeyed an identical trend. A repeated test with an intentionally blunted tip showed excellent agreement with a JKR model modified for flat tips. However, the Tabor parameter was only approximately 0.7, far away from the JKR regime. The aforementioned approach of Johnson [60], where partial slip is taken into account, could resolve this paradox; although not in the JKR regime, the contact area over which friction acts closely follows the JKR curve when partial slip is taken into account. Thus, Johnson's model [60] and the experiments agree very well.

In Situ Measurement of Contact Radius Using Repeated Pull-off Tests

Intermittent pull-off tests have been used to detect changes in tip shape over very long sliding distances [76]. For probe-based data storage applications, tip durability is critical for long-term reliability of the device; thus, accurate characterization of wear over long sliding distances is needed. In this and other related studies [77–79], repeated pull-off tests taken frequently provide a real-time measure of whether and how the sliding tip is changing over time. According to JKR and DMT predictions, the pull-off force is directly proportional to tip radius and to the work of adhesion (i.e., materials in contact). Therefore, this provides a quick and useful metric to verify that the tip is staying constant over time. In contrast, a steady increase in pull-off force is an indication of increasing tip radius due to wear of the probe tip, and a volatile (or even decreasing) pull-off force is an indication of a change in work of adhesion due to tip fouling by contamination or surface material. Once again this demonstrates that, in certain situations, trends identified by contact mechanics can be directly and usefully applied to nanoscale AFM contacts.

Observation of Atom-by-Atom Wear Governed by Stress-Mediated Chemical Reactions

Several investigations of nanoscale wear have used aspects of continuum contact mechanics to explain results [77, 79–81]. In Refs. [80, 81], the rate of wear is directly compared to the stress in the contact as predicted using continuum contact mechanics. In Refs. [77, 79], the rate of wear itself is inferred from tip radius measurements calculated using pull-off tests; a result arising directly from JKR and DMT predictions. The results produced physically reasonable values. It should be noted that authors differ about precisely which stress component is the most relevant to wear and how to analyze results for the various conditions under which their experiments were performed.

As a final note, while the atomistic computer simulation work that has been done does show differences in behavior from continuum predictions, these differences are not by orders of magnitude. Even the well-known paper entitled "The Breakdown of Contact Mechanics at the Nanoscale" predicts deviations from contact mechanics on the order of 10% for contact deformation and roughly 20–25% for contact radius [67]. While these differences are appreciable, they nevertheless verify that continuum mechanics provides at least the correct range of values. Other quantities, such as stress and lateral stiffness, do differ markedly from continuum predictions in some cases. As experiments improve in resolution and incorporate in situ methods to directly observe contact phenomena, as simulations become more accurate and approach the typically larger size of AFM experiments, and as theory and analysis methods improve, more convergence between all three will result. While many tools for analyzing contact mechanics for AFM exist, it is likely that very soon new levels of understanding and instrumentation will emerge that establish truly atomistic contact mechanics analysis.

ACKNOWLEDGMENTS

The authors would like to thank Ms. Desirae Johnson for assistance with obtaining permissions to reproduce figures. The authors acknowledge the support of the U.S. National Science Foundation, through grants DGE-0221664 (TDBJ) and CMMI-1200019 (RWC and KTT).

APPENDIX 2A SURFACE ENERGY AND WORK OF ADHESION

Different textbooks and scientific publications use different, and often conflicting, nomenclature to denote the surface energy and the work of adhesion of solid materials. In the present text, we adhere to the nomenclature used in Israelachvili's book [6]. First, note that all of these quantities are discussed here, and in most other publications, as energies per unit area of the surface or interface, even though they are called "work" or "energy." Hence, units are always energy per unit area, for example, J/m^2. Specifically, γ_{AB} is the "interfacial energy" of material A in contact with material B. This denotes the additional free energy required to increase the interface between A and B by one unit of area. The related quantity γ_A is the "surface energy" of material A and can be thought of as the interfacial energy of A in contact with vacuum. (Note that, for most materials, the value does not change significantly when in contact with air or other vapors.)

In contrast, the work of adhesion, denoted $W_{adh,A-B}$ or more commonly just W_{adh}, is the energy per unit area required to separate two materials from equilibrium contact to infinite separation (or large enough such that they no longer interact). In the initial state, material A and material B are in contact (with contact area A), thus the total energy of the initial state is $U_{initial}$. In the final state, materials A and B are each in contact with vacuum, and the total energy is U_{final}. If we assume that the path from the initial state to the final state was slow, quasi-static, and reversible, then the work required (or additional energy needed) per unit area is equal to the energy difference between the two states, that is,

$$W_{adh,\, A-B} = \frac{U_{final} - U_{initial}}{A} = (\gamma_A + \gamma_B) - (\gamma_{AB}) \tag{2A.1}$$

In the case where the two bodies being separated are composed of the same material (assuming identical surfaces, i.e., perfect, clean surfaces with the same surface orientation), then the above situation still holds, except that we replace material B with material A and replace all subscripts accordingly. Therefore,

$$W_{adh,\, A-A} = \frac{U_{final} - U_{initial}}{A} = (\gamma_A + \gamma_A) - (\gamma_{AA}) \tag{2A.2}$$

Since there is no energy penalty to create additional "interface" between a material and itself, then the quantity $\gamma_{AA} = 0$. Therefore, the work of adhesion of a material with itself is equal to twice the surface energy of that material:

$$W_{adh,A-A} = 2\gamma_A \tag{2A.3}$$

Since the two materials are identical, this is referred to as the "work of cohesion." It is assumed here that the structure of the contact interface is identical to the bulk structure; if the two materials have surface oxides or contaminants, or steps or other defects, or have different crystallographic orientations, then Eq. (2A.1) should be used instead. Values for the surface energy, interfacial energy, and work of adhesion of various materials and material pairs are given in Ref. [6].

REFERENCES

1. Hertz, H. J. *Reine Angew Math* **92** (1881): 156–171.
2. Timoshenko, S. P., and Goodier, J. N. *Theory of Elasticity*, McGraw-Hill: New York, 1987.

3. Johnson, K. L. *Contact Mechanics*, University Press: Cambridge, 1987.
4. Hamilton, G. M., and Goodman, L. E. *Journal of Applied Mechanics* **33** (1966): 371–376.
5. Lawn, B. R. *Fracture of Brittle Solids*, 2nd ed., Cambridge University Press: New York, 1993.
6. Israelachvili, J. N. *Intermolecular and Surface Forces*, 2nd ed., Academic: London, 1992.
7. Muller, V. M., Yushchenko, V. S., and Derjaguin, B. V. *Journal of Colloid and Interface Science* **77** (1980): 91–101.
8. Girifalco, L. A., and Good, R. J. *Journal of Physical Chemistry* **61** (1957): 904–909.
9. Garcia, R., and Perez, R. *Surface Science Reports* **47** (2002): 197–301.
10. Joyce, S. A., and Houston, J. E. *Review of Scientific Instruments* **62** (1991): 710–715.
11. Jarvis, S. P., Yamada, H., Yamamoto, S.-I., and Tokumoto, H. *Review of Scientific Instruments* **67** (1996): 2281–2285.
12. Ashby, P. D., Chen, L. W., and Lieber, C. M. *Journal of the American Chemical Society* **122** (2000): 9467–9472.
13. Mate, C. M. *Tribology on the Small Scale: A Bottom up Approach to Friction, Lubrication, and Wear*, Oxford University Press: Oxford, 2008.
14. Derjaguin, B. V., Muller, V. M., and Toporov, Y. P. J. *Colloid Interface Science* **53** (1975): 314–326.
15. Maugis, D. *Journal of Colloid and Interface Science* **150** (1992): 243–269.
16. Derjaguin, B. V. *Kolloid Zeits* **69** (1934): 15–164.
17. Johnson, K. L., Kendall, K., and Roberts, A. D. *Proc R Soc Lond A* **324** (1971): 301–313.
18. Tabor, D. *J Coll Interf Sci* **58** (1977): 2–13.
19. Carpick, R. W., Ogletree, D. F., and Salmeron, M. *Journal of Colloid and Interface Science* **211** (1999): 395–400.
20. Greenwood, J. A. *Philosophical Magazine* **89** (2009): 945–965.
21. Schwarzer, N. *Philosophical Magazine* **86** (2006): 5179–5197.
22. Maugis, D. *Langmuir* **11** (1995): 679–682.
23. Lin, Y.-Y., and Chen, H.-Y. *Journal of Polymer Science Part B–Polymer Physics* **44** (2006): 2912–2922.
24. Sneddon, I. N. *Proceedings of the Cambridge Philosophical Society* **42** (1946): 29–39.
25. Persson, B. N. J. *Wear* **254** (2003): 832–834.
26. Gao, H. J., and Yao, H. M. *Proceedings of the National Academy of Sciences of the United States of America* **101** (2004): 7851–7856.
27. Tang, T., and Hui, C. Y. *Journal of Polymer Science Part B-Polymer Physics* **43** (2005): 3628–3637.
28. Grierson, D. S., Liu, J., Carpick, R. W., and Turner, K. T. *Journal of the Mechanics and Physics of Solids* **61** (2012): 597–610.
29. Carpick, R. W., Agraït, N., Ogletree, D. F., and Salmeron, M. *Journal of Vacuum Science & Technology B* **14** (1996): 1289–1295.
30. Carpick, R. W., Agraït, N., Ogletree, D. F., and Salmeron, M. *Journal of Vacuum Science & Technology B* **14** (1996): 2772.
31. Zheng, Z., and Yu, J. *Journal of Colloid and Interface Science* **310** (2007): 27–34.
32. Grierson, D. S., Flater, E. E., and Carpick, R. W. *Journal of Adhesion Science and Technology* **19** (2005): 291–311.
33. Wortman, J. J., and Evans, R. A. *Journal of Applied Physics* **36** (1965): 153–156.
34. Willis, J. R. *Journal of the Mechanics and Physics of Solids* **14** (1966): 163–176.
35. Swanson, S. R. *International Journal of Solids and Structures* **41** (2004): 1945–1959.
36. Turner, J. R. *International Journal of Solids and Structures* **16** (1980): 409–419.
37. Lakes, R. S. *Viscoelastic Solids*, CRC Press: Boca Raton, FL, 1999.
38. Wahl, K. J., Stepnowski, S. V., and Unertl, W. N. *Tribology Letters* **5** (1998): 103–107.
39. Maugis, D., and Barquins, M. *Journal of Physics D (Applied Physics)* **11** (1978): 1989–2023.
40. Mangipudi, V. S., and Tirrell, M. *Rubber Chemistry and Technology* **71** (1998): 407–448.
41. Unertl, W. N. *Journal of Adhesion* **74** (2000): 195–226.
42. Giri, M., Bousfield, D. B., and Unertl, W. N. *Langmuir* **17** (2001): 2973–2981.

43. Haiat, G., Barthel, E., and Huy, M. C. P. *Journal of the Mechanics and Physics of Solids* **51** (2003): 69–99.
44. Shull, K. R. *Materials Science & Engineering R: Reports* **36** (2002): 1–45.
45. Greenwood, J. A., and Johnson, K. L. *Journal of Colloid and Interface Science* **296** (2006): 284–291.
46. Tranchida, D., Kiflie, Z., Acierno, S., and Piccarolo, S. *Measurement Science & Technology* **20** (2009): 095702.
47. Killgore, J. P., Yablon, D. G., Tsou, A. H., Gannepalli, A., Yuya, P. A., Turner, J. A., Proksch, R., and Hurley, D. C. *Langmuir* **27** (2011): 13983–13987.
48. Yablon, D. G., Gannepalli, A., Proksch, R., Killgore, J., Hurley, D. C., Grabowski, J., and Tsou, A. H. *Macromolecules* **45** (2012): 4363–4370.
49. Fischer–Cripps, A. C. *Surface & Coatings Technology* **200** (2006): 4153–4165.
50. Oliver, W. C., and Pharr, G. M. *Journal of Materials Research* **19** (2004): 3–20.
51. Chen, J., and Bull, S. J. *Vacuum* **83** (2009): 911–920.
52. Wang, M., Liechti, K. M., White, J. M., and Winter, R. M. *Journal of the Mechanics and Physics of Solids* **52** (2004): 2329–2354.
53. Hsueh, C. H., and Miranda, P. *Journal of Materials Research* **19** (2004): 2774–2781.
54. Reedy, E. D., Jr. *Journal of Materials Research* **21** (2006): 2660–2668.
55. Reedy, E. D., Jr. *Journal of Materials Research* **22** (2007): 2617–2622.
56. Chandross, M., Lorenz, C. D., Stevens, M. J., and Grest, G. S. *Langmuir* **24** (2008): 1240–1246.
57. Hill, I. J., and Sawyer, W. G. *Tribology Letters* **37** (2010): 453–461.
58. Johnson, K. L. *Proceedings of the Royal Society of London Series A—Mathematical and Physical Sciences* **230** (1955): 531–548.
59. Popov, V. L. In *Contact Mechanic and Friction: Physical Principles and Applications*, Springer: New York, 2010.
60. Johnson, K. L. *Proceedings of the Royal Society of London, Series A (Mathematical, Physical and Engineering Sciences)* **453** (1997): 163–179.
61. Kim, K. S., McMeeking, R. M., and Johnson, K. L. *Journal of the Mechanics and Physics of Solids* **46** (1998): 243–266.
62. Sørensen, M. R., Jacobsen, K. W., and Stoltze, P. *Physical Review B (Condensed Matter)* **53** (1996): 2101–2113.
63. Briscoe, B. J., and Evans, D. C. B. *Proceedings of the Royal Society of London Series a—Mathematical and Physical Sciences* **380** (1982): 389–407.
64. Szlufarska, I., Chandross, M., and Carpick, R. W. *J Phys D Appl Phys* **41** (2008): 123001.
65. Carpick, R. W., and Salmeron, M. *Chemical Reviews* **97** (1997): 1163–1194.
66. Luan, B., and Robbins, M. O. *Physical Review E (Statistical, Nonlinear, and Soft Matter Physics)* **74** (2006): 026111.
67. Luan, B., and Robbins, M. O. *Nature* **435** (2005): 929–932.
68. Wahl, K. J., Asif, S. A. S., Greenwood, J. A., and Johnson, K. L. *Journal of Colloid and Interface Science* **296** (2006): 178–188.
69. Carpick, R. W., Ogletree, D. F., and Salmeron, M. *Applied Physics Letters* **70** (1997): 1548–1550.
70. Lantz, M. A., O'Shea, S. J., Hoole, A. C. F., and Welland, M. E. *Applied Physics Letters* **70** (1997): 970–972.
71. Lantz, M. A., O'Shea, S. J., Welland, M. E., and Johnson, K. L. *Physical Review B (Condensed Matter)* **55** (1997): 10776–10785.
72. Piétrement, O., Beaudoin, J. L., and Troyon, M. *Tribology Letters* **7** (2000): 213–220.
73. Piétrement, O., and Troyon, M. *Surface Science* **490** (2001): L592–L596.
74. Piétrement, O., and Troyon, M. *Langmuir* **17** (2001): 6540–6546.
75. Hues, S. M., Draper, C. F., and Colton, R. J. *Journal of Vacuum Science & Technology B* **12** (1994): 2211–2214.

76. Lantz, M. A., Wiesmann, D., and Gotsmann, B. *Nature Nanotechnology* **4** (2009): 586–591.
77. Bhaskaran, H., Gotsmann, B., Sebastian, A., Drechsler, U., Lantz, M., Despont, M., Jaroenapibal, P., Carpick, R. W., Chen, Y., and Sridharan, K. *Nature Nanotechnology* **5** (2010): 181–185.
78. Lantz, M. A., Gotsmann, B., Jaroenapibal, P., Jacobs, T. D. B., O'Connor, S. D., Sridharan, K., and Carpick, R. W. *Advanced Functional Materials* **22** (2012): 1639–1645.
79. Gotsmann, B., and Lantz, M. A. *Physical Review Letters* **101** (2008): 125501.
80. Park, N. S., Kim, M. W., Langford, S. C., and Dickinson, J. T. *Journal of Applied Physics* **80** (1996): 2680–2686.
81. Sheehan, P. E. *Chemical Physics Letters* **410** (2005): 151–155.
82. Maugis, D. *Journal of Adhesion Science and Technology* **10** (1996): 161–175.

Chapter 3

Understanding Surface Forces Using Static and Dynamic Approach–Retraction Curves

Sudharsan Balasubramaniam, Daniel Kiracofe, and Arvind Raman

Department of Mechanical Engineering, Purdue University, West Lafayette, IN

The AFM is unique in its ability to measure surface forces with sub-nanonewton force resolution and nanometer spatial resolution [1, 2]. These surface forces include van der Waals, electrostatic, magnetic, sample viscoelasticity, capillary forces, chemical forces, hydration forces, magnetic forces, electrostatic forces, and the like and are directly correlated to the local composition of the sample within the small volume of the sample material probed by the AFM tip. The measurement of these forces as a function of the separation of the nanoscale tip of an AFM cantilever and the surface atoms is called force "spectroscopy." Essentially, force spectroscopy is a one-dimensional analog of AFM imaging where the tip approaches and then retracts from the surface in single-point measurements, and the surface forces are measured during this approach–retract time. Force spectroscopy is usually performed by monitoring some observables such as the cantilever's deflection (measured in static force curves) or oscillation amplitude and phase (measured in dynamic force curves) as the cantilever base is made to approach and then retract from the sample. Figure 3.1 shows examples from prior literature of these typical observables as a function of Z, the separation of the sample from the base of the cantilever with an example static force curve measurement shown in Figure 3.1(a) and dynamic force curve measurements of amplitude and phase shown, respectively, in Figure 3.1(b) and 3.1(c).

Scanning Probe Microscopy in Industrial Applications: Nanomechanical Characterization, First Edition. Edited by Dalia G. Yablon.

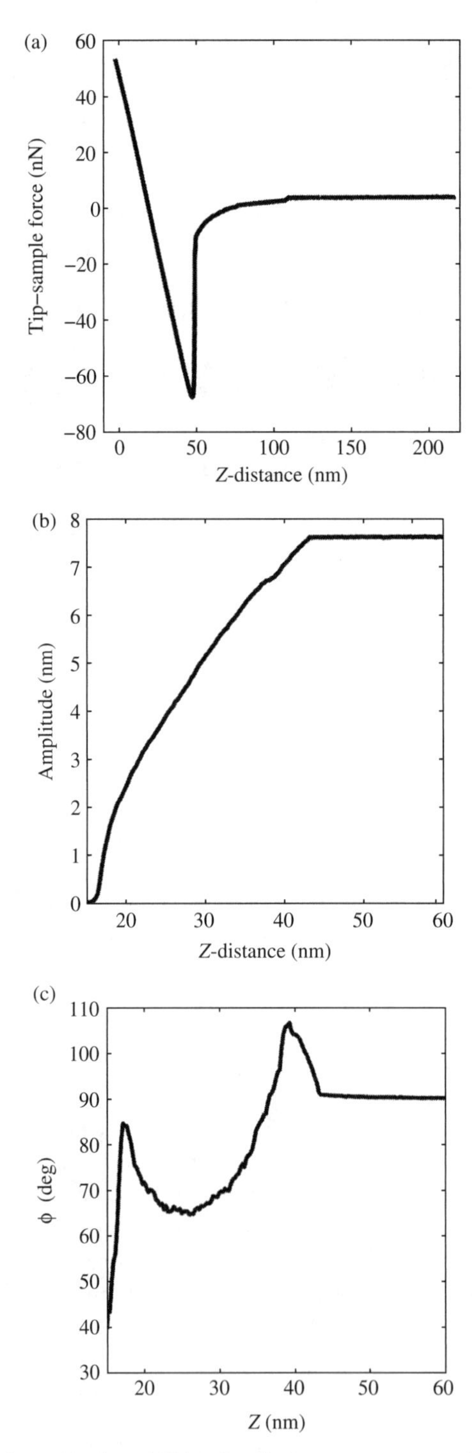

Figure 3.1 Typical observables in an AFM surface force spectroscopy experiment using a cantilever of stiffness 22.3 N/m, quality factor of 403, and resonance frequency of 276.6 kHz measured on a silicon sample under ambient conditions. (a) Static *F–Z* curve—Force measured against *Z*, and dynamic approach curves showing (b) Amplitude measured against *Z*.

Schematics of the most common forms of force spectroscopy are shown in Figure 3.2 and involve the static force–displacement curve [Fig. 3.2(a)] and the dynamic approach–retraction curve [Fig. 3.2(b)]. In the latter the cantilever is driven at a frequency close to its resonance and made to approach and retract from the sample while the cantilever oscillation amplitude and phase are allowed to evolve depending on the tip–sample interaction forces. The observables in the former case are tip–sample interaction force (F_{ts}) and Z separation between the undeflected cantilever tip and surface (also called Z piezo displacement) while in the latter are amplitude (A) and phase lag (Φ) versus Z. The characteristics of these two methods are shown in Table 3.1. Other more advanced techniques such as bimodal AFM [3] aim to monitor the amplitude and phase of two separate cantilever eigenmodes as a function of Z. Force spectroscopy refers to the extraction of meaningful information about the nature of forces acting between the tip and sample from the data measured in an AFM experiment. Converting the observables into tip–sample forces, that is, F_{ts} versus the tip–sample gap d is one such objective of force spectroscopy. The difference between Z (tip–sample separation) and d (tip–sample gap) is explained in detail in Section 3.2 and relies on the actual parameter that is measurable (tip–sample separation) and the parameter of interest (tip–sample gap, which is uncontrolled since this parameter includes the cantilever deflection that occurs in response to the interaction forces). The F_{ts} versus d

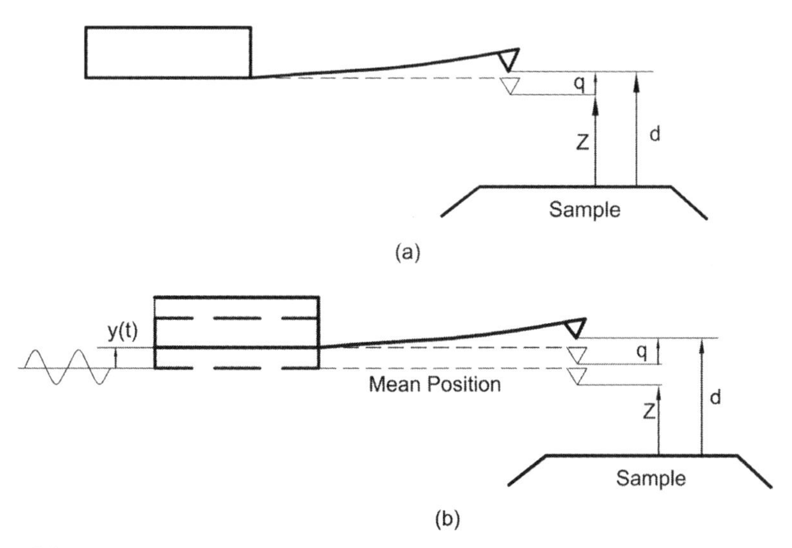

Figure 3.2 Schematic depicting the two main methods of force spectroscopy using AFM: (a) Static force–displacement curves showing the key notations: d is the tip–sample gap, Z is distance between the undeflected position of the cantilever, and the sample q is the deflection of the end of cantilever so that $d=Z+q$. (b) In the dynamic mode: $d(t)$ is the tip–sample gap, Z is the distance between the mean position of undeflected cantilever tip and sample, $q(t)$ is the cantilever deflection measured at its free end, and $y(t)$ is the base excitation such that $d=Z+q+y$.

Table 3.1 Characteristics of Static and Dynamic Modes

	Static Mode	Dynamic Mode
Cantilever Oscillated	No	Yes Direct/indirect
Observables	Deflection, Z distance	Amplitude, phase, frequency
Intermediate Observables	Tip–sample distance, interaction force	Virial, dissipation
Parameter of Interest	Adhesion, elastic modulus	Adhesion, elastic modulus, Hamaker constant, charge density, contact potential
Advantages	Converting intermediate observables to quantitative forces is easier	There is no snap-in phenomenon
Difficulties	Inability to reconstruct full force curve due to snap-in/pull-off phenomenon, high lateral forces during scanning, cannot distinguish hysteretic/ non-hysteretic attractive surface forces	Cantilever dynamics are to be understood, force spectroscopy is more complicated
Techniques	Force–volume mode, peak-force tapping, contact mode, etc.	Amplitude modulation, frequency modulation, phase modulation, etc.

graphs shown in Figure 3.3 contain rich information about the tip–sample interaction forces.

Force–displacement curves provide information about adhesion forces and Young's modulus among other important surface properties. Since the adhesion forces depend on the relative humidity, this influence has been extensively studied [4, 5]. Adhesion maps are consequently an important resource from which the topography of the surface can be derived. Larger adhesion is imparted in grooves due to a larger surface area and vice versa. Dynamic F–Z curves are used to understand hydrophobicity/hydrophilicity of surfaces through the effects of capillary forces [6] on the observables. Additionally, electrostatic double-layer forces have been measured using F–Z curves for varying pH of the medium [7, 8]. In this chapter, the elastic modulus of the sample has been extracted from experimental F–Z curves to highlight its importance.

In this chapter we aim to review the key features of these curves, the mathematics required to convert the results into quantitative results, the sources of error and uncertainties, and how to simulate these using the software VEDA (Virtual Environment for Dynamic Atomic Force Microscopy).

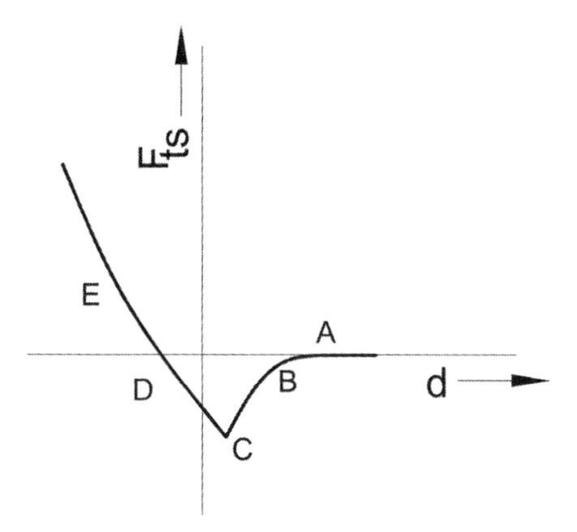

Figure 3.3 Typical force versus tip–sample gap curve showing the attractive regime *A-B-C* and the repulsive regime *C-D-E*. The zero line is the region of *d* values beyond point *A* where no attractive forces are active between the tip and the sample.

It is important to point out that there are many other more sophisticated forms of force spectroscopy possible, for example, frequency modulation techniques [9, 10] where the frequency shift and dissipation as a function of *Z* are observed or higher harmonic spectroscopy where higher harmonics of vibrations are measured versus *Z* [11]. These methods, however, are beyond the scope of the present chapter.

This chapter is divided into four sections: the first section reviews the different tip–sample interaction force models, the second section provides an insight into static force–displacement curves, the third section explains the theory behind dynamic approach curves, and the final section deals with the procedure to understand approach–retraction curves using VEDA, an open-source AFM simulator. Examples from literature have also been inserted within sections to understand behaviors that accompany measurements.

It should be noted that as a convention, the force–*distance* curves refer to the force F_{ts} measured as a function of the tip–sample gap *d*, and the force–*displacement* curves refer to the force F_{ts} measured as a function of *Z*. Figure 3.1 shows the "force-displacement" curve and as seen before, Figure 3.3 depicts the force–distance curve.

3.1 TIP–SAMPLE INTERACTION FORCES

A typical force–distance curve showing the interaction forces between a few hundred atoms on the tip of the AFM cantilever and the sample surface is seen in Figure 3.3. When the tip and sample are far apart (point *A*), there is no interaction between the tip and sample. As the distance *d* between the tip and sample is reduced, attractive forces (negative value of force) are experienced and the force becomes stronger when the distance is further decreased (point *B*). Beyond point *C* the tip starts to indent the sample and repulsive forces begin to show (point *D*). Further decrease in distance causes the repulsive forces to dominate over the attractive forces due to the inherent resistance of the sample to deform. The curve C–E starting from the point of contact and showing a relative increase in the force value is called the repulsive wall. Another term that is often used is the zero line. The zero line is the region where the force experienced by the cantilever is zero.

An important consideration is the concept of a conservative or dissipative tip–sample interaction force model. A conservative interaction force model is one where the tip–sample interaction force F_{ts} is a function of the tip–sample gap d alone. A dissipative force on the other hand is a function of both the tip–position d and velocity $\dot{d}$ relative to the sample surface. This represents the energy losses due to plastic deformation, viscoelasticity, and acoustic losses, which occur as the tip interacts with the sample. As a result, a dissipative tip–sample interaction force model can be recognized if it prescribes a different value when the tip approaches the sample, that is, when $\dot{d} < 0$ compared to when $\dot{d} > 0$, leading to the formation of a hysteretic loop in tip–sample force when the tip–sample gap d is made to go through a cycle of motion. The area of the loop measures the energy dissipated during tip–sample interaction. In what follows we will consider both conservative and dissipative forces.

Before describing force spectroscopy, it is useful to review the key tip–sample interaction models that are used in the analysis of *F–Z* curves.

3.1.1 Piecewise Linear Contact

The piecewise linear contact model to model interaction of two materials [Fig. 3.4(a)], the simplest of the models described here, is often applied to small indentations of suspended microelectromechanical systems (MEMS), small shell-like structures such as viral capsids, hollow microtubules, or carbon nanotubes. This model is applicable when van der Waals or electrostatic forces are negligible (e.g., in high ionic concentration buffer solutions). This model is especially useful for simple simulations and to compare sample stiffness directly to cantilever stiffness. For a contact stiffness (force gradient) k_{ts}^{rep} and tip–sample gap d, the piecewise linear contact model is mathematically represented as

$$F_{ts}(d) = \begin{cases} 0 & d > 0 \\ -k_{ts}^{rep} d & d < 0 \end{cases} \tag{3.1}$$

where $d<0$ indicates indentation and $d>0$ indicates no contact. This is a conservative force model since the force depends only on the gap d.

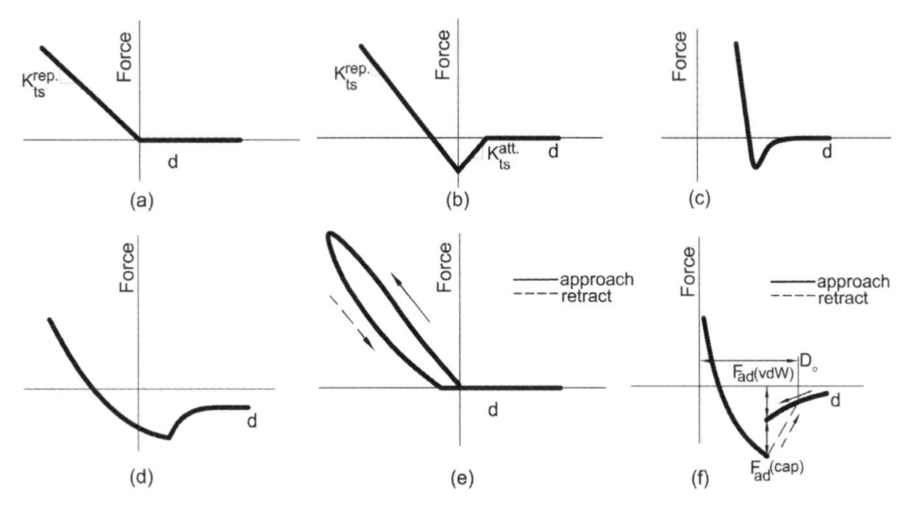

Figure 3.4 Schematics of various tip–sample interaction force models considered in this chapter: (a) piecewise linear contact, (b) piecewise linear attractive/repulsive contact, (c) Lennard-Jones potential, (d) DMT contact + van der Waals, (e) viscoelastic forces, and (f) capillary forces.

3.1.2 Piecewise Linear Attractive–Repulsive Contact

The piecewise linear attractive–repulsive model is the next more complex model shown in Figure 3.4(b). A linear attractive gradient is added to the repulsive gradient. This model is useful for simple simulations and to compare sample stiffness directly to cantilever stiffness, when van der Waals or electrostatic forces are not completely negligible. For a contact stiffness (force gradient) k_{ts}, gap between the tip and sample d, attractive gradient k_a, and maximum adhesion force F_{ad}, the piecewise linear attractive–repulsive contact model is

$$F_{ts}(d) = \begin{cases} 0, & d \geq L_0 \\ k_a(d - L_0), & 0 < d < L_0 \\ F_{ad} - k_{ts}d, & d \leq 0 \end{cases} \tag{3.2}$$

where $d = 0$ represents the sample surface. The adhesion force is given by $F_{ad} = K_a L_0$, where L_0 is the length of attractive gradient. This is also a conservative force model since the force depends only on the gap d. The presence of adhesion (negative interaction force) is not correlated to a force being conservative or dissipative.

3.1.3 Lennard-Jones Potential

The Lennard-Jones potential is commonly used to describe short-range interatomic forces [Fig. 3.4(c)] between a spherical tip and a flat surface. This model assumes the sample is infinitely stiff and the tip–sample repulsion arises from Pauli exclusion alone. In reality the tip and the sample actually contact in many AFM imaging modes, and many hundreds of atoms on the tip and sample surface interact. As a result this model is not a realistic representation of the forces when tip–sample contact occurs. The force is given by

$$F_{ts}(d) = -\frac{HR}{6d^2} + 12\frac{E_0}{r_0}\left[\left(\frac{r_0}{d}\right)^{13} - \left(\frac{r_0}{d}\right)^{7}\right] \tag{3.3}$$

where H is the Hamaker constant, R is the tip radius, E_0 is the binding energy, r_0 is the equilibrium distance, and d is the tip–sample gap. This is also a conservative force model.

3.1.4 Derjaguin–Müller–Toporov + van der Waals Model

The Derjaguin–Müller–Toporov (DMT) contact model [12] includes attractive, noncontact van der Waals forces [13, 14] combined with Hertz contact forces [13] and is valid for low adhesion relatively stiff contacts when operated in air or vacuum and under very dry conditions. For a tip–sample gap d, where $d = a_0$ (the intermolecular distance) is the sample surface, the DMT model can be written as

$$F_{ts}(d) = \begin{cases} -\dfrac{HR}{6d^2} & d > a_0 \\ -\dfrac{HR}{6a_0^2} + \dfrac{4}{3}E^*\sqrt{R}(a_0 - d)^{3/2} & d \le a_0 \end{cases} \tag{3.4}$$

where H is the Hamaker constant, R is the radius of the tip, E is the elastic modulus, and E^* is the effective elastic modulus between the tip and the sample system defined as

$$\frac{1}{E^*} = \left(\frac{1 - v_{tip}^2}{E_{tip}} + \frac{1 - v_{sample}^2}{E_{sample}}\right) \tag{3.5}$$

where v is the Poisson's ratio. For $d > a_0$ the tip experiences van der Waals forces. At $d \le a_0$ the van der Waals force saturates and Hertz contact forces begin. The adhesion force, that is, the maximum attractive force in the model is $HR/6a_0^2$. This model is shown in Figure 3.4(d). This is also a conservative force model.

3.1.5 Viscoelastic Forces

An elastic material is one where the stress (forces) at time t depends only on the strain (displacements) at time t and does not depend on the rate of strain, or on any past history of the strain. For example, a steel spring could be considered as an elastic body. A viscous material, on the other hand, is one in which the stresses are dependent only on the strain rate (velocity). For example, most common liquids such as water or oil are viscous. Many materials, however, display some mixture of viscous and elastic response. That is, the stresses may depend both on the current strain as well as the strain rate and the past history of the strain. This phenomenon is called viscoelasticity where examples of viscoelastic materials include polymers and probably one of the most famous viscoelastic materials, Silly Putty. There are several different continuum models for viscoelasticity, which can be combined with the tip–sample interaction models above in a few different ways.

Here we briefly discuss the Kelvin–Voigt viscoelastic dissipation theory, which assumes a stress relaxation that can be modeled as an elastic element in parallel with a damper. This model is applicable when the relaxation time of the material is very short compared to the tip–sample contact time.

This linear viscoelastic constitutive law can be combined with Hertzian contact mechanics in a self-consistent way, and the forces between the spherical tip and flat surface (without attractive forces) can be written as

$$F_{\text{ts}}(d) = \begin{cases} 0 & m = 0 \\ \frac{4}{3}E^*\sqrt{R}(-d)^{3/2} - \frac{2}{1-\nu}\eta\dot{d}\sqrt{-Rd} & m = 1,\ \dot{d} < 0 \\ \frac{4}{3}E^*\sqrt{R}\left(-d - \frac{1}{2}\frac{\eta}{G}\dot{d}\right)^{3/2} & m = 1, \dot{d} > 0 \end{cases} \tag{3.6}$$

where m is a mode variable that keeps track of the contact between the tip and sample; $m=0$ implies that there is no contact; $m=1$, $\dot{d}<0$ implies that the indentation is decreasing; and $m=1$, $\dot{d}>0$ implies that the indentation is increasing. The tip first contacts the sample ($m=0$ to $m=1$ transition) the first time that $\dot{d}<0$, and contact is broken ($m=1$ to $m=0$ transition) when either $\dot{d}>0$ or $-d-(1/2\times\eta/G)\,\dot{d}<0$. This is a dissipative interaction force, and this model is relevant for representing the contact physics of an AFM tip interacting with a flat polymer surface with low surface energy (adhesion). A van-der-Waals-like attractive force model can be easily added to this description. An example force curve is shown in Figure 3.4(e).

3.1.6 Capillary Forces

When a tip comes into contact with a sample in the presence of finite relative humidity, water molecules condense in the nanometric gap between the tip and the sample leading to the formation of a capillary neck. When the tip retracts, the capillary neck stretches and then breaks, releasing energy. This process causes a hysteretic, dissipative force superposed additively on any of the tip–sample interaction models described earlier. This additive force is given by

$$F_{\text{ts}}(d) = \begin{cases} 0 & m = 0 \\ \dfrac{2\Delta E}{D_0^2}(d - D_0) & m = 1, d > a_0 \\ F_{\text{ad,cap}} & m = 1, d < a_0 \end{cases} \tag{3.7}$$

where $d = a_0$ is the DMT intermolecular distance (set to 0 for the Hertz model), $d = D_0$ is the distance at which the neck breaks, E is the energy dissipated per hysteretic cycle, $F_{\text{ad,cap}}$ is the force jump when the capillary neck forms [Fig. 3.4(f)]. Here $F_{\text{ad,cap}} = 2\Delta E / D_0^2 (d - D_0)$, and F_{ts} is a state variable defined as follows: When $m = 0$, the capillary neck is "off" and when $m = 1$ the capillary neck is "on." When $d < a_0$, then m is set to 1. When $d > D_0$, then m is set to 0. This is clearly a dissipative force interaction.

In summary, the tip–sample interaction force models can range in complexity depending on the physics relevant to the material being probed. In one form or another, these models are fitted to experimental force spectroscopy results to extract material properties.

3.2 STATIC *F–Z* CURVES

As discussed above, the main goal of force spectroscopy is to reconstruct the F_{ts} versus d graph and then extract material properties by fitting it to one of the force models discussed in Section 3.1. Ideally, one would vary d and measure the force F_{ts} between the nanoscale tip and the surface. In practice in AFM, however, the tip–sample gap d is uncontrolled since the cantilever itself bends in response to the interaction forces. On the other hand, in an AFM it is the Z piezo displacement that is controllable. Thus the most common form of force spectroscopy measures the static F–Z curves or the interaction force versus Z piezo displacement.

Static F–Z curves can be used to directly quantify the tip–sample interaction forces. The force on the tip is assumed to act at a timescale much larger than the natural timescale of the cantilever oscillation. This force (F_{ts}) is calculated from the deflection of the cantilever (q) using

$$F_{\text{ts}}(d) = k_c q \tag{3.8}$$

where k_c is the stiffness of the cantilever. The deflection of the cantilever can be obtained by the optical lever method, the interferometry method, or the electron tunneling method. Of these, the optical lever method is most commonly used. A laser is directed to the tip of the cantilever, and a position-sensitive detector (essentially a photodiode) that contains four quadrants subtracts the voltage between the top two quadrants and bottom two to calculate the deflection. As a result, the photodiode measures the bending angle, which is proportional to the cantilever deflection.

Clearly, the *F–d* and the *F–Z* curves are different since Z and d must differ by the cantilever deflection q. Thus the *F–d* curve must be converted into an *F–Z* curve and vice versa to understand the link between the two. Experimentally, *F–Z* curves are what are measured and they must be converted to *F–d* curves. This is discussed further in Section 3.2.1.

Conversion of the *F–d* to the *F–Z* curve is conveniently understood through a graphical approach as shown in Figure 3.5. The plot in Figure 3.5(a) shows F_{ts} varying with d following the DMT interaction model (explained in the previous section). The tip–sample force equals the force experienced by the cantilever according to the Eq. (3.8). This can be further broken down into

$$F_{ts}(d) = k_c(d - z) \tag{3.9}$$

Equation (3.9) provides a way to determine d for any given value of displacement Z. For example, for a specific value of Z, the corresponding value of d must be such that that the right-hand side equals the left-hand side. Plotted as a function of d, the right-hand side of Eq. (3.9) is a straight line of slope k_c with an x-axis intercept of Z. The left-hand side of Eq. (3.9), F_{ts}, on the other hand comes from the standard *F–d* curve. The intersection of these two curves provides the values of d that correspond to the Z displacement (Z). The difference between d and Z denotes the deflection q.

Figure 3.5 shows the process of obtaining the Z displacements (Z_1, Z_2,…) graphically at different instances of the cantilever position from the sample. As we approach the sample (from 1 through 5), the straight line having the slope equaling cantilever stiffness (k_c) intersects the force curve at points A, B, C, D, and so on. At instance 4, there are two intersection points (D and B′). If the tip is brought closer, the intersection point jumps from the attractive part of the curve to the repulsive (point A′ in 5). This phenomenon is called snap-in. Following the retraction curve (5 back through 1) in Figure 3.5 and using the same procedure, the resulting points of intersection are A′–B′–C′–D′–A. We can see a similar jump from D′ to B, called the pull-off that is larger than the jump during approach. This happens due to the inherent nature of forces. There is a range of d values corresponding to the snap-in and pull-off locations that cannot be accessed while the cantilever is being pulled into or away from the sample. In this region, the exact force cannot be determined since the tip is not in stable equilibrium. This is because the gradient of the tip–sample interaction force is greater than the stiffness of the cantilever. Thus, although the *F–d* curve we analyzed here was conservative in nature, the corresponding *F–Z* curve shows hysteresis.

The approach shown in Figure 3.6 can be used on all the *F–d* models described earlier to determine the corresponding *F–Z* curve. For example, Figure 3.6(a) shows the force between the tip and an infinitely hard rigid wall. Once the tip comes into contact with the sample, there is no indentation possible; and following the earlier procedure, the corresponding *F–Z* curve would have to be a straight line having a slope equal to the cantilever stiffness. Similarly the *F–Z* curves corresponding to the *F–d* curves in Figures 3.6(b) and 3.6(c) can be deduced.

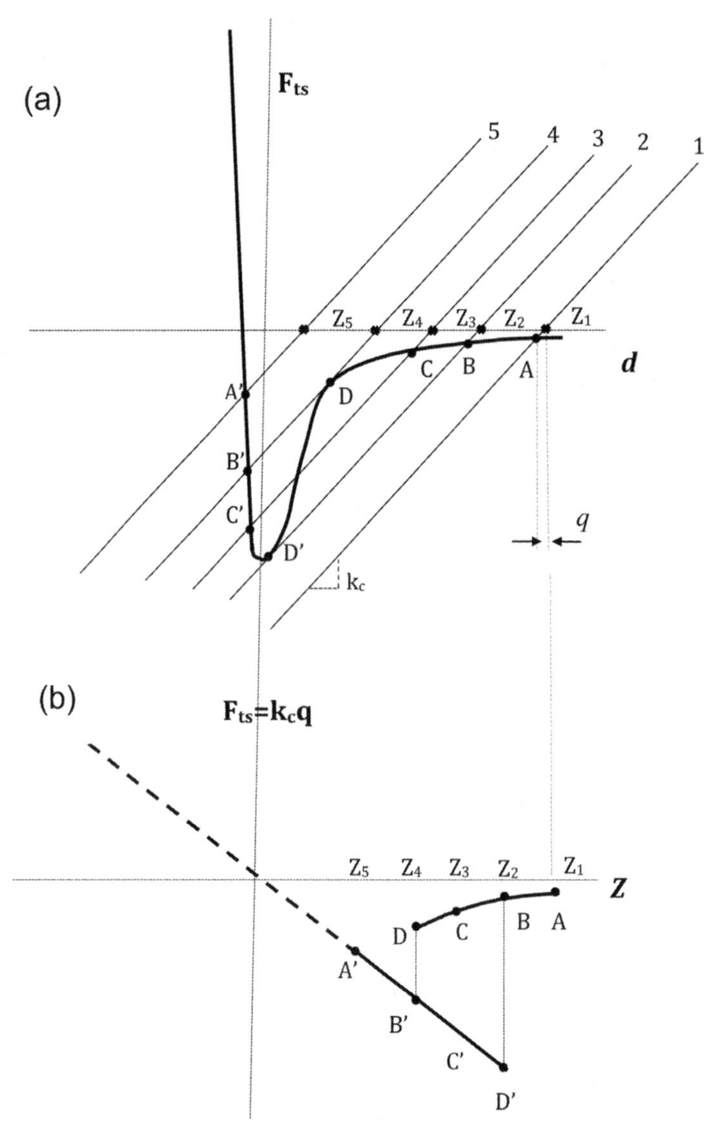

Figure 3.5 (a) Schematic representing the graphical interpretation of Eq. (3.8) in the text, which allows one to determine the value of the tip–sample gap d for given values of Z. The intersection of the F–d curve with the cantilever force $F_{ts} = k_c q$ provides the d value and force at five Z values. The converted F–Z curve is shown in (b). The Z values for this force are also marked. Snap-in occurs from D to B′ during approach (1-2-3-4-5) at instant 4 while retracting (5-4-3-2-1), a pull-off occurs from D′ to B at instant 2.

Figure 3.6(d) shows the Johnson–Kendall–Roberts (JKR) model, which is a dissipative interaction. The F–d curve during approach and retraction does not overlap leading to a true hysteresis. This leads to a hysteresis in the corresponding F–Z curve, which is qualitatively similar to the F–Z curve for the DMT model.

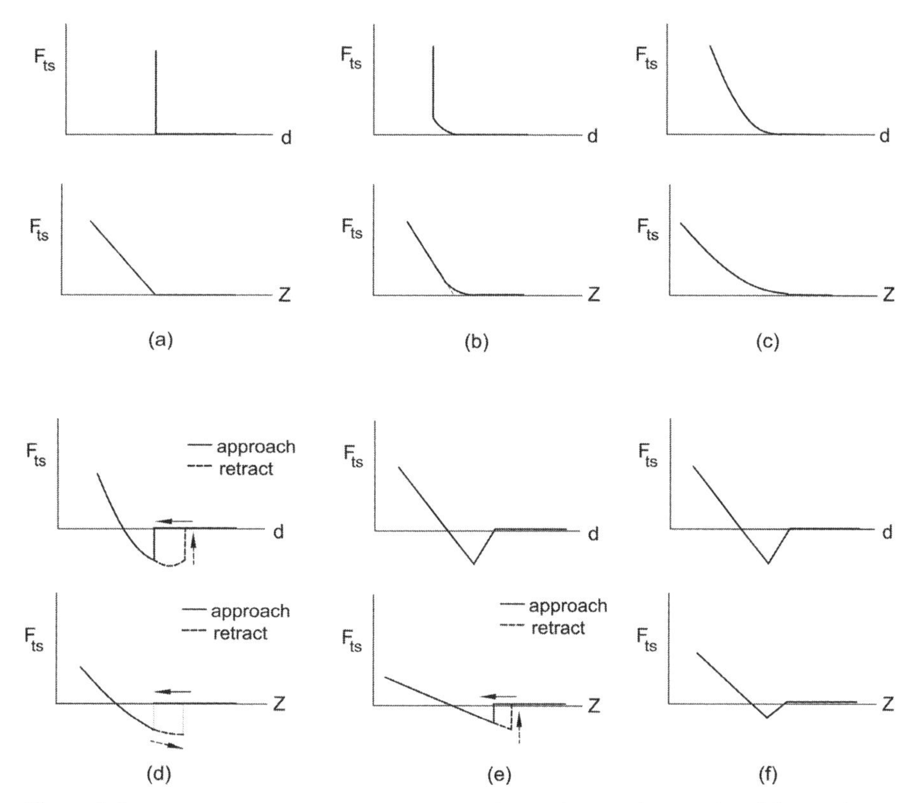

Figure 3.6 Schematics showing the conversion of F–d curves into F–Z for a number of tip–sample interaction models. (a) F–d curve on a rigid wall, and its corresponding F–Z curve, (b) F–d curve for an infinitely hard wall with long-range repulsion static AFM and its equivalent F–Z curve, (c) Hertz contact model and its corresponding F–Z curve, (d) JKR contact model and its corresponding F–Z curve, (e) linearly varying attractive and repulsive contact and the F–Z curve obtained using a soft cantilever, and (f) linearly varying attractive and repulsive contact model and its F–Z curve obtained using a stiff cantilever.

Clearly, a hysteretic F–d model must necessarily lead to a hysteretic F–Z curve. However the converse is not always true as described below.

Figures 3.6(e) and 3.6(f) show a linearly varying attractive and repulsive force between the tip and sample. For a soft cantilever, hysteresis can be observed in the F–Z curve even for this conservative interaction model, but if we choose a cantilever [Fig. 3.6(f)] that has a stiffness greater than the slope of the linear attractive regime, we do not observe hysteresis. It is clear that the nature of the F–Z curve obtained from experiments is closely correlated to the cantilever that is chosen.

If we compare Figures 3.6(d) and 3.6(e), there is a hysteretic F–Z curve for both nonconservative and conservative force models. This leads us to a very important observation that is often a source of misinterpretation in the literature. Hysteresis in an F–Z curve is not indicative of a dissipative interactive force or of the presence of hysteresis in the F–d curve. On the other hand, a dissipative tip–sample interaction force or hysteretic *F–d* curve must lead to a hysteretic *F–Z* curve. This will be discussed in more detail in Examples 1 and 2 in Section 3.4.

In summary, we have discussed how theoretical tip–sample force interaction models can be converted into *F–Z* curves. In experimental settings, however, the process is reversed, that is, one acquires an *F–Z* curve first, these being the experimental observables, and then extracts the *F–d* curve from the measured data. This is discussed next.

3.2.1 Conversion of *F–Z* Curves into *F–d* Curves

From AFM experiments we observe the cantilever deflection (*d*) or force F_{ts} measured against Z. However, the real physics and surface properties lie in the *F–d* curves; therefore, we need to convert the experimental data into force spectroscopy plots.

Figure 3.7 shows the deflection q plotted against Z, in which the regions of snap-in and pull-off can be observed. For the moment we choose the actual location of $Z=0$ arbitrarily. To convert them to *F–d* curves, we need to make the following transformations at every point:

$$F_{ts}(d) = k_c q \tag{3.10}$$

$$d = Z + q \tag{3.11}$$

Since $Z=0$ is arbitrary, so too is the definition of $d=0$. Often the *F–d* curve is offset along the d axis so that the largest adhesive force during the pull-out is labeled as $d=0$. There is no uniformly agreed-upon approach to choosing $d=0$ though. For example, if one assumes that a DMT model is representative of the contact mechanics, then the minimum adhesive force should in fact be $d \neq 0$, that is, $d=a_0$, the intermolecular distance. There is a third approach that is recommended where the choice of $d=0$ is treated as a parameter in a theoretical model, which is usually fit to the experimental data. This will be discussed later.

Often it is convenient to plot a F_{ts} versus d graph in terms of F_{ts} versus δ where $\delta = -d$; δ is the indentation of the tip into the sample. So δ is >0 when $d<0$. From Figure 3.7, there is an inaccessible region between the two areas of the plot (a to a′ or b to b′) where the force cannot be measured. This region could result from two possibilities: The force itself could be hysteretic (i.e., dissipative), or it could be a snap-in due to attractive force gradient exceeding the cantilever spring constant.

The importance of this conversion from *F–Z* to *F–d* or *F–*δ is emphasized in Figure 3.8. In contact, the tip–sample distance d is zero, or to be more exact it equals the interatomic distance. Beyond this point, the tip indents the sample. From the force–indentation plots, we can extract many parameters of the tip–sample force. For example, if one regards the DMT model as being representative of the tip–sample forces, then one could fit the extracted *F–d* curve to the DMT model as follows:

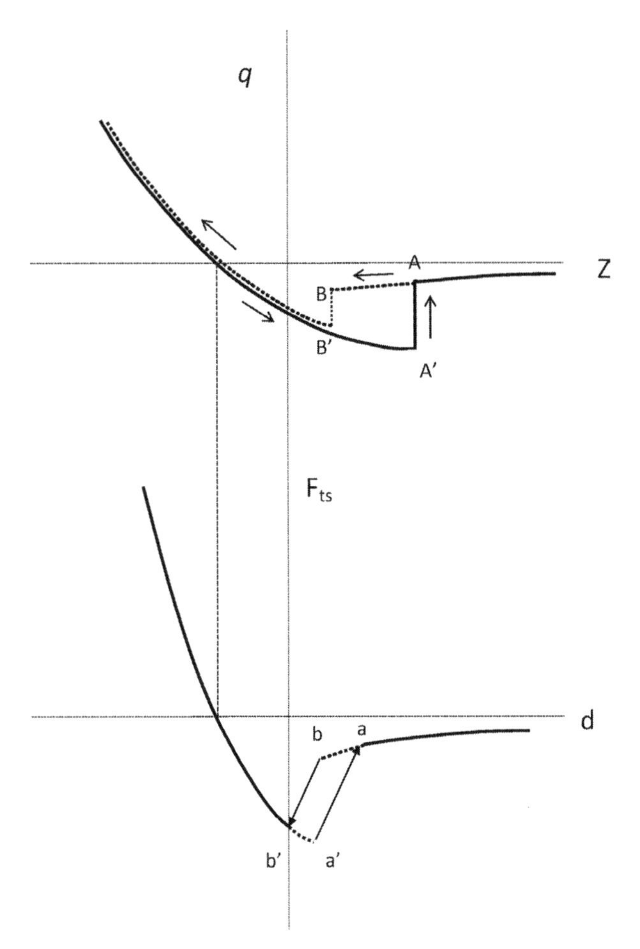

Figure 3.7 Schematic showing conversion of *F–Z* curve into an *F–d* curve using the transformation equations (3.10) and (3.11) in this chapter.

$$F_{ts} - F_{adhesion} = \frac{3}{4}E^*\sqrt{R}(\delta - \delta_0)^{3/2} = \frac{3}{4}E^*\sqrt{R}(d_0 - d)^{3/2} \tag{3.12}$$

In fitting this model to the acquired F_{ts} versus δ plot, the tip radius R needs to be known. The fitting parameters are the effective Young's modulus E^*, the adhesion force $F_{adhesion}$, and δ_0 or d_0, which define the location of $\delta=0$ or $d=0$. This way the $d=0$ is not chosen but rather made a fit parameter to the data. Using the equation for E^*, we can fit the value of E_{sample}^{fit} since we know E_{tip} and can assume the Poisson ratio of the tip and the sample. Young's modulus was fit for small indentations because the contact mechanics theory (DMT) models the interaction physics correctly in this regime. This theory is inaccurate for large indentations.

(a)

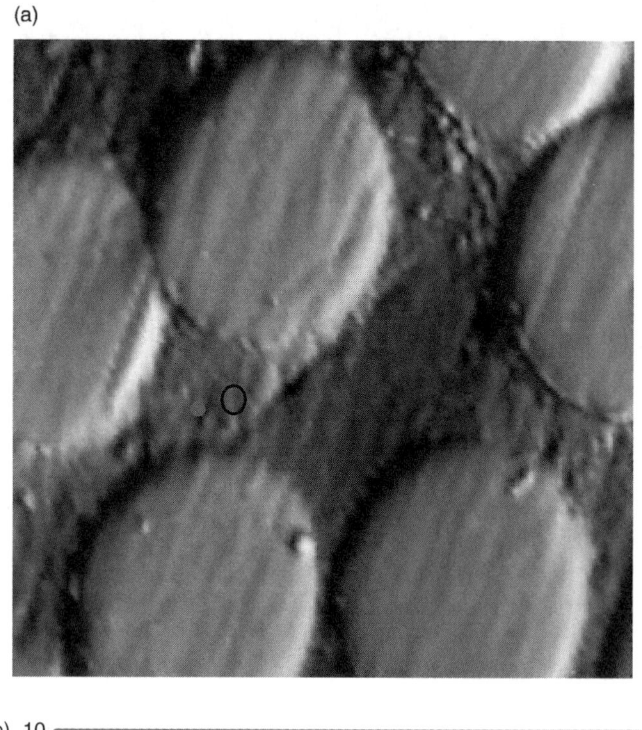

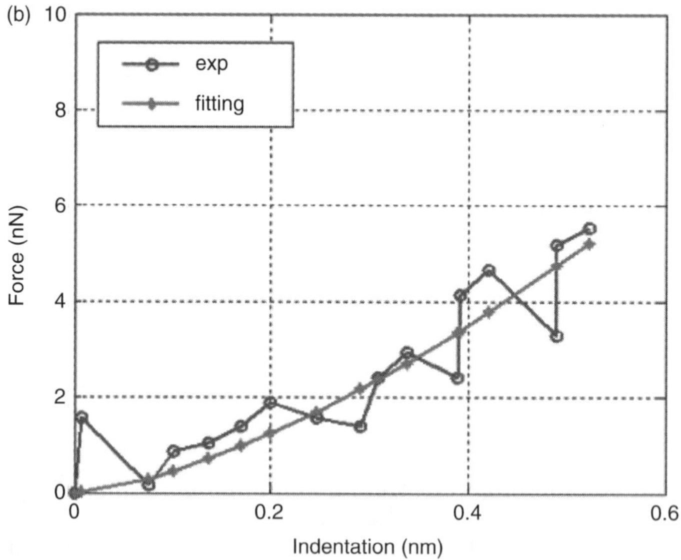

Figure 3.8 (a) 15 by 15 μm scan of a carbon fiber–epoxy sample showing the carbon fiber on an epoxy resin. (b) Force vs. indentation curve data acquired at the point *O* marked in the topography image on the epoxy. In (b) when the data is fit to small indentations, we get a more realistic value of $E_{\text{sample}}^{\text{fit}}$ sample = 3.23 GPa.

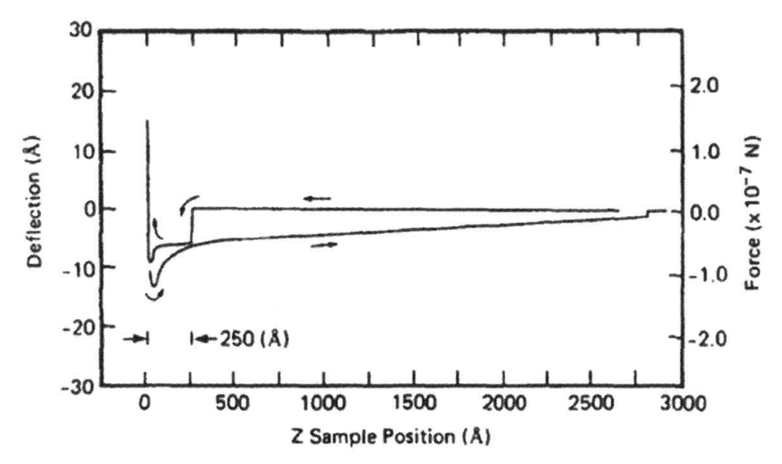

Figure 3.9 Example of *F–Z* curves on complex material interfaces. (a) Plot of the cantilever deflection versus *Z* on a perfluoropolyether polymer liquid film on silicon. Two distinctive pull-offs are seen when the forces are 120 and 10 nN. Adapted from [15].

3.2.2 Examples from Literature

Now we consider some experimental results on *F–Z* curves that have been published in the literature. Figure 3.9 shows the *F–Z* curve on a perfluoropolyether polymer liquid film on silicon [15]. The arrows indicate the direction in which the cantilever is moved. Initially at *Z* displacement of 2500 Å, there is no interaction between the tip and sample and therefore no deflection observed. When the displacement (*Z*) is close enough (250 Å), a sudden jump in the cantilever deflection is seen due to the attractive force exerted by the liquid film (see previous section). When the *Z* distance is further decreased, the force remains fairly constant as the tip penetrates the liquid film. Another small jump is observed at 10 Å, which is due to the attractive forces between the silicon surface and the tip in the liquid medium. Further reduction in *Z* displacement will lead to repulsive forces, and the observed deflection begins to increase. The key difference between this plot and those discussed before is that there are two snap-in and pull-off contact points seen here due to the layer of polymer film. Clearly *F–Z* curves can show more than one pair of snap-in/pull-off events depending on the nature of forces encountered by the tip.

The *F–Z* curves measured in liquids show a variation compared to measurements in air. In a liquid medium the cantilever tip and the surface tend to become charged either due to the dissociation of surface groups or by the adsorption of ions onto the surface. The result being that an electrostatic force acts between the charged surfaces (this force is often modeled using the DLVO (named after Derjaguin, Landau, Verwey, and Overbeek) theory [16]). The curves shown in Figure 3.10 were obtained for different KCl concentrations with a silicon nitride tip on mica [16]. At 0.5 mM concentration a repulsive force is seen due to the electrostatic forces between the tip and sample. As the concentration of KCl is increased, the magnitude of repulsive forces decrease, its decay becomes more rapid and the van der Waals forces show more prominence.

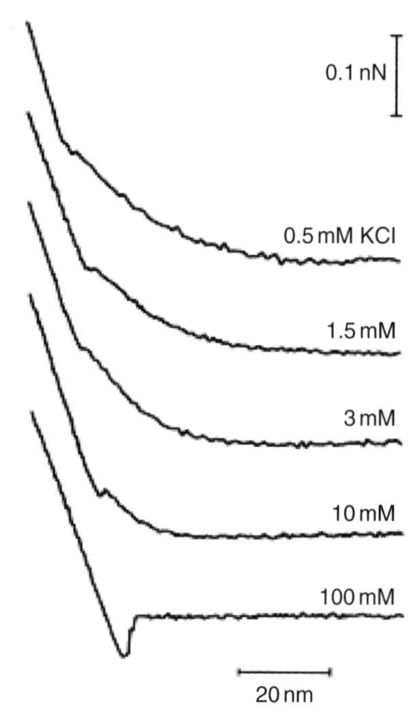

Figure 3.10 Force–displacement curves between a silicon nitride tip and mica in varying concentrations of KCl electrolyte. Presence of electrostatic forces can be seen with lower electrolyte concentrations. The attractive forces become distinct as the concentration of KCl gets higher. Adapted from [16].

At 100 mM concentration, the electrostatic forces are no longer visible and the van der Waals forces start to dominate. In this region, the attractive force gradient exceeds the force gradient of the spring force, and cantilever snap-in/snap-off can occur while using soft cantilevers. Also, at this concentration the decay length is so large that the decaying electrostatic forces cannot be distinguished from the zero force line.

3.2.3 Uncertainties and Sources of Error

Uncertainties in Input

One of the main uses of the F–Z curve is to extract local properties such as local elasticity and adhesion. Many sources of uncertainty and error need to be evaluated when extracting these properties from F–d or F–Z curves. In addition to finding uncertainty through multiple replicates of F–Z curves, a systematic study of uncertainties due to calibration (Z piezo sensitivity, photodiode sensitivity, cantilever stiffness) is required. This can help to quantify the confidence in the extracted surface properties [17]. A detailed review of various artifacts and uncertainties in all AFM measurements, including in force curves, is provided in Chapter 6, but some of the main ones associated with static force curves are included below.

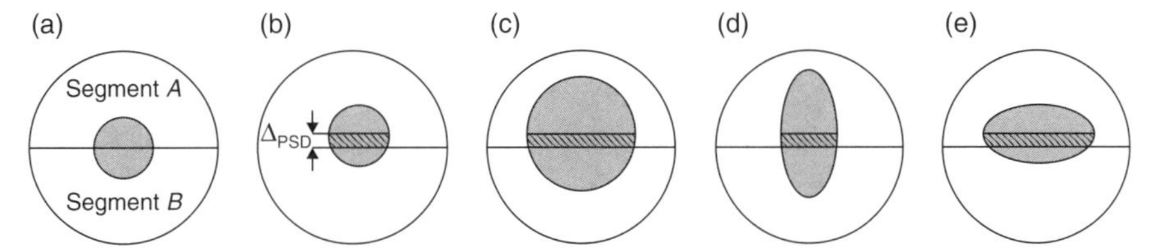

Figure 3.11 Shape of the laser spot on the photodiode produces an output $(A-B)/(A+B)$: (a) A centered laser spot produces a zero photodiode signal; (b) a small deflection of the cantilever causes a relatively higher change in the area between A and B, making it highly sensitive; (c) and (d) have poor sensitivity since the relative change in the area between the two segments is not much and hence the output from the photodiode does not show a marginal increase; and (e) an oval shape also has the tendency to produce higher sensitivity in this case. Adapted from [20].

Effect of Working Environment

During experiments, the zero line during retraction may or may not coincide exactly with that of the approach, depending on the environment in which the experiment is carried out. The reason is that the viscous drag of the surrounding environment bends the cantilever upward during approach and toward the sample during retraction [18, 19]. This leads to a hysteresis over the entire Z range of the experiment. To reduce this artifact, it may be necessary to reduce the approach–retraction speed or alternately reduce the frequency at which the curves are being performed.

Position and Size of Laser Spot on Photodiode

The position of the laser spot on the photodiode quadrant detector and its size both affect the sensitivity of the measurements (deflection in the static case). If the laser spot is not centered on the photodiode, the difference signal (output voltage from the photodiode) will have a net initial value. Further deflection measurements will be with respect to this value and as a result, the net F–Z plot in the measured forces and the zero line will be offset. Reading exact values can prove to be difficult if this is the case. In contrast, the size of the laser spot on the photodiode does not offset the measured forces but rather can reduce sensitivity of photodiode voltage output. There is a reduced difference in light received in the upper quadrants compared to the lower quadrants of the photodiode, and the relative area does not change much even though the center of the spot itself has shifted [13]. Oval-shaped laser spots can produce a lower or higher sensitivity depending on its orientation, as shown in Figure 3.11. More artifacts related to the laser spots have been discussed in [21].

Piezoelectric Scanner Effects Piezoelectric scanners can create discrepancies in the F–Z scurves due to three main reasons:

- *Hysteresis*: The piezoelectric element of a scanner consists of a large number of domains primarily polarized along a given poling direction through application of a large voltage. With time, the degree of poling decreases and the domains get randomly oriented either through the nucleation of random domains or through the movement of domain walls. The reoriented domains do not return to their original state when the voltage is dropped to zero [22, 23]. As a result there is a residual stress in the piezoelectric element. Therefore, the voltage required by the *Z* piezo to move the cantilever a certain distance toward the sample is not the same as the voltage required to move through the same distance away from the sample. This is a common problem in most AFM systems, and it depends on the piezoelectric material used. One method of reducing the *Z*-piezo hysteresis is through an iterative learning control (ILC) algorithm, which is used to learn the input required to position the probe tip over the sample to compensate for the distortions [24]. This method has been shown to reduce the error up to an order of the noise level. While imaging, hysteresis effects (along x and y direction) can be reduced by scanning in the same direction always. Design and compensation methods have been proposed by using the Preisach model to describe hysteresis [25].

- *Thermal Drift*: Even when there is no voltage applied, the piezo material continues to move (drift). This depends on the warming-up time, environmental conditions, as well as the thermal expansion coefficient of the piezoelectric material. Various drift compensation techniques (Kalman filters [26]) and algorithms have been developed to reduce this piezo drift [27–30]. The resulting *F*–*Z* plot is translated along the *Z*-distance axis.
- *Creep*: When a certain voltage is applied to the piezo scanner and left to remain, the piezo material continues to expand for some time after the voltage has stopped changing. The resulting plot will show values for the force that is actually at a different *Z* distance than the required *Z* distance. This can be countered by using feedback and feedforward mechanisms [31]. Feedforward mechanisms use the classic Preisach model [32] to compute inputs. Also, by employing charge control instead of voltage control, the effects of hysteresis and creep are known to be reduced. This comes at the cost of reducing the effective displacement of the piezo scanner [33]. Flexure stages that are common in nanopositioning have also been used to allow for independent *X*, *Y*, and *Z* motions. These stages provide a high degree of accuracy while handling larger samples [34].

Optical Interference

When the laser spot is not focused entirely within the cantilever, part of the beam reflects back from the sample surface leading to an interference pattern between the laser beam reflected off the back of the cantilever and the laser beam reflected from the sample surface. At certain tip–sample gaps the two reflected waves interfere constructively. For other tip–sample distances, there is a destructive interference. An undulation in the zero line occurs due to this phenomenon when performing approach–retraction curves over large *Z* ranges. It is for this reason that some instruments incorporate superluminescent diodes having lower coherence than lasers in order to reduce interference.

In summary, we have discussed many details of static *F–Z* and *F–d* curves, material property extraction, and sources of error and uncertainty in static force distance curves. We now discuss dynamic approach and retract curves.

3.3 DYNAMIC AMPLITUDE/PHASE–DISTANCE CURVES

3.3.1 Theory

As described before, there are two key disadvantages of force spectroscopy using static force–displacement curves, namely: (a) part of the *F–d* curve cannot be accessed due to snap-in and pull-off instabilities, and (b) it is not possible to distinguish an energy dissipation in tip–sample contact due to genuine hysteresis in tip–sample forces from the apparent hysteresis in the *F–Z* curves, which can happen from a purely conservative attractive force gradient. Dynamic AFM methods can provide significant advantages in this regard. In this chapter, we will focus on the study of amplitude and phase variations of a resonant cantilever probe with changing *Z* displacement. Here, the cantilever is initially far from the sample and is oscillated at its natural frequency. It is then brought closer to the sample by reducing the *Z* displacement. When the tip encounters an attractive–repulsive force from the sample, the resonance frequency is detuned from its natural resonance and there is added dissipation from the tip sample forces. Together these cause the amplitude and phase to change since the drive frequency is kept fixed.

We shall consider the connection between the amplitude–phase relations analytically. The origin of amplitude reduction can be best explained through a perturbation method that clearly defines two important quantities: the virial and dissipation [35, 36]. We begin with the point-mass model of the cantilever excited at a frequency Ω and excitation amplitude F_0 interacting with the sample:

$$m_c\ddot{q}+c_c\dot{q}+k_cq=F_0\cos(\Omega t)+F_{ts}(Z+q) \tag{3.13}$$

where m_c, c_c *and* k_c are the "effective mass," "effective damping," and "effective stiffness," respectively, of the cantilever probe. The relationship between these effective quantities and the material properties and geometric shape of the cantilever probe can be rigorously established [37]. It can be shown for rectangular cantilever probes that $m_c = m/4, m$ being the actual mass of the cantilever and k_c is the cantilever stiffness calculated from the thermal calibration method. In particular m_c and k_c must be such that $\omega_0 = \sqrt{k_c / m_c}$ where ω_0 is the fundamental natural frequency of the probe. The tip–sample interaction can be conservative or dissipative. A conservative force means that the force depends solely on $d(t)=Z(t)+q(t)$, while dissipative forces depend on both d and the tip velocity $\dot{d}$ or $\dot{q}$. Divding Eq. (3.13) by k_c we get

$$\frac{\ddot{q}}{\omega_0^2}+q+\frac{1}{\omega_0 Q}\dot{q}+=\frac{1}{k_c}\left[F_0\cos(\Omega t)+F_{ts}(Z+q)\right] \tag{3.14}$$

where $Q=m_c\omega_0/c_c$ is the quality factor. In this chapter we shall consider the case where the excitation equals the natural frequency of the cantilever, that is, $\Omega=\omega_0$ The general case can be derived in a similar manner and is described in Ref. [38]. It is well known that under most commonly occurring situations the oscillations are dominated by a harmonic component at the drive frequency. Accordingly, we write the harmonic response assumption of the probe:

$$q(t) = A\cos(\omega_0 t - \phi) \tag{3.15}$$

$$d(t) = Z + q(t) = Z + A\cos(\omega_0 t - \phi) \tag{3.16}$$

$$\dot{d}(t) = \dot{q}(t) = -A\omega_0 \sin(\omega_0 t - \phi) \tag{3.17}$$

where ϕ is the phase lag between the cantilever motion and the drive signal, and A is the steady-state operating setpoint amplitude. Now substitute Eqs. (3.16) and (3.17) into (3.14):

$$-\frac{1}{Q}\sin(\omega_0 t - \phi) = \frac{1}{k_c A}\{F_0\cos(\omega_0 t) + F_{\mathrm{ts}}[Z + A\cos(\omega_0 t - \phi)]\} \tag{3.18}$$

By multiplying Eq.(3.18) with $\sin(\omega_0 t-\phi)$ and $\cos(\omega_0 t-\phi)$ separately, followed by integration over a period $2\pi/\omega_0$ and grouping of terms, we, respectively, get two important quantities in terms of the observables (A and ϕ, which are the typical measured quantities in dynamic AFM):

$$E_{\mathrm{diss}} = \pi A F_0 \sin\phi - \pi k_c A^2\left(\frac{1}{Q}\right) \tag{3.19}$$

$$V_{\mathrm{t-s}} = -\frac{1}{2} A F_0 \cos\phi \tag{3.20}$$

Where

$$E_{\mathrm{diss}}(Z,A) = -\int_0^{2\pi/\omega_0} [-A\,\omega_0 \sin(\omega_0 t - \phi)]\cdot F_{\mathrm{ts}}[Z + A\cos(\omega_0 t - \phi)]\,dt \tag{3.21}$$

$$V_{\mathrm{t-s}}(Z,A) = \frac{\omega_0}{2\pi}\int_0^{2\pi/\omega_0} A\cos(\omega_0 t - \phi)\cdot F_{\mathrm{ts}}[Z + A\cos(\omega_0 t - \phi)]\,dt \tag{3.22}$$

The integrand in Eq. (3.21) is the product of the tip–sample interaction force and the tip velocity, which when integrated over an oscillation cycle represents the work done by the interaction force on the cantilever. Therefore, E_{diss} is the energy dissipated from the oscillating probe into the sample. This expression gives the information about the energy dissipated over a cycle. The virial V_{t-s} is the product of tip–sample force and tip–position and is a measure of the conservative forces between the tip and the sample. Ultimately, the E_{diss} and V_{t-s} depend on the Z displacement and the oscillation amplitude in addition to the local surface properties that are captured in the F_{ts} terms.

To show that E_{diss} measures the dissipative interactions and V_{t-s} measures the conservative interactions, consider the interaction force F_{ts} to consist of $F_{ts}^{cons} + F_{ts}^{diss}$ where F_{ts}^{cons} refers to the conservative part of the force and F_{ts}^{diss} refers to the dissipative part. The conservative part of the force depends on the position alone, and, since the position is considered to be harmonic, F_{ts}^{cons} must also be in phase with the tip motion while F_{ts}^{diss} must be out of phase with the tip motion by 90° due to its dependence on tip velocity. From Eqs. (3.23) and (3.24) we can infer that

$$E_{diss}(Z,A) = -\int_0^{2\pi/\omega_0} [A\,\omega_0 \sin(\omega_0 t - \phi)] \cdot F_{ts}^{diss}[Z + A\cos(\omega_0 t - \phi)]\,dt \tag{3.23}$$

$$V_{t-s}(Z,A) = \frac{\omega_0}{2\pi} \int_0^{2\pi/\omega_0} A\cos(\omega_0 t - \phi) \cdot F_{ts}^{cons}[Z + A\cos(\omega_0 t - \phi)]\ dt \tag{3.24}$$

To convert the amplitude of driving excitation F_0 in the above equations in terms of the observables, the transfer function of a simple harmonic oscillator excited at a frequency Ω is used. The transfer function is defined as the ratio of the measured response of the system to the input given to the system. This is usually a function of excitation (input) frequency. The output in this case is given by the product of the cantilever stiffness and amplitude, while the input is the magnitude of the driving force.

$$\frac{k_c A_0}{F_0} = \frac{1}{\sqrt{\left[1-\left(\frac{\Omega}{\omega_0}\right)^2\right]^2 + \left(\frac{\Omega}{\omega_0 Q}\right)^2}} \tag{3.25}$$

where A_0 is the amplitude when there is no interaction between the cantilever tip and sample; A_0 is sometimes called the "free amplitude" or the unconstrained amplitude, so that A/A_0 is called the operating setpoint for tapping mode AFM. In the commonly used tapping mode AFM, also called amplitude modulated AFM (AM-AFM), the Z piezo changes during a scan to maintain a constant setpoint amplitude. The changes in Z required to keep amplitude constant are rendered as the topography of the sample. Equation (3.25) reduces to $F_0 = k_c A_0/Q$ when $\Omega = \omega_0$.

This allows us to simplify the expressions for energy dissipation and virial derived in Eqs. (3.19) and (3.20) as follows:

$$E_{\text{diss}} = \frac{\pi k_c A A_0}{Q} \sin\phi - \pi k_c A^2 \left(\frac{1}{Q}\right) \tag{3.26}$$

$$V_{\text{t-s}} = -\frac{1}{2}\frac{k_c A A_0}{Q} \cos\phi \tag{3.27}$$

The above equations show that the virial and dissipation can be easily measured by monitoring the operating amplitude A and phase lag ϕ, so long as the probe is driven exactly at resonance and its stiffness k_c, quality factor Q, and free amplitude A_0 are known.

In a static AFM experiment, the observables are deflection (and hence the force) measured against Z. In the dynamic mode, the amplitude and phase are measured against the Z distance. These are converted into intermediate quantities E_{diss} and $V_{\text{t-s}}$ using Eqs. (3.26) and (3.27). This calculation has been performed in Figure 3.12 for the data presented in Figure 3.1 where a resonant cantilever was made to approach a Si surface. The shape of the E_{diss} and $V_{\text{t-s}}$ is formed by the combined effect of the amplitude and phase changes. The $V_{\text{t-s}}$ in the attractive regime ($\phi > 90$) is positive and in the repulsive regime ($\phi < 90$) is negative. The E_{diss} value is always going to be greater than zero if we look at Eq. (3.26). Virial and dissipation values are plotted till the cantilever no longer oscillates ($A = 0$); beyond this point the virial and dissipation formulas are no longer meaningful.

The E_{diss} and $V_{\text{t-s}}$ are important quantities that tell us about the forces of interaction. However, it is very hard to convert these measurements into tip–sample interaction forces since no simple analytical formulas exist for this. This is why simulations are essential and one such simulator, VEDA, is discussed in the final section.

3.3.2 Interpreting the Virial

Here, an insight into the implication of virial is presented. The virial is mathematically defined as the product of the tip–sample force and the tip–position over a cycle of cantilever oscillation. The physical meaning of the virial can be understood in two ways.

The first way, as described by Giessibl [39], rewrites Eq. (3.24) through integration by parts as the following expression:

$$V_{\text{t-s}}(Z, A) = -\frac{A}{\pi} \int_{q=-A}^{A} k_{\text{ts}}(Z+q)\sqrt{1-\left(\frac{q}{A}\right)^2}\, dq \tag{3.28}$$

where the gradient of the tip–sample interaction is $k_{\text{ts}} = \partial F_{\text{ts}}^{\text{cons}} / \partial d = -\partial F_{\text{ts}}^{\text{cons}} / \partial q$. Thus the virial can be considered as a weighted measure of tip–sample force gradients

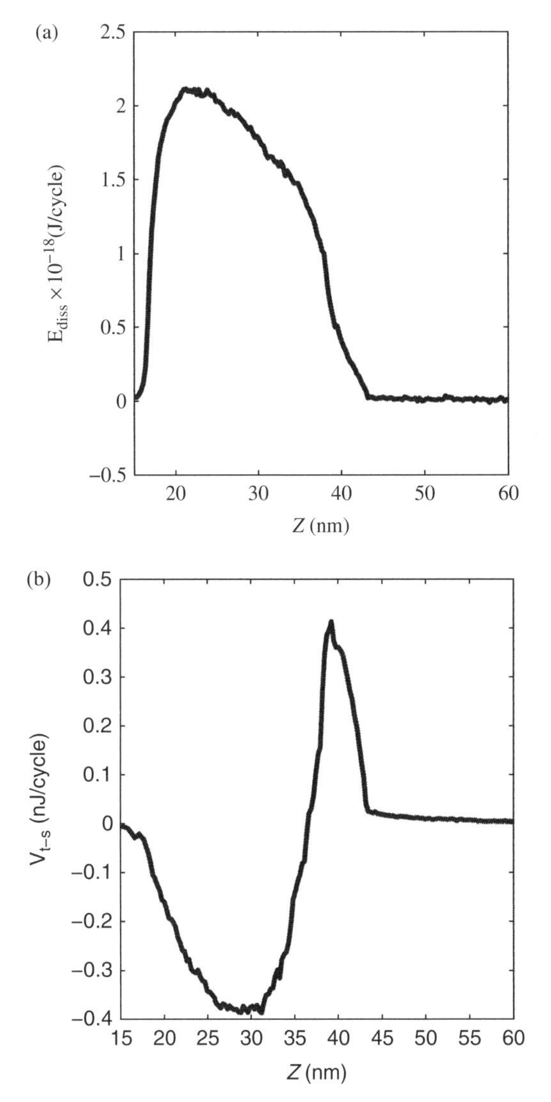

Figure 3.12 Tip–sample energy dissipation (a) and (b) virial plotted as a function of Z from the amplitude and phase data presented in Figure 3.1 (b, c) for a cantilever of stiffness $k = 22.3$ N/m, quality factor $Q = 403$, $\omega_0 = \Omega = 276.6$ kHz, and free amplitude $A_0 = 7.6$ nm.

over the oscillation cycle. The weight function is given by the semicircular kernel $\sqrt{1-(q/\mathrm{A})^2}$, which is maximum at the mean position of the cantilever and zero at the extremes of oscillation. It can also be shown that in the limit of small amplitudes, the expression can be approximated by [39]

$$V_{t-s}\left(Z, A\right) \sim -\frac{k_{ts}(Z)A^2}{2} \tag{3.29}$$

This shows that in the limit of infinitesimally small oscillation amplitudes, the virial can be regarded as the negative of the maximum potential energy stored in the tip–sample interaction. Figure 3.13(a) provides a graphical interpretation of Eq. (3.28). The area under the k_{ts} versus d plot is marked by the hatched region. The shaded region depicts the semicircular weight function $\sqrt{1-(q/A)^2}$.

An alternative way to interpret the virial arises by rewriting the virial [Eq. (3.24)] as follows:

$$V_{t-s}(Z, A) = \frac{1}{\pi} \int_{q=-A}^{A} F_{ts}^{cons}(Z+q) \frac{q}{\sqrt{A^2 - q^2}} dq \tag{3.30}$$

Equation (3.30) suggests that V_{t-s} is the weighted work done by the conservative forces as the tip moves from its lowest position to its highest position. The weight function $q/\sqrt{A^2-q^2}$ favors the work done at the extreme ends of the oscillation. Figure 3.13(b) depicts the above expression. The area under the oscillation range (from $-A$ to $+A$) of the F_{ts} versus d plot weighted by the shaded region $q/\sqrt{A^2-q^2}$ provides a measure of the virial.

3.3.3 Physics of Amplitude Reduction

As a cantilever driven at resonance is brought closer to the sample, its amplitude reduction is caused by both the conservative interaction forces as well as the dissipative forces. The final amplitude expression is arrived at by using the trigonometric identity $\sin^2\phi + \cos^2\phi = 1$, and eliminating ϕ from Eqs. (3.32) and (3.33):

$$A = \frac{A_0/Q}{\sqrt{(1-\Omega_{eff}^2)^2 + \left(\frac{1}{Q_{eff}}\right)^2}} \tag{3.31}$$

Equations (3.32) and (3.33) are obtained by rearranging Eqs. (3.26) and (3.27) as follows:

$$\frac{A_0}{QA} \sin\phi = \left(\frac{1}{Q_{eff}}\right) = \frac{1}{Q} + \frac{\omega_0}{\pi k_c A^2} E_{diss} \tag{3.32}$$

$$\frac{A_0}{QA} \cos\phi = \Omega_{eff}^2 - 1 = -\frac{2}{k_c A^2} V_{t-s} \tag{3.33}$$

Recall that the natural frequency of the probe is ω_0. However, when the cantilever oscillates close to the sample the tip–sample interaction potential modifies the natural frequency to (Ω_{eff} times ω_0); and Ω_{eff} is the nondimensional effective resonance

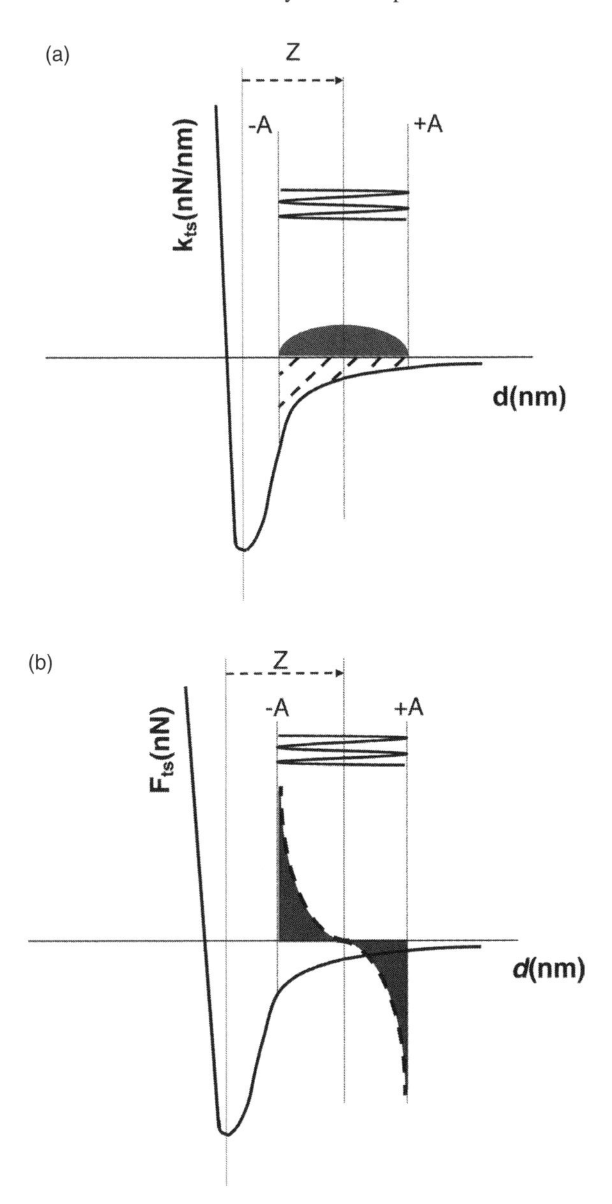

Figure 3.13 (a) Plot showing the physical meaning of virial understood through Eq. (3.28). The hatched area is the area under the k_{ts} vs. d plot and the shaded region is the weighting function. The virial is the area under the F–d curve covered by the tip oscillation weighted by the kernel $\sqrt{1-(q/A)^2}$. (b) Plot showing the physical meaning of virial understood through Eq. (3.30). The shaded region (bounded by the dashed curve) is the weight function. The virial is the area under the F_{ts} vs. d plot covered by the tip oscillation weighted by the shaded region $q/\sqrt{A^2-q^2}$. In both diagrams, the Z position is shown along with the harmonic tip oscillation of amplitude A. Adapted from [40].

frequency shift of the probe. When the tip encounters attractive forces, Ω_{eff} becomes less than 1 since V_{t-s} takes a positive value here. Similarly, in the repulsive regime Ω_{eff} is greater than 1 and the resonance frequency becomes higher than the initial case. Recall from Eqs. (3.23) and (3.24) that both V_{t-s} and E_{diss} depend on both the Z and the amplitude A at which they are measured. So too then must the frequency shift Ω_{eff} and the effective quality factor Q_{eff}, where Q_{eff} measures the change in effective Q factor due to the additional dissipative interactions.

The results from these analytical expressions show that when one scans over a sample with constant amplitude in the tapping mode, the consequent image is not a contour of constant force (which is the case for static AFM) or of force gradient (as in frequency-modulated AFM) but is the contour that combines tip–sample dissipation and conservative force gradients.

3.3.4 Attractive and Repulsive Regimes of Interaction

We derived an expression for the amplitude in terms of the conservative and dissipative interactions in Eq. (3.31). The denominator can take the same value for two different cases of Ω_{eff}; for values of $\Omega_{eff} < 1$ (attractive regime) and for $\Omega_{eff} > 1$ (repulsive regime). These two situations usually arise at two different Z displacements. Thus the dynamic approach–retraction curves exhibit the behavior of bistability wherein the same sample amplitude can be achieved at two different Z positions.

This phenomenon was observed by several authors [35, 41, 42]. Three typical amplitude curves observed from their experiments are shown in Figure 3.14 and are obtained on mica for a cantilever with free amplitude $A_0 = 16$ nm and free resonance frequency $f_0 = 326$ kHz. When the tip is far away from the sample, the amplitude remains constant as there are no interactions with the sample. The occurrence of these three types of curves can be explained by considering the denominator of the right-hand side term in Eq. (3.31). Figure 3.14(a) occurs for small free amplitudes A_0 when the tip is entirely in the attractive regime. Figure 3.14(b) is attributed to the tip moving from a region of strong attractive force to a region of strong repulsive force. Figure 3.14(c) is caused by the interplay of attractive and repulsive forces for large free amplitudes. To understand the factors that control the amplitude reduction and the formation of peaks, the amplitude curves were studied for a specific cantilever [41] on a surface for varying free amplitudes. The results show that the spike in the amplitude curves evolve from the tip transition from long-range attractive region to a region dominated by repulsive forces.

The main practical effect of this bistability is that if an amplitude setpoint is chosen such that two different Z values provide the same amplitude, then the controller can switch from one state to the other during a scan leading to jumps in Z (or topography) during a scan. To understand what probes and operating conditions to choose to avoid this instability, it becomes very important to be able to simulate approach curves using VEDA.

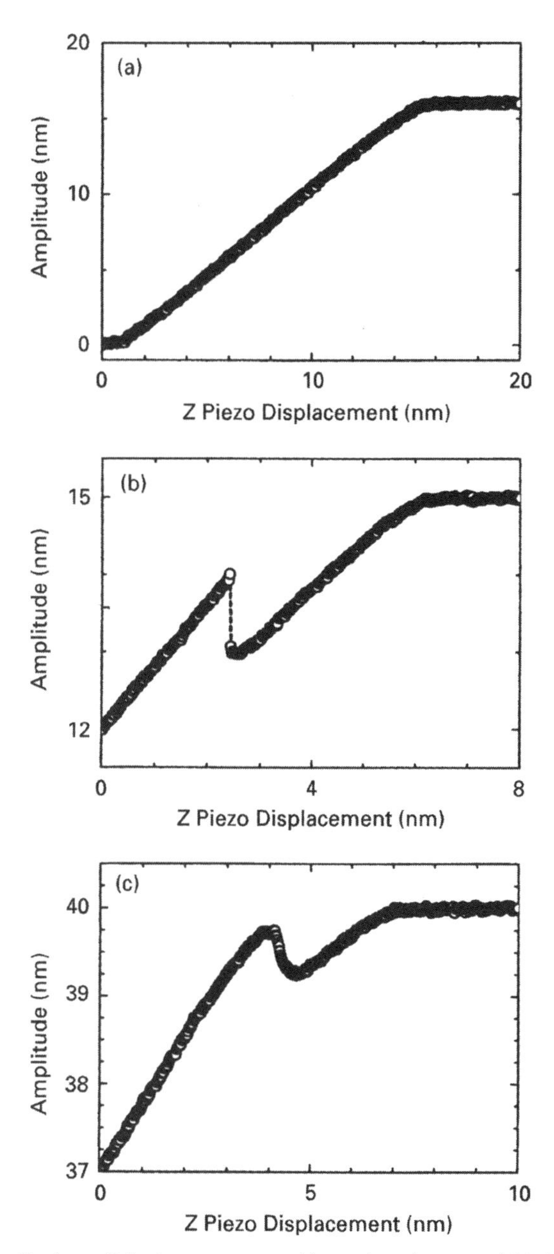

Figure 3.14 Amplitude vs. *Z*-displacement curves (a) on mica substrate and (b), (c) InAs/GaAs substrate. The free amplitudes are (a) 16 nm, (b) 15 nm, and (c) 40 nm. The nature of the amplitude curves depend on the setpoint value chosen as well as the nature of interactions. Adapted from [41].

3.3.5 Reconstruction of Forces

The observables from the dynamic AFM experiment (A, ϕ) have been used to calculate the virial and dissipation as a function of Z in the previous sections. The aim of force spectroscopy, however, is the measurement of the tip–sample forces against tip–sample gap. The most common methods involve the force reconstruction using the frequency modulation AFM technique where the drive frequency is locked to the nonlinear effective frequency of the probe [43]. We discuss briefly here the methods to reconstruct interaction forces using A versus Z and ϕ versus Z curves, which are acquired at a fixed drive frequency.

The inversion of the integral formulation of the steady-state equations of dynamic AFM was proposed first in Ref. [44] to reconstruct the interaction forces from A versus Z and ϕ versus Z curves. This was generalized in Ref. [45] wherein Chybechev polynomials were used to approximate the interaction force and the polynomial coefficients were estimated from the A versus Z and ϕ versus Z curves. Experiments were performed on common polymers to recreate the interaction forces. In spite of these methods, force reconstruction is most commonly performed when the drive frequency is locked into the resonance frequency of the probe [43]. This process has been described in many papers and texts on the subject of frequency modulation AFM and is not discussed further here.

3.4 BRIEF GUIDE TO VEDA SIMULATIONS

As described earlier, the analytical description of dynamic AFM allows one to convert the observables into the energy dissipation and the virial. However, understanding the connection between energy dissipation, virial, sample material properties, cantilever properties, and the operating conditions is quite challenging and nonintuitive [46, 47]. Where intuition fails, simulations can provide deep insight.

Virtual Environment for Dynamic Atomic Force Microscopy (VEDA) is an open-source suite of sophisticated simulation tools that can compute the response of an AFM probe in a variety of modes, on diverse material surfaces, and in different environments—air, vacuum, or liquids. The tool is deployed using the cyber-infrastructure of the nanohub (www.nanohub.org), which means that users need not download the software but rather they can run it from any Java-enabled web browser, have the computations performed in the cloud, and returned to the user. The tool can be manually located by selecting the Tools section from the Resources menu (from the nanohub.org home page), and browsing through the list of tools followed by selecting the Atomic Force Microscopy tag. Once that is done, click on VEDA 2.0 (Virtual Environment for Dynamic AFM) and hit Launch Tool.

The VEDA environment consists of a primary Applications drop-down menu that has the list of various tools. Presently, our interest is in the Force Distance curves and Amplitude Modulated Approach Curves. Within these tools are lists of parameters grouped into tabs according to their relevance. The tabs themselves will be located right below the Applications menu once the chosen application is loaded. The three most common tabs are Operating Conditions and Cantilever Properties, Tip–Sample Interaction Properties, and Simulate. The number of tabs vary depending on the chosen tool.

VEDA is designed in such a way that when the tool is loaded, a default set of parameters appears that lead to physically reasonable simulations. Also, there are a set of examples (located in the Example Loader box) that load predetermined parameter values directly when chosen. These values are chosen to guide the first-time user and aid the understanding of experimental results. It is strongly encouraged that users input values manually instead of loading the example to get a feel for the environment.

With this background, the following sections will guide you through the process of simulating static force–displacement curves and dynamic amplitude–phase distance curves. For users interested in additional tools, a manual is provided in the list of supporting documents attached to the VEDA simulator.

3.4.1 *F–Z* Curve Tutorial

Once VEDA is launched, choose Force Distance Curves under the Application menu. After the tool is loaded, the first tab (right under the header), which is named as Operating Conditions and Cantilever Properties, is automatically displayed. The other tabs appear dim when the current tab is displayed; however, the user can switch to any tab as and when required to change its values. The final tab is the Simulate tab, which runs the tool for the current set of values.

EXAMPLE 1 *Approaching and Retracting from a Sample Modeled Using DMT Contact*

Start with the Operating Conditions and Cantilever Properties tab and enter the values from Table 3.2. If the value of a particular parameter is not known (for, e.g., the Hamaker constant of a particular sample), it is best to retain the default value. Once the values in the first tab have been entered, click the button on the bottom right of the window that says Tip Sample Interaction Properties. This displays the Tip Sample Interaction Properties tab where information regarding the tip and sample are required. Notice that from the DMT Calculation Options, choose the Enter Adhesion Force, Auto-calculate Hamaker Constant and Intermolecular Distance menu for conveniently autocalculating the Hamaker constant and intermolecular distance. The user is required to input only the parameters relevant to the model. Jumping between tabs (to change values/rerun simulations) is achieved by directly clicking on the dimmed tab name.

Continue to input the parameters in each tab. The final tab is called Simulate. Selecting this will start the simulation. A progress bar appears that denotes the percentage of simulation completed. In certain cases, when spurious inputs are given, the Simulate tab displays an error displaying the reason. In certain other cases, the transient phase does not stabilize. This means that the controller does not reach the required setpoint for the given set of controller gains. As the tool is used more frequently, debugging becomes easier.

After the simulation is finished, plots are displayed within the Simulate tab. Tabs can once again be switched to check/change parameters. To complete the first example, click back on Operating Conditions and Cantilever Properties and change Initial *Z*-Separation to −5 and Final *Z*-Separation to 10. Click Simulate. This simulates the retraction curves as can be seen from

Table 3.2 Parameter Values for Examples 1, 2, and 3

Parameter	Example 1	Example 2	Example 3
Operating conditions and cantilever properties tab			
Number of eigenmodes	1	1	1
k_i (N/m)	0.87	0.87	0.1
Q_i	33	33	100
f_i (kHz)	44	44	50
Initial Z separation (nm)	10	10	5
Final Z separation (nm)	–5	–5	–5
Tip–sample interaction properties tab			
Tip–sample interaction model	DMT	JKR	Hertz
Tip radius (nm)	10	10	10
Young's modulus of tip (GPa)	130	130	130
Hamaker constant (J)	N/A	N/A	N/A
Adhesion Force (nN)	1.42	1.42	—
Young's modulus of sample (GPa)	1	1	0.5

the Z values. By changing the input parameters and clicking on Simulate; the old results do not disappear. Instead, a slider bar shows up below the plot indicating which simulation is currently being viewed. To view all results, click on All; which is to the left of the slider bar. If none of the parameters are changed, clicking the Simulate will not rerun the code, instead it displays the previously computed values. Different plots can be viewed by pulling down the Result menu and selecting the desired plot. To download the plot or to store the values to the hard drive, use the Download button to the far right of the Result menu. These values can later be processed in the user's favorite software (e.g., Matlab or Excel). Instead of running the simulation twice by interchanging the initial and final Z values, the user can directly choose the Approach and Retract option from the Choose Operating Mode menu that is the first label under Operating Conditions and Cantilever Properties.

RESULTS AND ANALYSIS Figure 3.15(c) shows the computed deflection of the cantilever versus Z distance. This is the readily available quantity in an AFM experiment. The force versus tip–sample gap is calculated within the program. Considering the F_{ts} versus d plot, there is a discontinuous region in d arising due to the snap-in and pull-off phenomenon. This region cannot be accessed since the cantilever is snapped into the sample. Stiffer cantilevers are needed to avoid snap-in instabilities. ■

EXAMPLE 2 ***Approaching and Retracting from a Sample Modeled Using JKR Theory***

This example helps to understand force spectroscopy with a JKR contact model and contrasts it with the DMT contact model. Follow the procedure below to run the simulation using VEDA. If you had simulated Example 1, it is necessary that the Clear button is clicked to erase previous results. Pull down Example 2: Approaching and Retracting from a Sample Modeled Using JKR

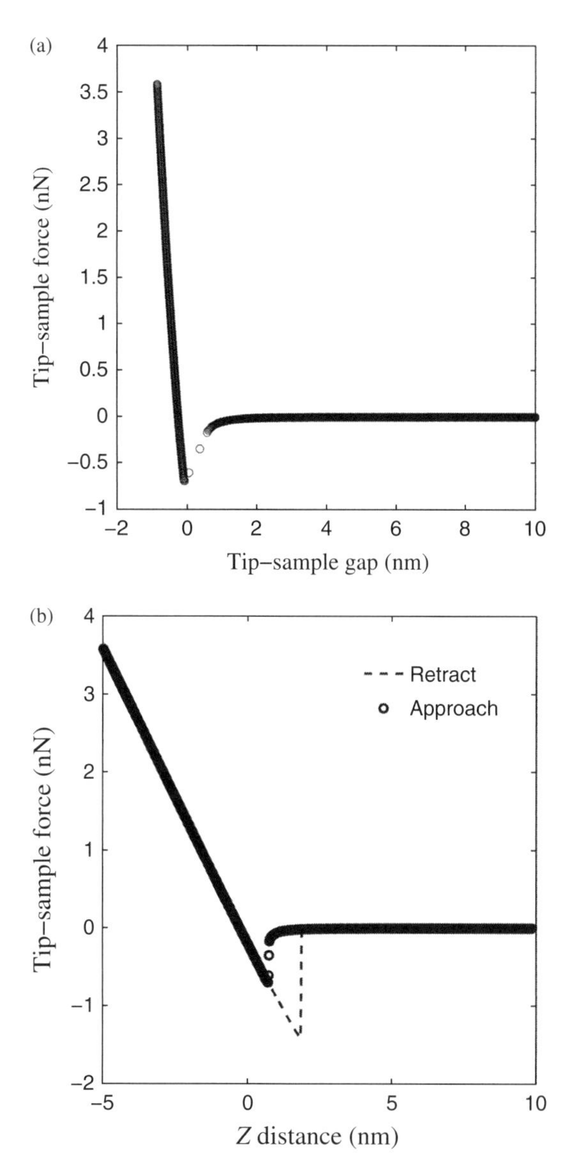

Figure 3.15 VEDA simulation of an F–Z curve using a cantilever of $k_c = 0.87$ N/m, $Q = 33$, and natural frequency of 44 kHz with DMT contact model: (a) F_{ts} vs. d, (b) F_{ts} vs. Z shows the snap-in phenomenon at 1 nm during approach. The retraction curve shows hysteresis, and the pull-off occurs at 2 nm, and (c) the observed deflection vs. Z displacement as would be displayed in the experiment.

from the Example Loader present in the Operating Conditions and Cantilever Properties tab. Doing this changes the values automatically without having to navigate from one tab to another. Alternately, you can enter the values manually from the table. Once the parameters are entered, hit the Simulate tab to run the first part of the simulation. From the values of initial and final Z

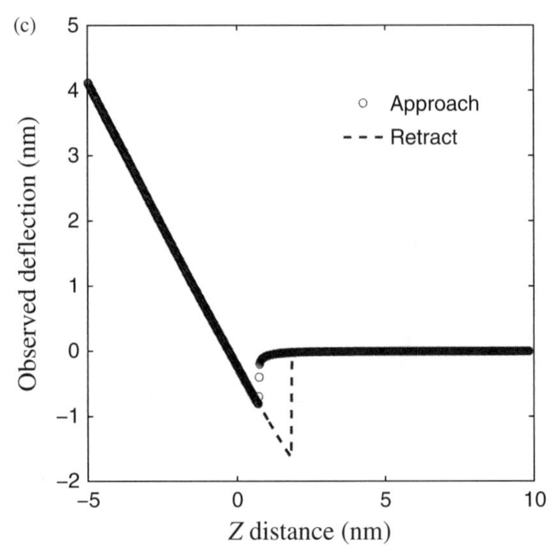

Figure 3.15 *(cont'd)*

displacement, it is obvious that the approach curve is simulated. To obtain the retraction curve, go back to Operating Conditions and Cantilever Properties tab and interchange the initial and final Z-separation values and now click on Simulate. The approach and retraction curves can both be seen simultaneously by clicking the All button next to the slider bar beneath the plot.

RESULTS AND ANALYSIS The primary difference between this example and the previous example is the type of model used. When the JKR model is used, it can be seen that the tip snaps into the sample only after it touches the sample. This is because there are no attractive forces in the JKR model. There is a discontinuity in the model that causes the cantilever to snap-in when $d < 0$ as seen in Figure 3.16(b). On retraction, if we follow the same procedure that we used to convert F–d curves to F–Z, we find the pull-off occurring at a greater distance. Once again, if we look at the F–d curves, we find a region of inaccessibility due to the snap-in. ■

The above two examples illustrate the fact that hysteretic F–Z curves are obtained from two different models, one that is conservative (DMT) and the other that is dissipative (JKR). This emphasizes the fact that no conclusion can be made on the presence of dissipative forces acting between the tip and sample just by looking at F–Z curves.

EXAMPLE 3 *Choosing the Right Cantilever for Experiments*

If one is required to perform an experiment on a biological sample having modulus of 0.5 GPa with the knowledge that a force of 200 pN or more will irreversibly damage it, which cantilever would be ideal from a set of 0.1, 1, and 10 N/m stiffness values?

To answer this question via simulations, after the force–distance curve tool is loaded, from the Operating Conditions and Cantilever Properties tab type in 0.1 N/m for the cantilever stiffness. Choose a quality factor of 100 (which is close to the actual value in air) and a natural frequency of

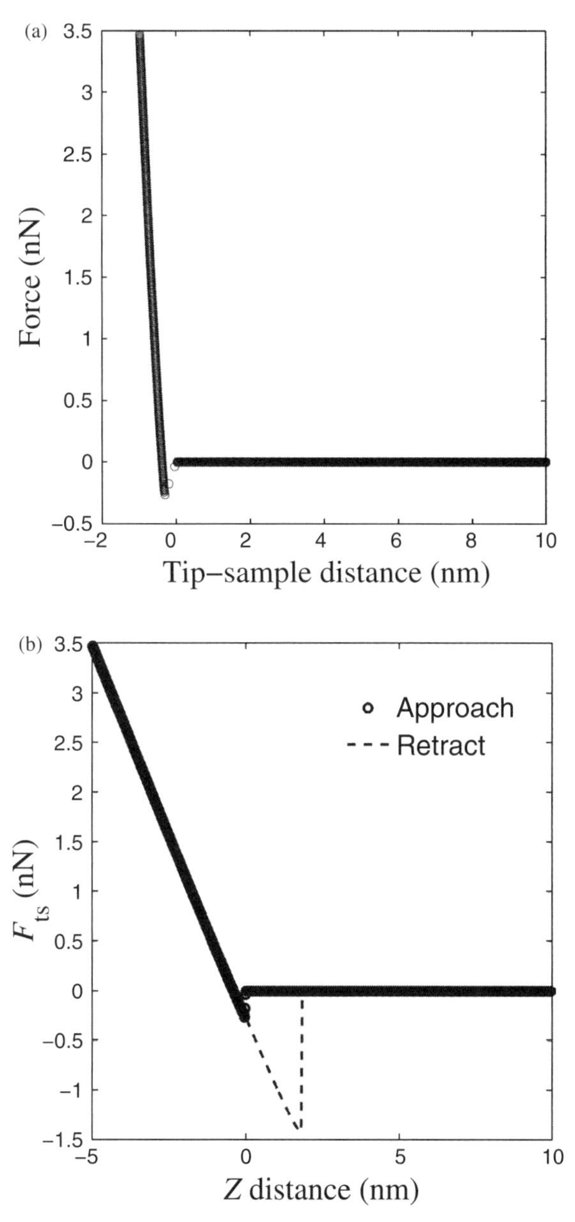

Figure 3.16 VEDA simulation of an F–Z curve using a cantilever of $k_c = 0.87$ N/m, $Q = 33$, and natural frequency of 44 kHz with JKR contact model: (a) F_{ts} vs. d, (b) F_{ts} vs. Z shows the snap-in phenomenon and hysteresis, and (c) the observed deflection vs. Z displacement as would be displayed from the experiment.

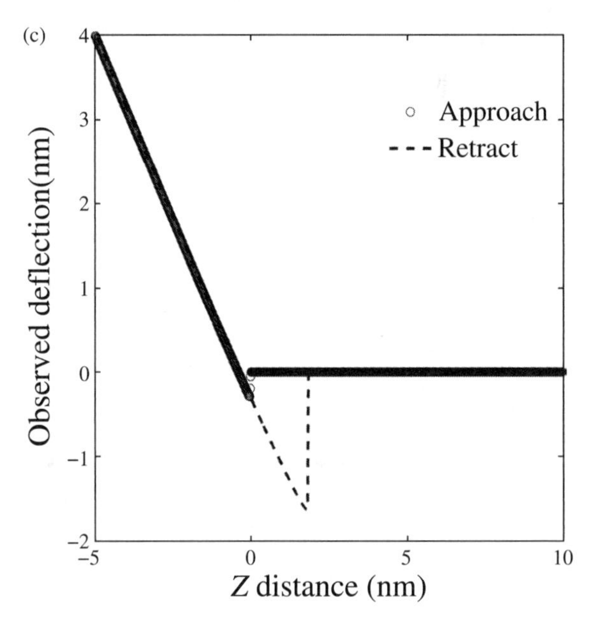

Figure 3.16 (*cont'd*)

50 kHz. Type an initial Z separation of 5 nm and a final Z separation of −5 nm. From the Tip Sample Interaction Properties tab choose a Hertz contact model and specify the stiffness (modulus) of the sample to be 0.5 GPa. The other parameter values are left unchanged from the default values. Now hit the Simulate button. Repeat the procedure by going back to the Operating Conditions and Cantilever Properties tab and changing the stiffness values to 1 and 10 N/m.

RESULTS AND ANALYSIS Figure 3.17 shows the force exerted by the tip for varying Z displacements. The cantilever having stiffness of 0.1 N/m exerts the minimum force for a given Z displacement. This cantilever is, therefore, ideal to perform F–Z curves since it can cover the maximum Z displacement of up to −2.2 nm without damaging the sample. On the other hand, we need to be extra careful when using the stiffer cantilevers because even a small step in the Z displacement exerts an appreciable force, and given the piezo creep, these cantilevers are most likely to damage the sample. ■

3.4.2 Amplitude/Phase–Distance Curves Tutorial

In VEDA, the amplitude-modulated approach curves are divided into two sections: basic and advanced. The basic tool consists of the minimum parameters required to carry out the experiment. For the more experienced user who is interested in carrying out experiments in liquid or who uses bimodal excitation (exciting at two different resonant frequencies of the cantilever), the advanced tool comes in handy. There are a number of examples for each tool. The manual [48] discusses these results individually. In this section, one example each from the basic and advanced tool is taken.

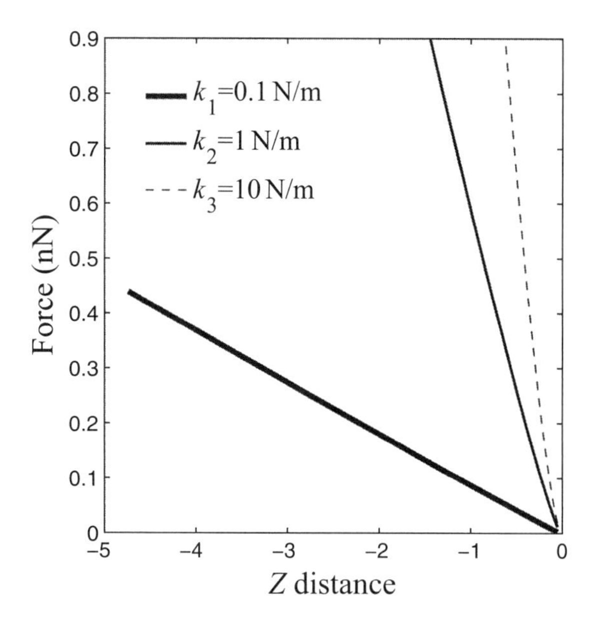

Figure 3.17 VEDA simulation of an *F–Z* curve using a cantilever of $k_c = 0.1$ N/m, $Q = 100$, and natural frequency of 50-kHz *F–Z* curves for spring stiffness of 0.1 N/m, 1, and 10 N/m on a biological sample of Young's modulus $E = 0.5$ GPa.

EXAMPLE 4 *Retraction Curves*

Example 4 consists of two parts. The first part simulates the retraction curves, and the second simulates the approach curves. This illustrates the use of dynamic approach curves in determining the attractive and repulsive regimes. This plot can be valuable if a scan has to be performed in tapping mode. Reasons for this will be discussed in the Results section.

To run the first part of the example, start by launching VEDA. From the Applications menu, choose Amplitude Modulated Approach Curves (Basic). Once this tool is loaded, from the pull-down menu titled Example Loader; select Example 2: Retraction Curves or manually enter the values from Table 3.3. The bold headings in the table correspond to the input tabs in VEDA that the user must choose to view the appropriate parameters. The parameters that are not mentioned in the table do not have to be changed from its default values. Similar to the previous examples, once the values are set, the Simulate tab is clicked to start the simulation. A progress bar shows the percentage completion. Bogus values will stop the simulation, resulting in the display of an error message. Notice that the initial *Z* displacement is 0 nm and the final is 35 nm (which is 5 nm higher than the unconstrained amplitude). This simulates the retraction curves.

To move on to the second part of the simulation, go back to the Operating Conditions and Cantilever Properties tab and swap the values for initial and final *Z* separation. Alternatively, you can load Example 1: Attractive and Repulsive Regimes of Oscillations from the Example Loader menu. This simulates the approach curves. Now both the approach and retract curves can be viewed as we have not cleared out the previous simulation. You will see a slider bar beneath the plot that indicates which simulation is currently in view. To view both results in the same plot click All.

Table 3.3 Parameter Values for Dynamic Approach/Retraction Curves

Parameter	Example 4	Example 5
Operating conditions and cantilever properties tab		
Number of frequencies	—	2
Unconstrained amplitude (nm)	30	17
k_i (N/m)	40	0.9
Q_i	400	225, 1000
f_i (kHz)	350	48.9, 306.6
f_d (kHz)	350	48.9, 306.6
Z approach velocity (nm/s)	200	100
Z range determination	Specify	Specify
Initial Z separation (nm)	0	20
Final Z separation (nm)	35	1.05
Tip–sample interaction properties tab		
Tip–sample interaction model	DMT	DMT
Tip radius (nm)	20	20
Young's modulus of tip (GPa)	130	130
Intermolecular distance (nm)	0.164	0.1
Hamaker constant (J)	7.1×10^{-20}	4×10^{-20}
Young's modulus of sample (GPa)	1	0.5

RESULTS AND ANALYSIS The default plot that is output is the first harmonic amplitude versus Z displacement. The next plot of importance is the variation of phase with Z displacement. This can be viewed by selecting the First Harmonic Phase from the results drop-down menu. These plots are shown in Figure 3.18. First consider the amplitude versus Z curve [Fig. 3.18(a)]. Initially, the cantilever is at a distance of 35 nm away from the sample (during the approach phase). At this distance, the cantilever tip does not interact with the sample, and therefore the unconstrained amplitude (30 nm) is observed. Since the cantilever is driven at resonance, the observed phase is 90°. As the Z displacement is reduced (moving from right to left in the plot), the tip starts to interact with the sample (at 30 nm), and we see a decrease in the amplitude. As the Z displacement is continually decreased, a spike is observed in the amplitude (at 27 nm). This denotes the transition from the attractive to repulsive regime. When the Z displacement is decreased below 27 nm, amplitude starts to decrease once again. These jumps in amplitude are accompanied by jumps in phase as discussed earlier.

If the user repeats the simulations for a different drive frequency (f_d) that is not equal to the natural frequency, the results will be slightly different. This is due to the detuning effects of the attractive and repulsive regimes accomplishing different results. The Mean Interaction Forces plot can also be used to determine the attractive and repulsive regimes. A negative force implies attractive regime and a positive force denotes repulsive regime oscillations.

Now let us consider the retraction curves in Figure 3.18. The initial Z displacement is set to zero. When the Z displacement increases, the amplitude also increases due to the same reason mentioned above. The important observation is that the spike is no longer observed at the same Z value leading to hysteresis. If an amplitude value of close to 27.5 nm is chosen, there are two possible values for Z satisfying this condition (one in the attractive branch and the other in the repulsive branch). The controller does not differentiate between these two

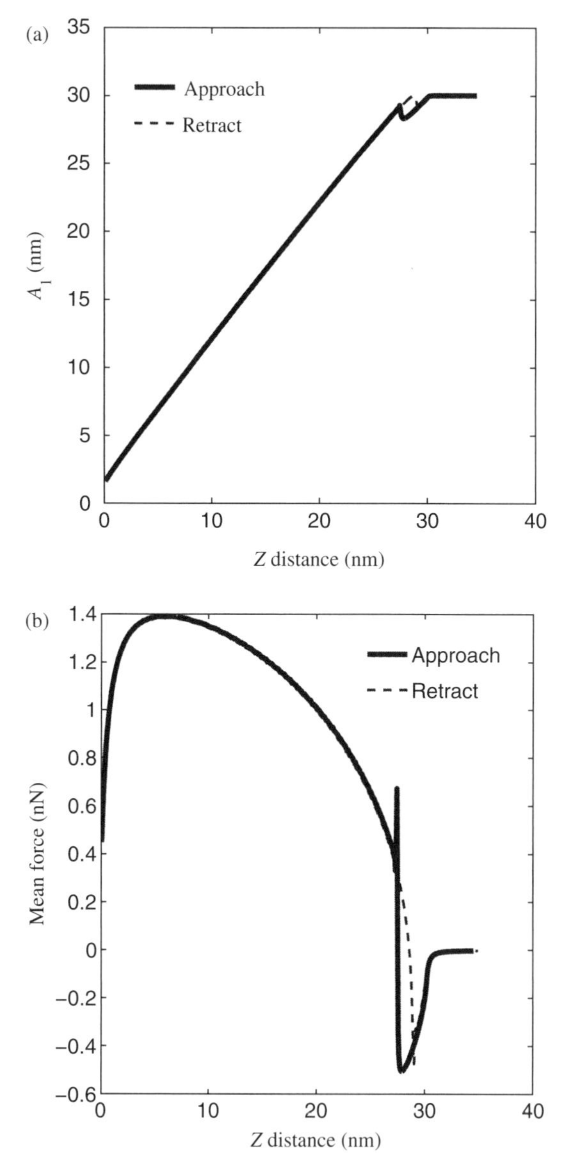

Figure 3.18 VEDA simulations of dynamic approach–retraction curves showing the results for Example 3 in the text: Approach and retraction curves for a cantilever of $k_c = 40$ N/m, $Q = 400$, $\omega_0 = \Omega =$ 350 kHz, a free amplitude of 30 nm: (a) Shows the amplitude variation with Z displacement. We can see a sudden increase in the amplitude when the cantilever enters the repulsive regime during approach. The jump from the repulsive to the attractive oscillation upon retraction occurs at a slightly different Z location. (b) The average interaction forces are initially negative in the attractive regime and suddenly become positive when the cantilever transitions into the repulsive regime. (c) The phase vs. Z displacement shows an increase (>90°) in the attractive regime and a dip (<90°) as the cantilever enters the repulsive regime.

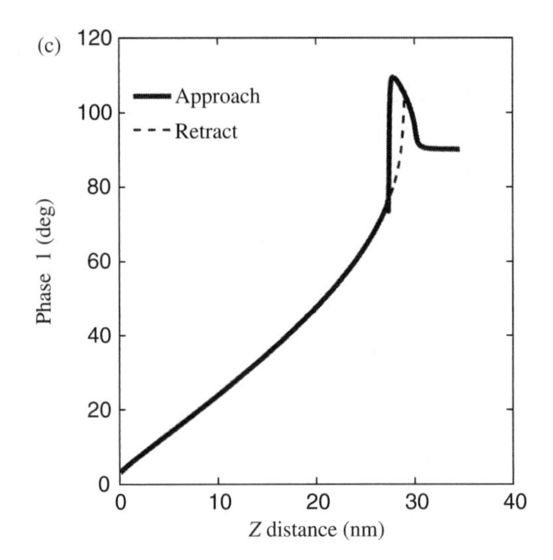

Figure 3.18 (*cont'd*)

values, and there is a high probability that the controller jumps from one regime to the other. This region is called the bistable region. ■

The above result is important in determining the setpoint ratio of amplitude-modulated scanning (tapping mode) because, if the setpoint amplitude is chosen in the bistable region, a jump into another branch is possible leading to dubious inversion in contrast.

Identifying the attractive and repulsive regimes can also help to avoid damaging the sample. If the sample is delicate it is recommended to image in the attractive regime.

3.4.3 Advanced Amplitude/Phase–Distance Curves Tutorial

In this section, we shall look at an example using the advanced feature of the amplitude-modulated approach curves available in VEDA. Example 5 uses the bimodal technique to study the contrast arising due to changing attractive forces.

EXAMPLE 5 ***Phase Spectroscopy in Bimodal AFM***

Bimodal excitation is one of the more advanced features in VEDA. This example simulates the bimodal excitation where the cantilever is excited at two frequencies (in this case, the frequency of the first and second modes) and observing the changes in results for varying conservative attractive forces (changes in Hamaker constant).

Enter the parameter values from Table 3.3 or select Phase Spectroscopy in Bimodal AFM from the example loader present in the Amplitude Modulated Approach Curves tool to automatically fit in the values required for this simulation. Once the values are entered, click on Simulate to start the simulation. If you look at the Simulation Parameters tab, you will see that

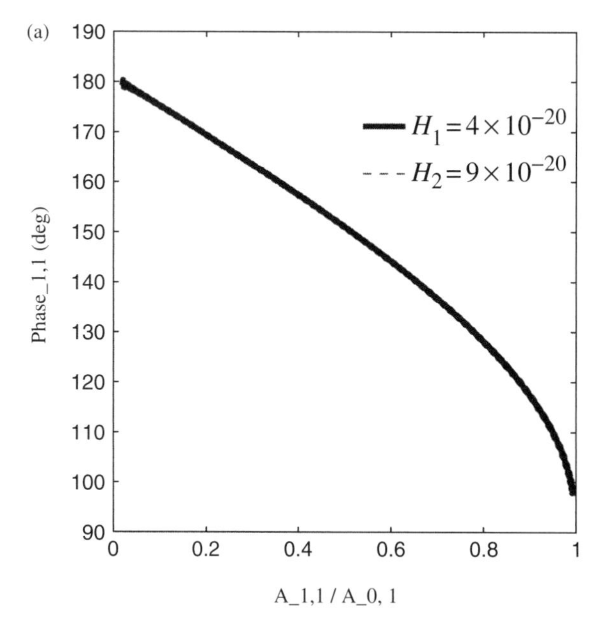

Figure 3.19 VEDA simulations showing the sensitivity of bimodal AFM observables to changes in conservative attractive forces. The first subscript refers to the harmonic and the second subscript to the mode. (a) First harmonic phase of the first mode plotted against the amplitude ratio of mode one, (b) first frequency phase plotted against the amplitude ratio of mode two, (c) second frequency phase plotted against amplitude ratio of mode one, and (d) second frequency phase plotted against the amplitude ratio of mode 2. The parallel mode representations [(a) and (d)] do not show a contrast for changes in Hamaker constant.

the x-Axis Variable is Amplitude Ratio. After the first run of the simulation is over, go back and change the x-Axis Variable to 2nd Frequency Amplitude Ratio and perform the simulation. After these two runs, change the Hamaker constant value (Tip–Sample Interaction Properties tab) from 4×10^{-20} to 9×10^{-20}, and repeat the above process to plot the results against both the Amplitude ratio and 2nd Frequency Amplitude Ratio. In total, four simulation runs should have been performed.

RESULTS AND ANALYSIS Material contrast can be achieved only in the cross-mode representation—plotting the amplitude phase of one mode with the phase/amplitude of the other mode. This was studied in [49] where the contrast was explained by combining parameters from both modes into the total virial of the system. The result showed that the parallel-mode representation (amplitude of one mode plotted against the phase of the same mode) is not sensitive to changes in Hamaker constant.

Figure 3.19 summarizes these results. On cross-mode representation [Figs. 3.19(b) and 3.19(c)], the contrast between the two material properties can be observed. On the other hand parallel-mode representation shows no such contrast.

The other advanced features include simulations in a liquid environment using multiple excitation schemes. The reader is encouraged to try out the examples that come along with the tool and interpret the results. For more insight into the results, the reader is directed to the VEDA manual where the theory of formulation for different models (cantilever, controllers, etc.) is also provided. ■

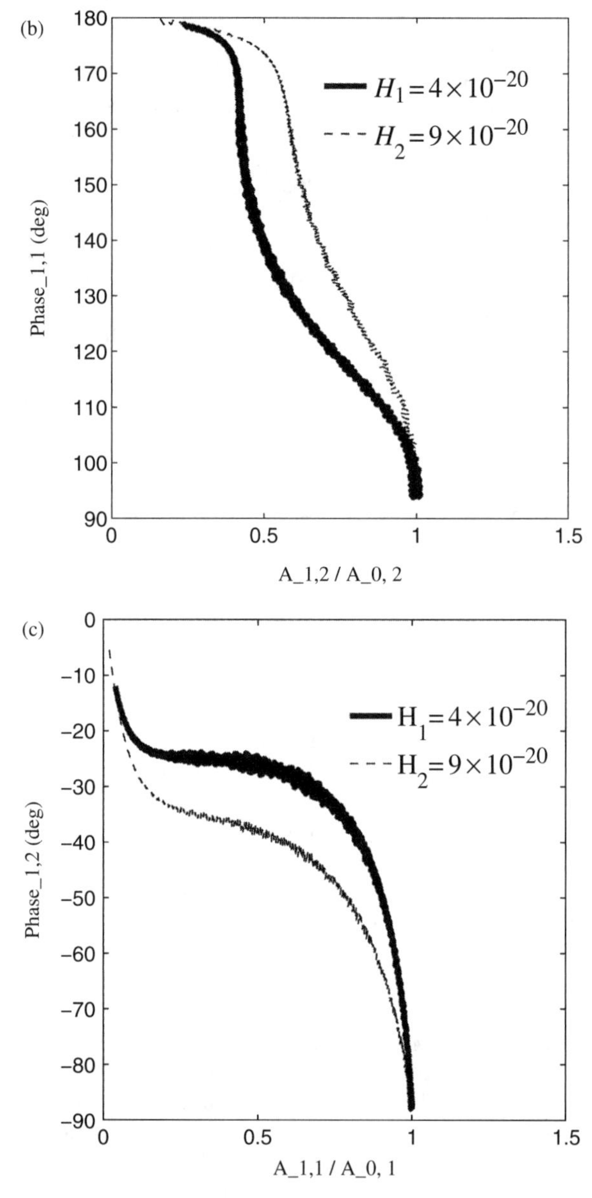

Figure 3.19 (*cont'd*)

3.5 CONCLUSIONS

In summary we have discussed in detail the relationships between the local material properties and the static and dynamic approach and retraction curves. While static force–distance methods provide a direct quantitative measurement of the force, they are unable to measure forces in some ranges of tip–sample gaps and cannot distinguish

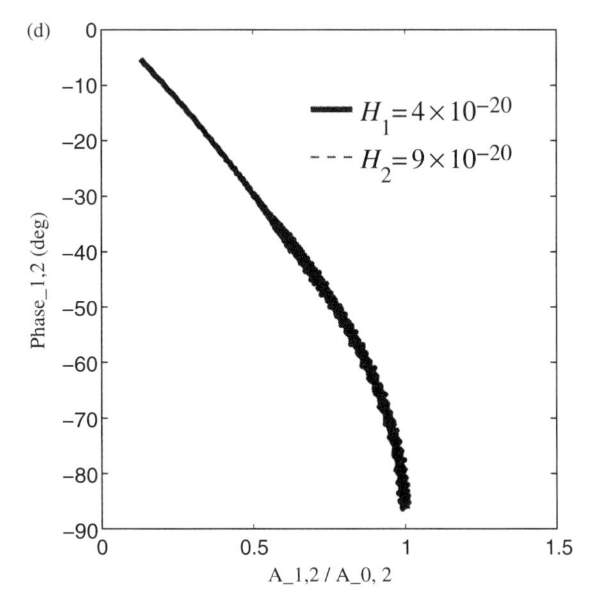

Figure 3.19 (*cont'd*)

between a true tip–sample hysteresis and one caused by tip pull-in an pull-out due to conservative forces. Dynamic approach curves either at fixed drive frequency or in the frequency modulation mode bypass these restrictions; however, converting the observables such as amplitude and phase into tip–sample forces is not easy and requires the use of simulations.

We also presented a brief guide to the use of VEDA simulations to provide insight into static and dynamic approach and retraction curves. VEDA simulations offer the opportunity to explore these links for a surprisingly diverse variety of sample materials, operating conditions, and spectroscopy modes.

GLOSSARY

F–Z Curve The tip–sample interaction force (F_{ts}) plotted against the *Z* displacement. The *Z* displacement is defined as the distance between the undeflected cantilever tip and the sample changed by moving the base of the cantilever, or the sample. This curve is typically obtained from an atomic force microscopy (AFM) experiment.

F–d Curve The tip–sample interaction force (F_{ts}) plotted against the tip–sample distance *d*. This curve is extracted from experimental *F–Z* curves. *F–d* curves contain direct information about surface physics.

Dissipative Force (Nonconservative Force) The force values are different depending on which direction the cantilever tip traverses. The *F–d* curve thus shows hysteresis, and the area under this loop equals the energy dissipated. Examples include viscoelastic force, capillary forces, and plastic deformation.

Conservative Force These forces are independent of the velocity, direction of the cantilever tip, and depend only on the position. The magnitude of the conservative force is the average value of force as the tip approaches and recedes from the sample with no loss in energy. Examples include van der Waals force and electrostatic force.

Tip–Sample Dissipation The energy dissipated by the tip into the sample over one cycle of oscillation. Alternatively, it is the irreversible work done in one oscillating cycle by the tip on the sample.

Tip–Sample Virial Mathematically, it is regarded as the integral of the force times tip position over a cycle of oscillation. The physical meaning of this term can be thought of as the weighted measure of the conservative tip–sample force gradients encountered by the tip over an oscillation cycle.

Dynamic Approach Curve The characteristic curves obtained when the "excited" or oscillated cantilever is brought closer to the sample. These include amplitude versus *Z*-displacement curves, phase versus *Z*-displacement curves, drive amplitude versus *Z*-displacement, and so forth depending on the mode of operation and parameter of interest.

Phase Both the cantilever excitation force and the cantilever response are sine waves at the same frequency. However, there will be some delay between when the excitation force is applied and when the cantilever responds. Therefore, the two sine waves will be shifted with respect to each other, and this is referred to as "phase." There are two possible sign conventions for phase, which are referred to as "lead" and "lag."

Point Mass Model The AFM cantilever can be reduced to a spring–mass–damper system for most calculations. This is called the point mass model, and the equivalent parameter values for the reduced model can be obtained through various methods that include the energy conservation principle.

Peak Force The maximum force that is applied by the cantilever tip on the sample during one oscillation cycle.

Attractive–Repulsive Regime In static mode, the attractive regime can be defined as the region where the gradient of the interaction force is attractive and vice versa. In dynamic mode, the attractive regime is characterized by a phase lag greater than 90° while the repulsive regime is identified by a phase lag that is lower than 90° (i.e., in the attractive regime the tip–sample virial is positive, the frequency shift is negative).

Static Force Spectroscopy The method of obtaining force–distance curves from force–displacement curves to understand local tip–sample interactions, and other surface properties is termed "static" force spectroscopy. If these force–distance curves are obtained from dynamic measurements such as amplitude and phase curves, this process is called "dynamic" force spectroscopy. The dynamic force spectroscopy is more advanced than the static spectroscopy.

REFERENCES

1. Aoki, T., Hiroshima, M., Kitamura, K., Tokunaga, M., and Yanagida, T. Non-contact Scanning Probe Microscopy with Sub-piconewton Force Sensitivity, *Ultramicroscopy* **70**(1–2) (1997): 45.
2. Rotsch, C., and Radmacher, M. Mapping Local Electrostatic Forces with the Atomic Force Microscope, *Langmuir* **13**(10) (1997): 2825.
3. Martínez, N. F., Lozano, J. R., Herruzo, E. T., García, F., Richter, C., Sulzbach, T., and García, R. Bimodal Atomic Force Microscopy Imaging of Isolated Antibodies in Air and Liquids, *Nanotechnology* **19** (2008): 384001.
4. Torii, A., Sasaki, M., Hane, K., and Okuma, S. Adhesive Force Distribution on Microstructures Investigated by an Atomic-Force Microscope, *Sensors and Actuators A–Physical* **44**(2) (1994): 153–158.
5. Thundat, T., Zheng, X. Y., Chen, G. Y., Sharp, S. L., Warmack, R. J., and Schowalter, L. J. Characterization of Atomic-Force Microscope Tips by Adhesion Force Measurements, *Applied Physics Letters* **63**(15) (1993): 2150–2152.
6. Eastma, T., and Zhu, D. M. Adhesion Forces between Surface-Modified AFM Tips and a Mica Surface, *Langmuir* **12**(11) (1996): 2859–2862.
7. Capella, B., Baschieri, P., Frediani, C., Miccoli, P., and Ascoli, C. Force-Distance Curves by AFM—A Powerful Technique for Studying Surface Interactions, *IEEE Engineering in Medicine and Biology Magazinze* **16**(2) (1997): 58–65.
8. Butt, H. J. Electrostatic Interaction in Atomic Force Microscopy, *Biophysical Journal* **60**(4) (1991): 777–785.
9. Albrecht, T. R., Grütter, P., Horne, D., and Rugar, D. Frequency Modulation Detection Using High-*q* Cantilevers for Enhanced Force Microscope Sensitivity, *Journal of Applied Physics* **69** (1991): 668.
10. Ueyama, H., Sugawara, Y., and Morita, S. Stable Operation Mode for Dynamic Noncontact Atomic Force Microscopy, *Applied Physics A—Materials Science & Processing* **66**(Part 1, S) (1998): S295.
11. Crittenden, S., Raman, A., and Reifenberger, R. Probing Attractive Forces at the Nanoscale Using Higher-Harmonic Dynamic Force Microscopy, *Physical Review B* **72** (2005): 235422.
12. Derjaguin, B. V., Muller, V. M.,and Toporov, Y. P. Effect of Contact Deformation on the Adhesion of Particles, *Journal of Colloid and Interface Science* **53**(2) (1975): 314.
13. Butt, H. J., Cappella, B., and Kappl, M. Force Measurements with the Atomic Force Microscope: Technique, Interpretation and Applications, *Surface Science Reports* **59**(1–6) (2005): 1.
14. Israelachvili, J. *Intermolecular and Surface Forces*, 2nd ed., Academic Press Limited: London, 1992.
15. Mate, C. M., Lorenz, M. R., and Novotny, V. J. Atomic Force Microscopy of Polymeric Liquid-Films, *Journal of Chemical Physics* **90**(12) (1989): 7550.
16. Butt, H. J. Measuring Electrostatic, Vanderwaals, and Hydration Forces in Electrolytesolutions with an Atomic Force Microscope, *Biophysical Journal* **60**(6) (1991): 1438.
17. Wagner, R., Moon, R., Pratt, J., Shaw, G., and Raman, A. Uncertainty Quantification in Nanomechanical Measurements Using the Atomic Force Microscope, *Nanotechnology* **22**(45) (2011): 455703.
18. Siedle, P., and Butt, H. J. Artifacts in Force Measurements with the Atomic Force Microscope Due to Digitalization, *Langmuir* **11**(4) (1995): 1065.
19. Hoh, H. J., and Engel, A. Friction Effects on Force Measurements with an Atomic Force Microscope, *Langmuir* **9**(11) (1993): 3310.
20. D'Costa, N. P., and Hoh, J. H. Calibration of Optical Lever Sensitivity for Atomic Force Microscopy, *Review of Scientific Instruments* **66**(10) (1995).
21. Schäffer, T. E. Calculation of Thermal Noise in an Atomic Force Microscope with a Finite Optical Spot Size, *Nanotechnology* **16**(6) (2005): 664.
22. Jona, F., and Shirane, G. *Ferroelectric Crystals*, Macmillan Co.: New York, 1962.
23. Lines, M. E., and Glass, A. M. *Principles and Applications of Ferroelectrics and Related Materials*, Clarendon: Oxford, 1977.

24. Leang, K. K., and Devasia, S. Design of Hysteresis-Compensating Iterative Learning Control for Piezo-positioners: Application to Atomic Force Microscopes, *Mechatronics* **16**(3–4) (2006):141.
25. Cruz-Hernandez, J. M., and Hayward, V. Phase Control Approach to Hysteresis Reduction, *IEEE Transactions on Control Systems Technology* **9**(1) (2001): 17–26.
26. Mokaberi, B., and Requicha, A. A. G. Towards Automatic Nanomanipulation: Drift Compensation in Scanning Probe Microscopes, IEEE International Conference on Robotics and Automation, Vol. 1–5 (2004): 416.
27. Huerth, S. H., and Hallen, H. D. Quantitative Method of Image Analysis When Drift is Present in a Scanning Probe Microscope, *Journal of Vacuum Science & Technology* **21**(2) (2003): 714.
28. Staub, R., Alliata, D., and Nicolini, C. Drift Elimination in the Calibration of Scanning Probe Microscopes, *Review of Scientific Instruments* **66**(3) (1995): 2513.
29. Woodward, J. T., and Schwartz, D. K. Removing Drift from Scanning Probe Microscope Images of Periodic Samples, *Journal of Vacuum Science & Technology* **16**(1) (1998): 51.
30. Yurov, V. Y., and Klimov, A. N. Scanning Tunneling Microscope Calibration and Reconstruction of Real Image—Drift and Slope Elimination, *Review of Scientific Instruments* **65**(5) (1994): 1551.
31. Leang, K. K., and Devasia, S. Feedback-Linearized Inverse Feedforward for Creep, *Hysteresis, and* Vibration Compensation in AFM Piezoactuators, *IEEE Transactions on Control Systems Technology* **15**(5) (2007): 927–935.
32. Schafer, J., and Janocha, H. Compensation of Fysteresis in Solid-State Actuators, *Sensors and Actuators A–Physical* **49**(1–2) (1995): 97–102.
33. Kaizuka, H., and Siu, B. A Simple Way to Reduce Hysteresis and Creep When Using Piezoelectric Actuators, *Japanese Journal of Applied Physics* **27**(Part 2, No. 5) (1988): L773–L776.
34. Kwon, J., Hong, J., Kim, Y. S., Lee, D. Y., Lee, K., Lee, S. M., and Park, S. I. Atomic Force Microscope with Improved Scan Accuracy, Scan Speed, and Optical Vision, *Review of Scientific Instruments* **74**(10) (2003): 4378–4383.
35. Wang, L. Analytical Descriptions of the Tapping-Mode Atomic Force Microscopy Response, *Applied Physics Letters* **73**(25) (1998): 3781.
36. Anczykowski, B., Gotsmann, B., Fuchs, H., Cleveland, J. P., and Elings, V. B. How to Measure Energy Dissipation in Dynamic Mode Atomic Force Microscopy, *Applied Surface Science* **140** (1999): 376.
37. Melcher, J., Hu, S., and Raman, A. Equivalent Point-Mass Models of Continuous Atomic Force Microscope Probes, *Applied Physics Letters* **91** (2007): 53101.
38. http://nanohub.org/resources/9961.
39. Giessibl, F. J. Forces and Frequency Shifts in Atomic-Resolution Dynamic-Force Microscopy, *Physical Review B* **56**(24) (1997): 16010–16015.
40. Giessibl, F. J. A Direct Method to Calculate Tip–Sample Forces from Frequency Shifts in Frequency-Modulation Atomic Force Microscopy, *Applied Physics Letters* **78** (2001): 123.
41. García, R., and San Paulo, A. Amplitude Curves and Operating Regimes in Dynamic Atomic Force Microscopy, *Ultramicroscopy* **82**(1–4) (2000): 79.
42. Lee, S. I., Howell, S. W., Raman, A., and Reifenberger, R. Nonlinear Dynamics of Microcantilevers in Tapping Mode Atomic Force Microscopy: A Comparison Between Theory and Experiment, *Physics Review B* **66** (2002): 115409.
43. Welker, J., Illek, E., and Giessibl, F. J. Analysis of Force-Deconvolution Methods in Frequency-Modulation Atomic Force Microscopy, *Beilstein Journal of Nanotechnology* **3** (2012): 238.
44. Lee, M., and Jhe, W. General Theory of Amplitude-Modulation Atomic Force Microscopy, *Physical Review Letters* **97**(3) (2006): 036104.
45. Hu, S., and Raman, A. Inverting Amplitude and Phase to Reconstruct Tip–sample Interaction Forces in Tapping Mode Atomic Force Microscopy, *Nanotechnology* **19** (2008): 375704.
46. Kim, J. H., Jang, J., and Zin, W. C. Estimation of the Thickness Dependence of the Glass Transition Temperature in Various Thin Polymer Films, *Langmuir* **16**(9) (2000): 4064.
47. Sun, C. T., and Zhang, H. Size-Dependent Elastic Moduli of Platelike Nanomaterials, *Journal of Applied Physics* **93**(2) (2003): 1212–1218.
48. https://nanohub.org/tools/adac.
49. Lozano, J. R., and García, R. Theory of Phase Spectroscopy in Bimodal Atomic Force Microscopy, *Physical Review B* **79** (2009): 014110.

Chapter 4

Phase Imaging

Dalia G. Yablon[1] and Greg Haugstad[2]

[1] *SurfaceChar LLC, Sharon, MA*
[2] *University of Minnesota, Minneapolis, MN*

4.1 INTRODUCTION

Phase imaging is perhaps one of the best known and most widely used modes in atomic force microscopy (AFM). There are several reasons for its popularity. First, it is a simple method in that it is a channel collected in tapping mode (also called amplitude modulation mode, intermittent contact mode) and requires no postprocessing. Second, it is one of the easiest to implement modes that is also sensitive to material properties, which is one of the key attractions of AFM aside from providing nanoscale topographic information. There has been a significant amount of research activity trying to relate the phase signal to actual material properties such as storage modulus, loss modulus, and loss tangent. However, phase imaging is rather complicated and interpretation of its meaning is still widely debated and researched [1–4].

Fundamentally, the phase shift measured in a phase image is a measure of the energy dissipated by the cantilever into the material. It is a measure of both conservative and nonconservative tip–sample interaction forces (see Chapter 3), which really depends on a number of factors including viscoelasticity, adhesion, capillary forces, and contact area. That being said, phase imaging is a very convenient and easy-to-use method to distinguish materials based on their material properties, as can be evidenced by numerous examples in the applications section.

Phase shift is always measured with respect to a reference and, in this case, to the driven cantilever. The cantilever is driven at a resonant frequency and interacts with the sample at a given oscillation amplitude set by the user. The phase shift (ϕ) is then induced by interacting with the sample, as illustrated in Figure 4.1. This phase shift is then mapped as a channel while the tip raster scans over the surface.

At resonance, the phase lag of a driven harmonic oscillator with damping is 90° with respect to the driving or excitation force, as shown in Figure 4.2 where a phase

Scanning Probe Microscopy in Industrial Applications: Nanomechanical Characterization, First Edition. Edited by Dalia G. Yablon.

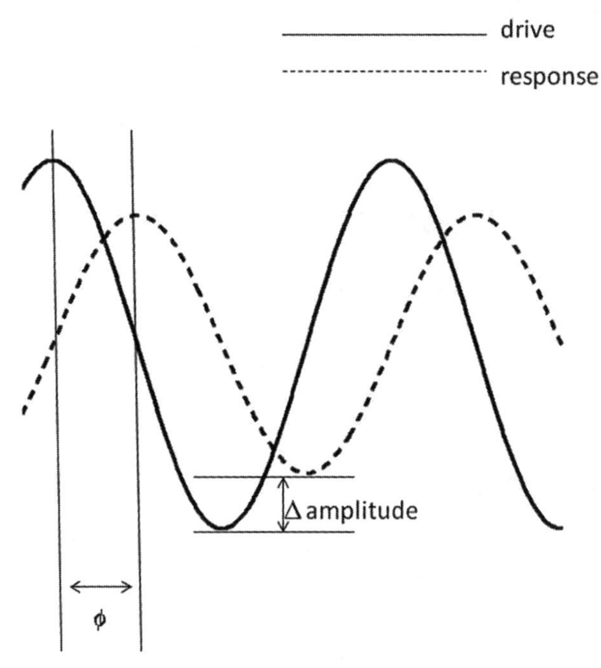

Figure 4.1 Illustration of the concept of phase lag. The cantilever is driven at a certain resonant frequency with a given oscillation amplitude (solid line). As it interacts with the surface, the response of the cantilever (dotted line) is an oscillation that shifts by amount (ϕ), the phase lag. The oscillation amplitude will also reduce as a result of interacting with the surface from the free air oscillation, as marked by the Δ amplitude.

lag of 90° is easily observed at the frequency resonance peak, F_0. To understand the concept of phase shift and a 90° shift at resonance, one can try the following experiment at home. Shake a flexible ruler. Once the ruler is at resonance, watch the motion of the wrist with respect to the peak of the ruler or string. The peak of one's wrist will be 90° out of phase with the peak of the ruler or string's motion at resonance.

4.1.1 Definition of Phase: Instrument (Vendor) Dependent

Note that in some instruments this 90° at resonance is labeled as "zero degrees" and has been made the reference value. This causes some confusion as different instrument vendors employ different conventions for phase, whether actually displaying the true phase (90° at resonance) or "reference" values (0° at resonance). A table of main AFM vendors and their convention for definition of phase is shown in Table 4.1.

Specific attention is now turned to the phase measurement in one of the most common commercial AFMs on the market, manufactured by Bruker (formerly Veeco and DI). The most common source of difficulty is the so-called Extender electronics package sold with certain first-generation commercial systems (Bruker), which nevertheless

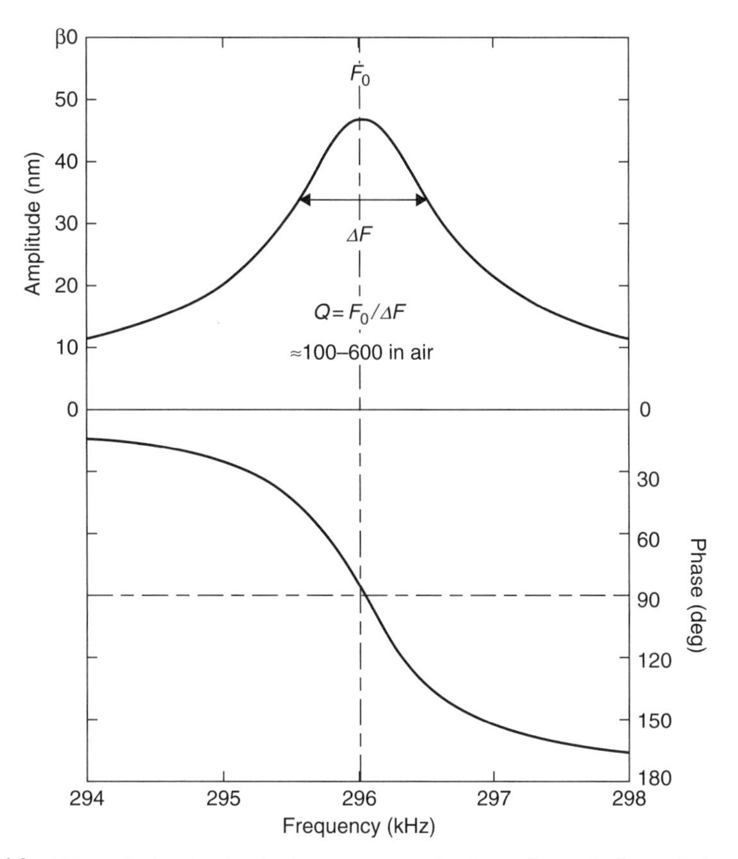

Figure 4.2 Schematic showing that (top), a resonance peak of a cantilever. On (bottom), the phase lag is at 90° on resonance.

remain in common use (i.e., remain valuable instrumentation). The Phase Extender unit is an add on to many of the early generation controllers (Nanoscope III–IV) and was designed for purposes other than to measure phase per se [5]. A measurement of *true phase* requires a lock-in amplifier or similarly sophisticated device to track the relative offset in time of two sine waves and, moreover, with a frequency range extending up to hundreds of kilohertz. But the Extender can be calibrated to provide numbers very close to true phase for most measurements of interest. A detailed explanation of this calibration is provided in Section 4.3.3 for those interested.

Now the artifacts and challenges associated with phase imaging are discussed.

4.2 BISTABILITY: ATTRACTIVE AND REPULSIVE MODE

Phase can play a critical *diagnostic* role in controlling topographic imaging. It is well established in the scientific literature that an oscillating AFM tip (and cantilever) interacting with a sample surface can be *bistable* [6–9]. This means that for a given

Table 4.1 Main AFM Vendors

Vendor	Phase at resonance	Phase change in net attractive regime	Phase change in net repulsive regime
Agilent	0	<0	>0
Asylum Research	90	>90	<90
Bruker (NS V controller)	0	<0	>0
Bruker (NS III and NS IV controller)[a]	0	<0	>0
NT-MDT	0	>0	>0
AIST	0	>0	<0

[a] phase in Bocek degrees - needs to be calibrated for true phase degrees.

(average) Z position of tip relative to sample, one of two possible states can arise. We refer to these as (net) attractive and (net) repulsive regimes. (Some authors use the terminology *high- and low-amplitude states*, for a given Z positional value [8].)This concept means that during AFM imaging (i.e., at a fixed setpoint amplitude) there are *two possibilities for height*: one for which the net tip–sample interaction is repulsive and the other for which the net interaction is attractive (i.e., tip does not reach as close to the surface). Bistable means that at any location the oscillating tip can settle into one state *or* the other, net attractive or repulsive. Enough oscillations occur at each pixel that a unique dynamic equilibrium can be reached; that is, the oscillator may stabilize in one regime or the other *in specific subregions of the image* in correspondence with local sample properties [10].

Artifacts induced by bistability are well known to practitioners of phase imaging and are responsible for many of the observed contrast reversal artifacts. When doing phase imaging, it is very important to understand whether the image is collected in the net-attractive or net-repulsive regime. Bistability and operation in the two different regimes is perhaps the most significant source of confusion in interpretation of tapping mode-phase images (height channel, phase channel, as described below). Fortunately, the phase signal provides a useful measure of the regime of operation.

In the net-attractive regime, the phase is greater than 90°; in the net-repulsive regime, the phase is less than 90°. Because different vendors define the phase at resonance at different values (e.g., 90° or 0°), the different regimes of net attractive and net repulsive for AFMs manufactured by various vendors is also included in Table 4.1. Both of these regimes are referred to in the net terminology because, for example, the net attractive regime can have some repulsive interactions as well, but the net interaction is attractive. Much less energy is dissipated in the attractive regime than in the repulsive regime. This can be easily understood in the attractive regime on typical hard surfaces (e.g., silicon) because this corresponds to noncontact imaging where the tip is not really making contact with the sample, hence no real pathway for energy dissipation. Additionally, this regime generally preserves the tip shape better as the tip wear is minimized. On the other hand, a larger tip or one that has worn will have more attractive forces than a sharper tip.

Fortunately, there are several experimental knobs that can tune between the net-attractive and net-repulsive mode of operation. Drive frequency is the first parameter, where drives slightly above resonance favor the attractive regime, while drives slightly below resonance favor the repulsive regime. Larger oscillation amplitudes (controlled by the drive voltage) mean more kinetic energy and thus also favor repulsive over attractive regimes. The amplitude setpoint is a final parameter to tune between the regimes where lower amplitude setpoint voltage will favor repulsive interaction and higher amplitude setpoint voltage (i.e., a setpoint closer to the free air oscillation setpoint) will favor attractive interaction.

The data in Figures 4.3 and 4.4 exemplify the issue of bistability of the attractive and repulsive modes on an ultrathin film of polyvinyl alcohol cast on cleaved mica from aqueous solution. These films contain three characteristic domains differing in properties [10, 11]. A first layer completely covers the substrate and is strongly adhered [11]. Two types of discontinuous second-layer domains differ in crystallinity; minority domains 1 nm thick apparently are highly crystalline (being brittle in contact mode and exhibiting friction that is independent of scan velocity). A weblike (~3-nm thick) majority component is more heterogeneous in its properties (i.e., friction, phase, adhesion). In Figure 4.3 we label these layers 1, 2, and 3, respectively. Height and phase images are shown for each of *three cases of free oscillation amplitude* at resonance: (a) 14, (b) 10, and (c) 4 nm. In all cases the setpoint amplitude was 70% of the free value, and the drive frequency was 315.16 kHz, that is, 0.30 kHz below the resonance frequency of 315.46 kHz (nominal spring constant 40 N/m). In each phase image the color scale was offset to provide desirable contrast.

The phase distribution plots in Figure 4.4 quantify the phase images of Figure 4.3 (in true phase). Figures 4.3 and 4.4 demonstrate a physically derived artifact in dynamic AFM: significant changes in *apparent* relative height, from domain to domain, the result of *switching* between attractive and repulsive regimes. A widely split distribution of phase values for case (b) indicates the coexistence of both dominant attractive and repulsive dynamic states within the image. The phase distributions for cases (a) and (c) are much more restricted and predominantly correspond to the attractive (c) *or* repulsive (a) regimes, not both. The brightest phase domains in (b) show "heights" that are *sunken by about 0.3 nm* compared to the adjacent regions; whereas in (a) and (c), these same domains *are instead elevated by about 1 nm* compared to the adjacent regions. Cases (a) and (c) are moreover consistent with observations in contact and force curve mapping modes (which for this polymer system are not problematic, i.e., do not generate deleterious shear forces). Clearly something is operationally "wrong" to give the height image in case (b): the *coexistence of two dynamic equilibrium states* across the image, specific to the type of material interacting with the tip. Obviously, this must

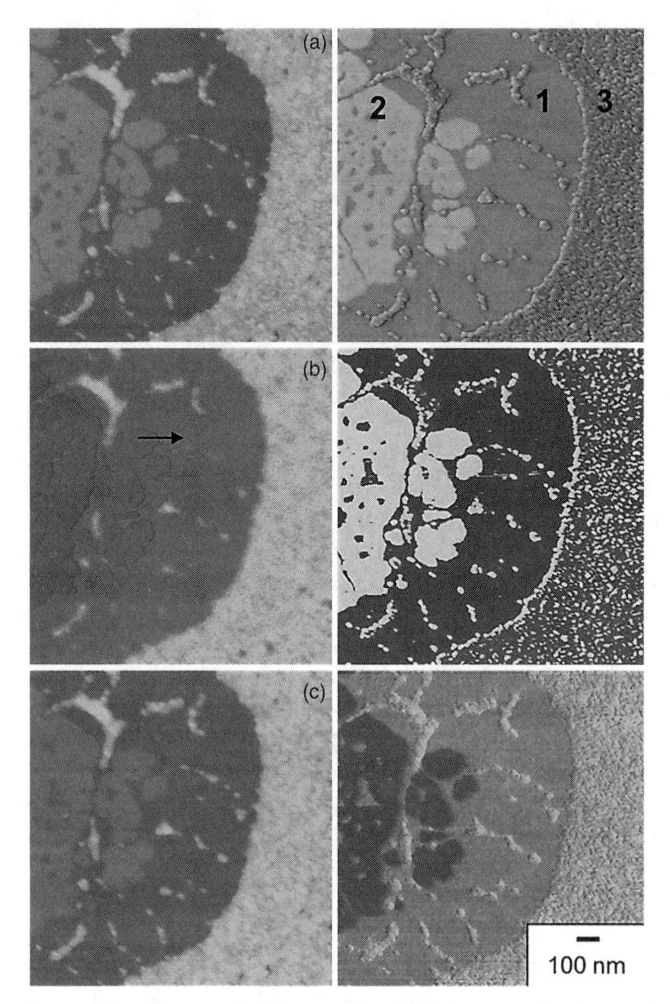

Figure 4.3 Height (left) and phase (right) images of a polyvinyl alcohol thin film containing three characteristic components. Three drive amplitudes produce free oscillation resonance amplitudes (A_0) of (a) 14, (b) 10, and (c) 4 nm corresponding to uniform repulsive, coexisting, and uniform attractive regimes, respectively. In height image, bright contrast refers to topographically high and dark contrast refers to topographically low. In phase image, bright contrast refers to high phase and dark contrast refers to low phase. For color details, please see color plate section.

be avoided, and one way to do this is to change the kinetic energy of the oscillator—per the free oscillation amplitude (via the drive amplitude)—while keeping the setpoint amplitude constant as a fraction of the free oscillation amplitude. In this example one can go either way, to larger or smaller kinetic energy (via larger or smaller drive amplitude) to *stabilize a single stable regime* across the image.

Region-specific interaction regimes—spatial domains of dominant attractive or repulsive interaction—do not necessarily result from heterogeneities in material *properties*. Surface shape also can cause this artifact. Even on a homogeneous

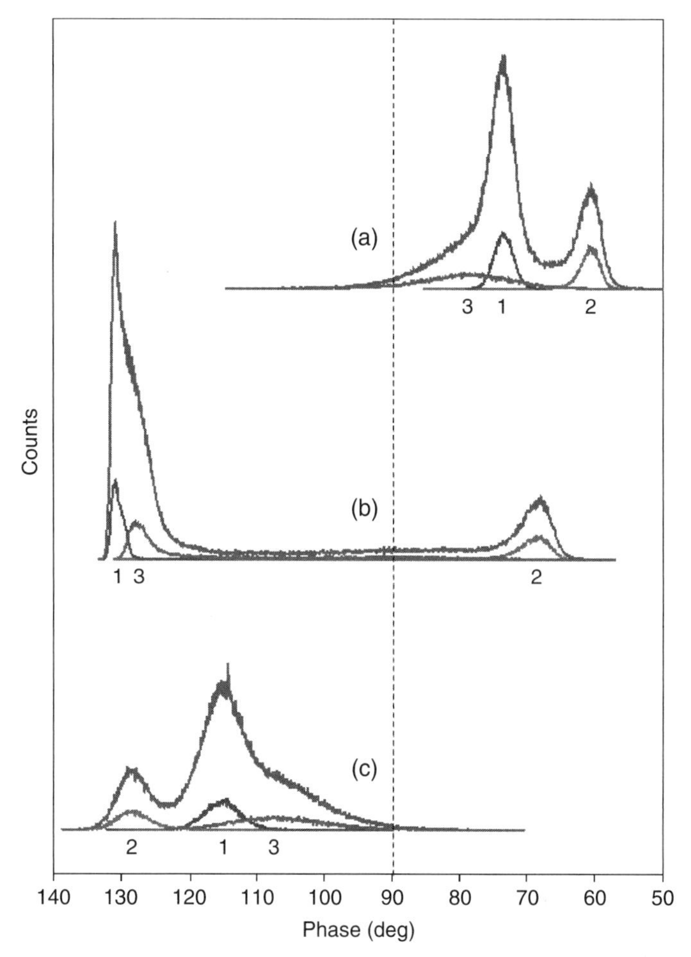

Figure 4.4 Distributional representations of phase data from the images in Figure 4.3 for film components 1–3. Multipeak histograms are from entire phase image; smaller single peaks derive from individual film components. For color details, please see color plate section.

(in properties) material, if there are valleys similar to the tip in *radius of curvature*, then attractive forces will be significantly greater in valleys than on flats or atop hills. Thus parameters selected to stabilize the repulsive regime may work *except in valleys*, where attraction is stronger; an oscillator stabilized in the repulsive regime on a flat surface may transition into the attractive regime in a valley. (Similarly, an oscillator stabilized in the attractive regime on a flat surface may transition into the repulsive regime on a hilltop, where attraction is weaker.) Figure 4.5 exemplifies this situation for a polycrystalline gold surface. In (a), parameters were chosen to stabilize the repulsive regime (by driving below resonance with a free-oscillation amplitude between 50–60 nm using a nominal $k = 40$ N/m cantilever). But in the resulting height–amplitude–phase images, at several narrow valleys the oscillator switched to the

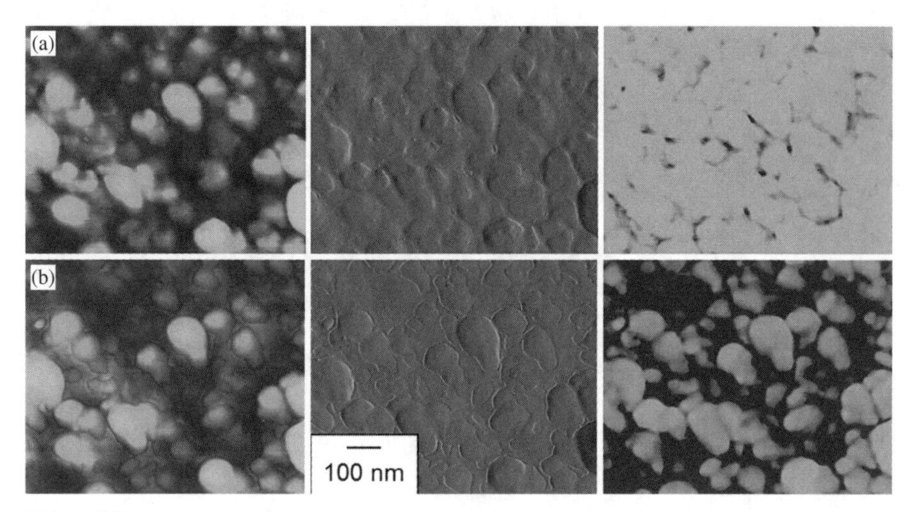

Figure 4.5 Coexisting attractive and repulsive imaging regimes on a polycrystalline gold surface, driving at frequencies below (a) and above resonance (b). In order from left to right, height, amplitude (error signal), and phase images reveal features characteristic of regime switching in both (a) below-resonance and (b) above-resonance cases.

attractive regime (dark contrast in the phase image). It was found that the regions of net attraction greatly expanded when driving the cantilever above resonance, case (b). Boundaries demarcating domains of net-attractive or net-repulsive interaction are also visible in the amplitude image.

It is also important to note that *long-range electrostatic forces* can have a significant effect on the interplay of attractive and repulsive forces in dynamic AFM. A voltage bias can be applied between tip and sample to intentionally produce such long-range electrostatic forces, related to the capacitance of the tip–sample system. Or an external bias can be chosen so as to *null intrinsic differences in surface potential* (i.e., work function for metals) in order to *remove* long-range forces. Thus the freedom to bias tip or sample can be important to ordinary dynamic AFM, *even if interested only in surface topography*—especially on highly conductive materials such as the gold of Figure 4.5. A long-range electrostatic force, if uncompensated, can lessen the stability of the repulsive regime. Indeed the presence of net attraction in the valleys in Figure 4.5(a) is in part due to electrostatic forces derived from different tip and sample potentials. If one is attempting to image within the attractive regime, the difference in surface potentials can stabilize this regime; but the longer range of these forces can result in a significantly increased "fly height" and thus a smearing of the topographic image. In deciding whether to invoke an external bias, the user needs to weigh advantages and disadvantages per the realities of the surface being imaged.

Arguably the most elucidating measurements to assess interaction regime are dynamic force curves or amplitude and phase versus *Z* curves (AP-*Z*). These single-point measurement curves are discussed in detail in Chapter 3. AP-*Z* curves are single-point force distance curves (*F*-*Z*) that are done while oscillating the cantilever at resonance, such that amplitude and phase are measured as a function of tip–sample gap.

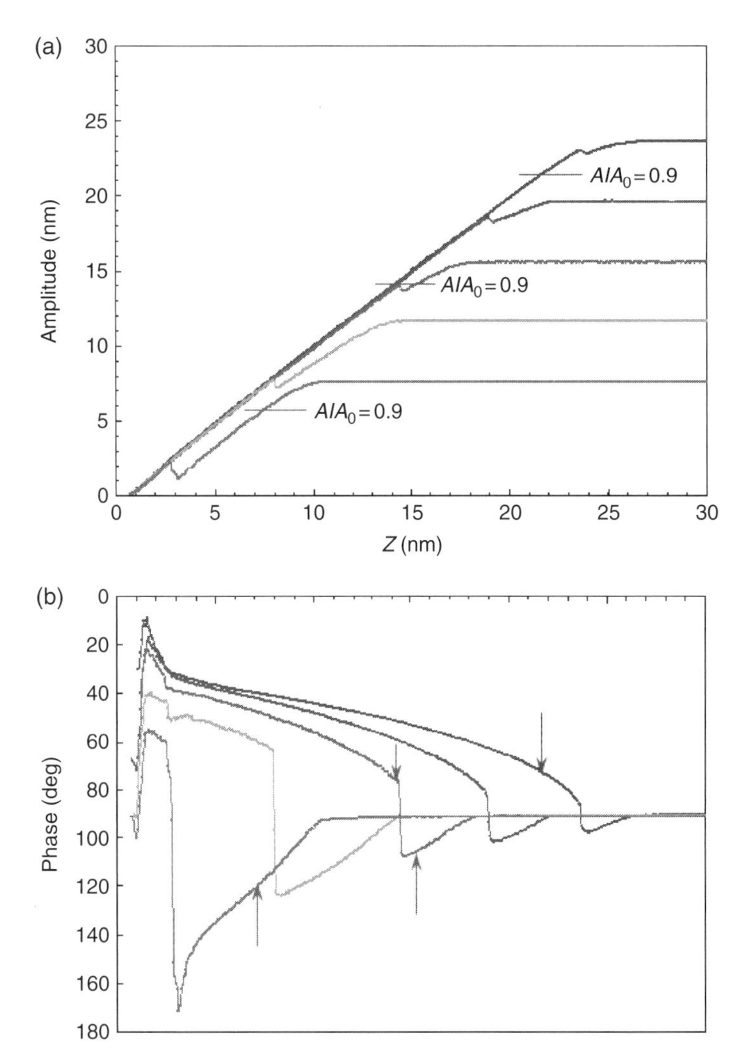

Figure 4.6 (a) Amplitude-Z and (b) phase-Z curves on a silicon wafer driving at resonance frequency and at five different drive amplitudes, producing free oscillation amplitudes A_0 ranging from 7.6 to 23.8 nm. Dashed horizontal lines denote typical setpoint amplitudes for repulsive-regime (top curve) and attractive-regime (bottom curve) imaging states at a setpoint amplitude ratio $A/A_0=0.9$. For color details, please see color plate section.

Jumps between attractive and repulsive regimes are easily observed in AP-Z curves. Five overlaid AP-Z are shown in Figure 4.6, which were obtained during an approach at five different free oscillation amplitudes A_0 ranging from 7.7 to 23.6 nm, on an untreated silicon wafer using silicon tip and rigid rectangular cantilever ($k \sim 40$ N/m, $Q \approx 400$). The oscillator was driven at its resonant frequency $f_0 = 315.46$ kHz such that the phase reading for Z beyond tip–sample interaction is 90°. During approach, the amplitude [Fig 4.6(a)] decreases upon tip–sample interaction; eventually the amplitude abruptly increases by ~1 nm, then decreases again with a somewhat steeper slope.

Correspondingly, the phase data [Fig 4.6(b)] shows a dramatic upward jump at the same value of Z at which the amplitude abruptly increases. This tells us that the transition to ~1-nm higher amplitude yields a transition from attractive to repulsive regime. (Thus the alternative nomenclature of *low- and high-amplitude states* [8].)

Among the five cases in Figure 4.6, one notes that the jumps occur at very different values of A/A_0. Specifically, the repulsive regime is dominant for a large range of amplitudes when the cantilever has a large amount of kinetic energy, and a small range of amplitudes when the cantilever has a small amount of kinetic energy. One could in principle choose a setpoint amplitude A that results in the net-attractive *or* net-repulsive regime. For a commonly used value of $A/A_0=0.9$, marked with horizontal dashed lines in Figure 4.6(a) for three cases, the attractive *or* repulsive regime results, as indicated by the corresponding phase values (marked with vertical arrows) to either side of the zero of raw phase. An attempt to image in the intermediate of these three cases would be problematic: The selected amplitude for $A/A_0=0.9$ corresponds to *two values of* Z, cases of net repulsion (lower Z) or net attraction (higher Z).

One must recognize that on a heterogeneous surface the AP-Z curves may depend on the (X,Y) location; thus the *value of amplitude at which the oscillator transitions from attractive to repulsive may vary with location.* One usually needs to choose a setpoint amplitude value that is not close to the transition, anticipating that the amplitude at which the transition takes place will vary, perhaps with local surface curvature, from location to location. Moreover, if the transition is inadvertently reached, the oscillator may remain "locked in" in the new state until a very different kind of location is reached. The "locking in" is due to *hysteresis* over an extension, then retraction, of Z, and is routinely observed in AP-Z curves when comparing data during approach to that during retraction [9].

Driving the oscillator somewhat below the resonance frequency is a common approach to stabilize the repulsive regime. Again, AP-Z curves can be used to explore this behavior, by varying the drive frequency from one data set to the next. Figure 4.7 overlays phase Z curves acquired during approach for 15 different drive frequencies in the vicinity of resonance. These data were acquired on a polyvinyl alcohol (PVA) film similar to that in Figure 4.3 (region 1), using a similar tip–cantilever. When driving at resonance frequency f_0, there is a jump discontinuity from attractive to repulsive regime as Z decreases, at $Z \approx 28$ nm. When driving at the lowest frequency f_-, about 0.25 kHz below resonance, a sudden decrease of phase is seen upon tip–sample interaction during approach, but phase then continues to increase as Z decreases. When driving at frequency f_+, phase decreases monotonically as Z decreases, lacking a transition to the repulsive regime. The behaviors at intermediate drive frequencies interpolate between the cases of f_+ and f_-.

An important question is whether any of the above strategies (selection of drive frequency, free oscillation amplitude, setpoint amplitude) will adequately stabilize a given regime *across the entire image*. For this reason, AP-Z curve mapping can be elucidating, to see the extent to which regime switching is taking place. AP-Z curve mapping can moreover be utilized (with custom algorithms) to generate a *corrective* height mapping that not only accounts for switching phenomena but also can quantify compliance (indentation) effects [9].

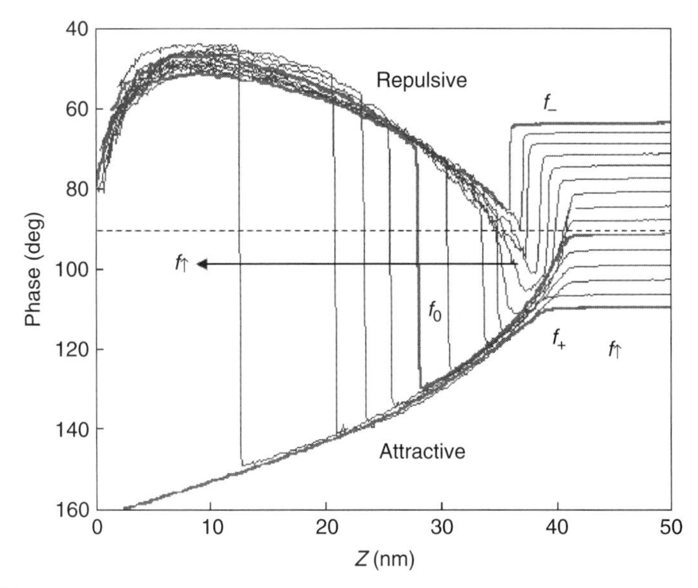

Figure 4.7 Phase-*Z* curves obtained at the 15 values of drive frequency. Arrows indicate the effect of increasing drive frequency on both the free oscillation phase value (right) and the discontinuous transition from attractive to repulsive regime (left). For color details, please see color plate section.

Our first example of AP-*Z* curve mapping is another ultrathin film of PVA similar to that in Figure 4.3. Figure 4.8 shows conventional height (a) and phase (b) images. The same three characteristic subregions have been identified: a first layer ~1 nm thick (1) and two types of second-layer domains exhibiting relatively low and high dissipation (2 and 3). The phase contrast in (b) is *too* strong: 80° separate brightest from darkest, with the brightest apparently corresponding to a net-repulsive shift of oscillator and the darkest correspond to a net-attractive shift. Figure 4.8(c) contains three pairs of raw AP-*Z* curves that are representative of the *behaviors* of regions 1–3. The trigger for retraction during the acquisition of AP-*Z* curves was a 1.25-V decrease of amplitude signal from the free oscillation value of 1.87 V. Examining the phase-*Z* curves, one sees that the *Z* displacement from the onset of interaction (steep drop in phase versus decreasing *Z*) to the ultimate approach point is in fact *different* among the three measurement locations. The point of steep drop is demarcated by the vertical dotted lines and occurs at approximately 23, 25, and 27 nm, for the three different locations spanning a distance of 4 nm. The distance between the steep drop to the ultimate approach is thus different in the three different areas: greatest at location 2 and smallest at location 3. One can also utilize this data cube to better understand the conventional height image in (a) acquired at a setpoint of 1.65 V. In (d) the AP-*Z* curves are shifted horizontally to force the amplitude curves to coincide at 1.65 V, corresponding to the feedback condition of (a)–(b). The relative shifts of the steep drop locations in the phase *Z* curves quantify the height artifacts contained at corresponding locations in

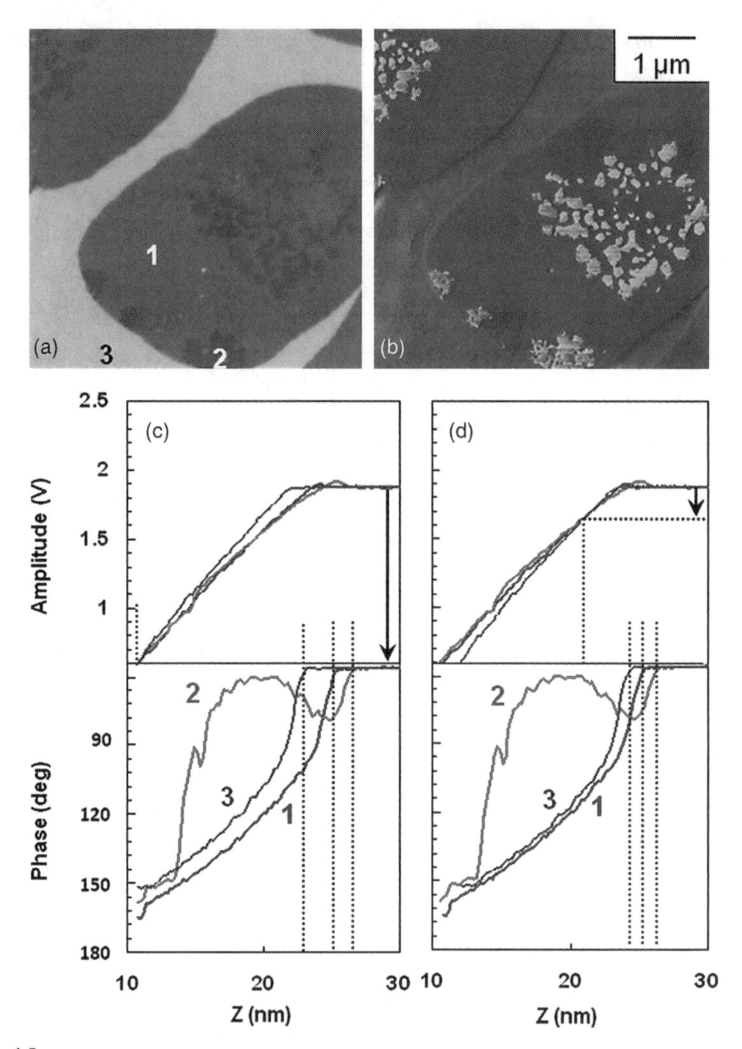

Figure 4.8 (a) Height and (b) phase images of a three-component PVA surface. (c) Raw amplitude-Z and phase-Z curves acquired at the three locations denoted in (a). (d) Curves from (c) Z shifted to achieve the same Z value at the amplitude value of 1.65 V. For color details, please see color plate section.

(a), roughly 1 nm too high (3) or too low (2) relative to region 1 [10]. Custom algorithms can allow one to create a mapping of these "corrections." What appears to be a recessed region 2 (by a fraction of a nanometer) is in reality approximately 1 nm *higher* than the surrounding region 1. One also finds that region 2 was imaged at a net oscillator shift of approximately zero relative to the free oscillation value (phase near 90°), whereas regions 1 and 3 were imaged "deep" in the attractive regime. Thus AP-Z curve mapping provides better insight into what is going on during conventional imaging.

4.3 COMPLICATIONS IN PHASE QUANTIFICATION

4.3.1 Overview

In this section we will see that the proper quantification of phase can be troublesome because of both instrumental and physical effects. We discuss five classes of issues affecting the collected phase data. These arise from:

1. An *apparatus phase shift* that is in series with the phase shift of interest (i.e., that between the base and tip ends of the cantilever)
2. Substantial change in oscillator behavior as a function of distance *beyond* tip–sample engagement (i.e., seen in coarse approach)
3. Thermal *drift*, potentially both intrinsic (machine warm-up) and extrinsic (lab temperature if many human bodies occupy a small space, e.g., in demos)
4. *Coupling* between topography and phase, primarily the influence of surface slope on measured phase
5. Issues of the "authenticity" of phase measurement (i.e., whether a true *lock-in amplifier* is being used or some other electronics)

Measuring the phase shift between a vibrational signal imparted to the base of the cantilever and the sinusoidal motion at the tip end is simpler in concept than in practice. An electrical alternating current (AC) driving signal must travel from a (sinusoidal) function generator device through electrical contacts to a vibrational element; then the vibration must propagate through materials and across interfaces until it reaches the base of the cantilever. Thus there can be additional phase shifts due to other elements, both mechanical and electrical (i.e., capacitive). At the measurement end there can be an input device phase shift as well. One makes no attempt to quantify each contributing phase shift; instead one *subtracts an aggregate apparatus phase shift to yield the mechanically correct phase lag* between cantilever base and tip ends at resonance, 90°. Often the software calls this "zeroing the phase" or "adjusting the phase offset." To ensure a more correctly zeroed phase, one typically performs some variant of the following (possibly automated) before executing the coarse approach, a procedure often called "cantilever tune":

1. Examine the response amplitude during a sweep of the drive frequency.
2. Locate the drive frequency producing a prominent peak.
3. Narrow the sweep range centered on this peak.
4. Zero the phase at this frequency by subtracting an apparatus phase shift (usually done by software).

Prior to coarse approach, one chooses a combination of drive frequency and drive amplitude to (hopefully) stabilize the attractive or repulsive regime, and set the cantilever's kinetic energy beyond tip–sample engagement. After the ensuing coarse approach to tip–sample engagement (at some reduced tip amplitude), an important question is whether the frequency-dependent response has substantially *changed*

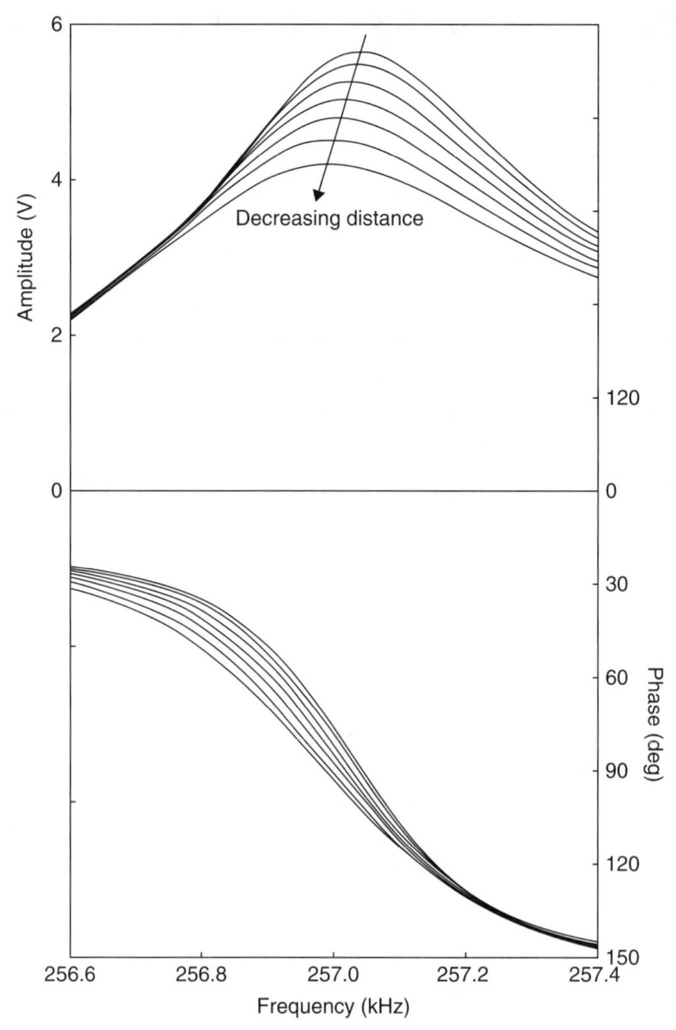

Figure 4.9 Frequency dependence of raw amplitude and phase signal at different distances actuated by the Z motor. Highest to lowest resonance amplitudes A_0 ranging from 75 to 54 nm were measured at individual distances of 19, 10, 5, 3.2, 2, 1.2, and 0.4 μm.

from that viewed far from the surface (the preceding step). One can slightly withdraw from the surface (say 1 μm instead of tens of micrometers) and repeat the frequency sweep measurement: Usually one finds that indeed the resonant response functions (amplitude and phase) have shifted to lower frequency and broadened. This behavior is shown in Figure 4.9 for seven distances above the substrate ranging from 19 μm down to 0.4 μm. The changes in the amplitude and phase sweeps are substantial, a result of *squeeze film damping* [12, 13]. The air drag on the cantilever increases as the distance between cantilever and sample surface decreases: Net work is done in pushing air out of the gap between the cantilever and the sample surface during each oscillation cycle. The distance over which this effect changes is comparable to the lateral dimensions of the cantilever.

Even a slight change in resonance frequency relative to a fixed drive frequency, as well as a slight change in phase offset, can translate into several degrees of change in phase. This can produce inconsistent results, from session to session or sample to sample, when steps 1–4 listed above are *not performed at the same cantilever–sample distance*. For detailed and quantitative work, one should carefully examine AP-Z curves while adjusting drive frequency and phase offset. A procedure is to first adjust the drive frequency until the amplitude is maximal at a Z value just beyond tip–sample engagement (i.e., nanometers), thereby finding the "correct" resonance frequency in free oscillation; then adjust the apparatus phase offset to zero the phase at this same value of Z.

The preceding adjustment also should be repeated during the course of a session to compensate for *thermal drift*. Properly zeroing the phase early in a lab session does not guarantee quantitatively correct or consistent results. Thermal warm-up generally may affect the resonant frequency and Q factor of the oscillator, the instrumental phase offset, as well as the amount of response amplitude for a given drive amplitude. After a cold start, and usually through the first hour and perhaps as long as several hours of operation, one may need to adjust the drive frequency and phase offset to account for this drift. Failure to do so may produce at least several degrees and perhaps tens of degrees of inconsistency in phase measurement.

Another, independent source of potential drift in free-oscillation amplitude is change in the resonance amplitude itself. That is, the entire amplitude–frequency curve may drift upwards or downwards; thus the energy of the driven oscillator may increase or decrease, perhaps as the mechanical coupling of the vibrational signal to the cantilever chip changes during warmup or with change of humidity (e.g., human bodies in the lab). Over the course of ~2 hours from a cold start, a monotonic increase in free-oscillation amplitude A_0 of 15–20% may be observed in some cases, corresponding to a substantial increase in oscillator kinetic energy. On soft materials, one may observe significant changes in both height and phase images due to this drift.

4.3.2 Coupling of Topography and Phase

Upon first examining a phase image, one may notice that topography data seems to be "leaking" into phase data. The phase image, in addition to containing material contrast, may resemble the height and/or amplitude image. Why? There are three principal mechanisms whereby phase and topography are coupled. The first was already mentioned: attraction increases in valleys with *curvature* comparable to tip and decreases atop hills with similar curvature opposite in sign. The second and third mechanisms of coupling between phase and topography relate to surface slope.

The first slope-derived effect is instrumental rather than physical, deriving from the imperfect tracking of topography. The operational amplitude varies (with respect to setpoint) with the slope of topography, meaning the value of Z has not been reactively positioned precisely to null the error signal. But generally we have seen in

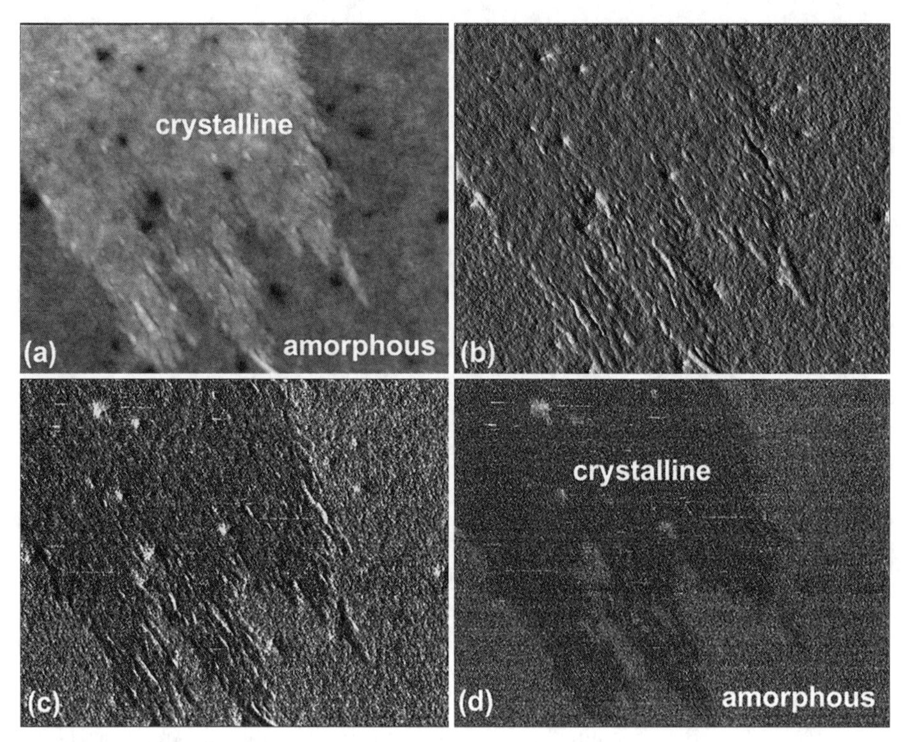

Figure 4.10 Height (a), amplitude (b), and phase (c) images of a polyacrylamide thin film containing a (1) highly crystalline island on a (2) more amorphous underlayer. Subtraction of the (scaled) amplitude data from the phase data results in a (d) "corrected" phase image that better contrasts materials differences between 1 and 2 by removing the effects of topographic slope (which causes the operational amplitude to deviate from setpoint).

AP-Z measurements that *phase also* varies when amplitude varies as a function of Z. This variation is often strong. An example is provided in Figure 4.10, for the case of attractive regime imaging of a semicrystalline polymer film. A highly crystalline island of polyacrylamide contrasts with a more amorphous underlayer. The amplitude (error) signal (b) relates simply to the slope of the height signal (a) and is clearly manifest in the phase image. Upon subtracting a scaled amplitude image from the phase image to produce a modified ("corrected") phase image (d), the presence of topographic slope is greatly reduced; material-derived phase contrast is starker.

It should be noted that a potential problem with the preceding type of adjustment, to obtain the image in Figure 4.10(d), arises when the amplitude–phase relationship is not approximately constant across the image on materially heterogeneous surfaces. Thus if one scales an amplitude image and subtracts from a phase image in the attempt to remove variations that derive from topographic slope, this subtraction *may provide a good correction at some locations but not others* [9].

The third mechanism by which topography affects phase images is another surface slope effect but dependent on cantilever orientation. At any point the local surface slope measured *parallel* to the cantilever long axis affects the energy dissipation during tip–sample interaction, which in turn affects the phase as a result of the cantilever tilt. A tilted cantilever confers a lateral component to the tip's oscillatory motion and thus produces shear dissipation when tip interacts with sample [14]. A surface slope that is downhill in the direction pointing from tip to base of cantilever often may exhibit higher dissipation, and thereby a change of phase signal, compared to a surface -location where this slope is zero or uphill in this direction.

Figure 4.11 compares height and phase traces along the slow-scan, long-cantilever axis direction (vertical) in attractive-regime phase images of water droplets hundreds of nanometers tall, containing slopes dZ/dY of nearly 0.4. The water droplets condensed on a micrometer-thick coating of a biodegradable commercial polymer Polyactive—the semicrystalline block copolymer poly(ethylene oxide-butlyene terephthalate) (PEO-PBT)—at 94% relative humidity imaging conditions. (This phenomenon is reversible, i.e., the droplets appear and disappear or grow and shrink with cycling of humidity.) Of interest to our discussion is the pronounced tilt in the phase cross section. Higher phase near the top edge of the imaged droplet (right side of droplet in cross section) corresponds to higher energy dissipation in the attractive regime. The effect is superimposed on an overall higher phase, and thus dissipation, atop the droplet, compared to the surrounding surface.

4.3.3 Phase Electronics and Its Calibration

The last difficulty with phase measurement treated in this chapter is very much dependent on the electronic equipment used to measure phase. The most common source of difficulty is the so-called Extender electronics package sold with certain first-generation commercial systems (Bruker/Veeco), which nevertheless remain in common use (i.e., remain valuable instrumentation). The Extender was designed for purposes other than to measure phase per se [5]. A measurement of *true phase* requires a lock-in amplifier or similarly sophisticated device to track the relative offset in time of two sine waves and, moreover, with a frequency range extending up to hundreds of kilohertz. But the Extender can be calibrated to provide numbers very close to true phase for most measurements of interest. Extender phase values within 50° of the (conventional) zero can be adjusted by simply multiplying by 0.9. But checking this calibration with an oscilloscope is recommended. Upon doing so and examining the numbers closely, one may empirically find that invoking a second fitting parameter produces a better calibration, for example, multiplying the Extender phase P by

$$c_1\{1+[(P/c_2)^2]^{1/2}\} \quad (4.1)$$

with fitting parameters c_1 (typically less than 0.9) and c_2 (100–250) [9]. (The square followed by square root handles both positive and negative Extender phase.) To demonstrate one result, in Figure 4.12 we plot phase Z curves in the attractive regime during retraction from polycrystalline gold, consecutively acquired using the Extender package and a lock-in amplifier. (A nominal $k=3$-N/m uncoated silicon

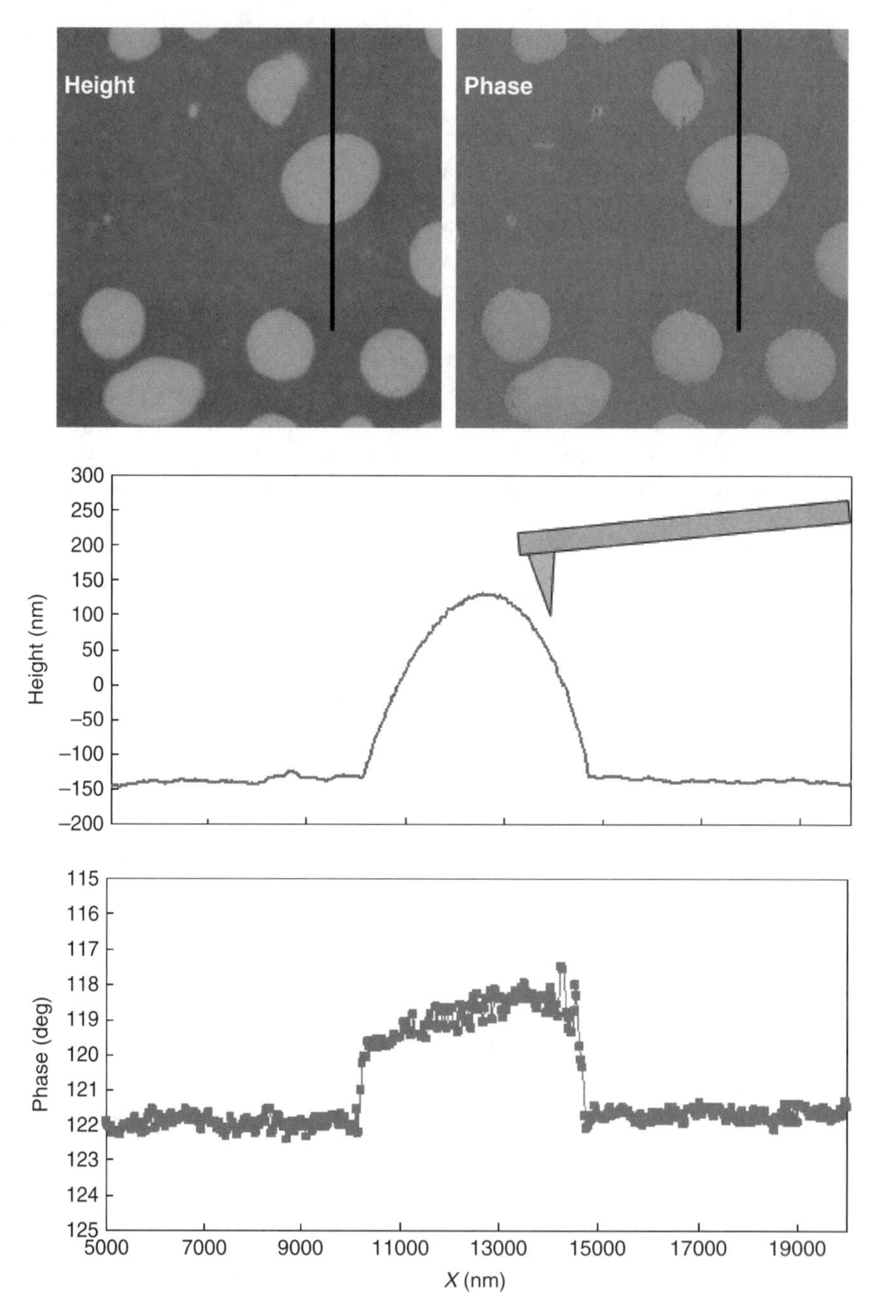

Figure 4.11 Effects of topographic slope (along the horizontal component of the cantilever's principal axis) on phase, with the fast-scan axis chosen to be perpendicular to cantilever. Height and phase cross sections (middle and bottom) derive from data along the vertical lines marked in the images at top.

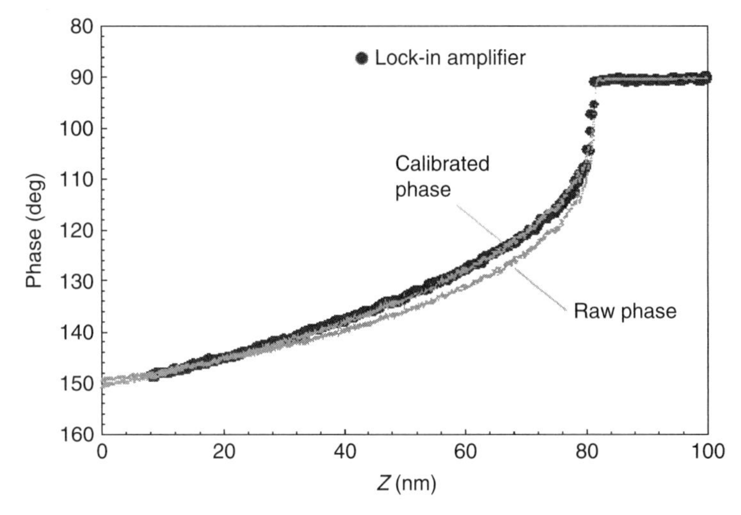

Figure 4.12 Phase-*Z* curve comparing phase measurements using a lock-in amplifier (circles) to raw and nonlinearly calibrated measurements (lines) using the Extender electronics. For color details, please see color plate section.

cantilever was used, with resonance frequency of about 64 kHz to enable the use of a common lock-in amplifier limited to 100 kHz range.) We also overlay Extender phase data recalibrated with the above two-parameter expression: Agreement is very good within 60° of the phase zero value.

As a final remark, those with access to a lock-in amplifier should not necessarily use it for routine acquisition, whether AP-*Z* curves or images: In addition to the likely limitation in maximum frequency (again, often 100 kHz), common lock-in amplifiers are quite limited in *measurement rate*. The phase *Z* curve in Figure 4.12 was acquired slowly, 2 seconds per approach–retract cycle, and using the smallest lock-in-amplifier time constant; yet some approach–retract hysteresis was observed in lock-in-acquired data while absent in Extender-acquired data (on a nonviscoelastic, i.e., not intrinsically time-dependent, surface). More notably the measurement delay can be apparent in images as a shift in the corresponding features of height and phase images along the fast-scan direction [9].

REFERENCES

1. Cleveland, J. P., Anczykowski, B., Schmid, A.E., and Elings, V. B. Energy dissipation in tapping mode atomic force microscopy, *Applied Physics Letters* (1998): **72**, 2613–2615.
2. Garcia, R., Tamayo, J., Calleja, M., and Garcia, F. Phase contrast in tapping mode scanning force microscopy, *Applied Physics A [Materials Science Processing]* (1998): **66** (suppl, pt 1–2), S309–312.
3. Garcia, R., Tamayo, J., and Paulo, A. S. Phase contrast and surface energy hysteresis in tapping mode scanning force microscopy, *Surface and Interface Analysis* (1999): **27**(5–6), 312–316.
4. Martinez, N. F., and Garcia, R. Measuring phase shifts and energy dissipation with amplitude modulation atomic force microscopy, *Nanotechnology* (2006): **17**, S167–S172.
5. Bocek, D. About "Bocek" Degrees used in the Extender Package on Digital Instruments Nanoscope Scanning Probe Microscopes, *Microscopy Today* (2001, June).

6. Gleyzes, P., Kuo, P. K., and Boccara, A. C. Bistable Behavior of a Vibrating Tip Near a Solid Surface, *Appl Phys Lett* (1991): **58**(25), 2989–2991.
7. Anczykowski, B., Kruger, D., and Fuchs, H. Cantilever Dynamics in Quasinoncontact Force Microscopy: Spectroscopic Aspects, *Phys Rev B* (1996): **53**(23), 15485–15488.
8. Garcia, R., and Perez, R. Dynamic Atomic Force Microscopy Methods, *Surf Sci Rep* (2002): **47**, 197–301.
9. Haugstad, G. *Atomic Force Microscopy: Exploring Basic Modes and Advanced Applications*, Wiley: Hoboken, NJ, 2012.
10. Haugstad, G., and Jones, R. R. Mechanisms of Dynamic Force Microscopy on Polyvinyl Alcohol: Region-Specific Non-contact and Intermittent Contact Regimes, *Ultramicroscopy* (1999): **76**, 77–86.
11. Haugstad, G. Contrasting static-to-kinetic friction transitions on layers of an autophobically dewetted polymer film using Fourier–analyzed shear modulation force microscopy, *Trib Lett* (2005): **19**, 49–58.
12. Hosaka, H., Itao, K., and Kuroda, S. Damping Characteristics of Beam-Shaped Micro-oscillators. *Sensors Actuators A* (1995): **49**, 87–95.
13. Dareing, D. W., Yi, D., and Thundat, T. Vibration Response of Microcantilevers Bounded by a Confined Fluid. *Ultramicroscopy* (2007): **107**, 1105–1110.
14. D'Amato, M. J., Marcus, M. S., Eriksson, M. A., and Carpick, R. W. Phase imaging and the lever-sample tilt angle in dynamic atomic force microscopy. *App Phys Lett* (2004): **85**(20), 4738–4740.

Chapter 5

Dynamic Contact AFM Methods for Nanomechanical Properties

Donna C. Hurley and Jason P. Killgore

National Institute of Standards and Technology, Boulder, CO

5.1 INTRODUCTION

This chapter focuses on two atomic force microscopy (AFM) methods for nanomechanical characterization: force modulation and contact resonance (CR). Usually called "force modulation microscopy" (FMM) to denote its imaging ability, force modulation techniques were developed over 20 years ago. Contact resonance methods are a newer, emerging set of techniques encompassing point spectroscopy, qualitative imaging, and quantitative imaging (here called contact resonance force microscopy or CR-FM). FMM and CR methods share several common features that distinguish them from other methods discussed in this book. Specifically, FMM and CR methods fall at the intersection of AFM methods that involve *dynamic excitation* with those that operate in *continuous contact*. Both of these characteristics present distinct measurement advantages. As a result, dynamic contact AFM methods enable more accurate, quantitative characterization of nanoscale mechanical properties.

5.1.1 Common Measurement Themes

Before discussing FMM and CR methods separately, we consider some of their common characteristics. These common themes are (a) operation in continuous contact, (b) dynamic cantilever excitation at relatively high frequencies but low amplitude, (c) frequency-specific detection, and (d) data analysis with contact

Scanning Probe Microscopy in Industrial Applications: Nanomechanical Characterization,
First Edition. Edited by Dalia G. Yablon.

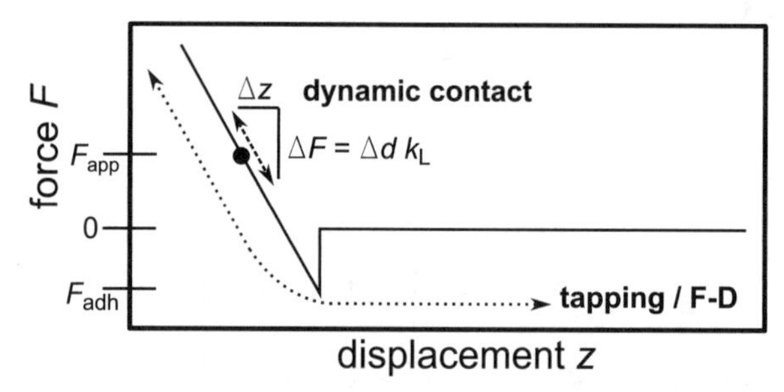

Figure 5.1 Concepts of dynamic contact AFM methods. The solid line indicates the force F versus displacement z for the retraction phase of force–distance spectroscopy. An attractive adhesion force F_{adh} exists between the tip and the sample. A quasi-static applied force F_{app} (filled circle) keeps the tip in the repulsive contact regime for dynamic contact methods. Lines with arrows indicate the relative portions of the curve traversed by dynamic contact methods (dashed line) and tapping mode or force–distance (F–D) spectroscopy mode (dotted line). The inset shows parameters used for the discussion of FMM analysis in Section 5.2.2: sample modulation amplitude Δz, cantilever response amplitude Δd, resulting dynamic force ΔF, and cantilever spring constant k_L.

mechanics. Figure 5.1 describes the concepts of dynamic excitation in continuous contact by considering the familiar force-displacement (F–D) relation for the AFM tip as it pulls away from a sample. The dotted line shows that in intermittent-contact (tapping) mode or F–D spectroscopy (static force curve mode), the majority of the force–distance curve is traversed. During each cycle in these modes, initially the tip is out of contact with the sample. The tip then approaches the sample and eventually enters the repulsive regime of interaction forces. During the final part of each cycle, the process is reversed, and the tip overcomes the adhesive force F_{adh} and breaks contact.

In contrast, during operation in FMM and CR modes, the AFM tip is kept in contact by a static force F_{app} applied normal to the surface of the sample. This constant force is maintained throughout the measurement through feedback on the cantilever deflection (voltage setpoint). As indicated in Figure 5.1, F_{app} is sufficiently high to ensure operation in the repulsive regime of tip–sample contact; typically, it is in the range from tens of nanonewtons to a few micronewtons. Operating in the repulsive contact regime ensures a linear force–displacement relation between the tip and the sample. It is much simpler to accurately model this linear contact rather than the highly nonlinear interaction involved in intermittent-contact (tapping) modes. The inset in Figure 5.1 defines parameters used for analysis of FMM measurements, which is discussed in Section 5.2.2.

In FMM and CR modes, dynamic excitation is performed to induce relative motion of the tip and the sample while the tip is in contact. As indicated by the short dashed line in Figure 5.1, dynamic excitation results in the tip traversing only a small section of the force–displacement curve about the constant (quasi-static) value F_{app}. It is important to note that the portion of the force–displacement curve traversed is much smaller than in intermittent contact modes such as tapping or F–D spectroscopy. Thus, unlike tapping mode or F–D spectroscopy, the tip and sample remain

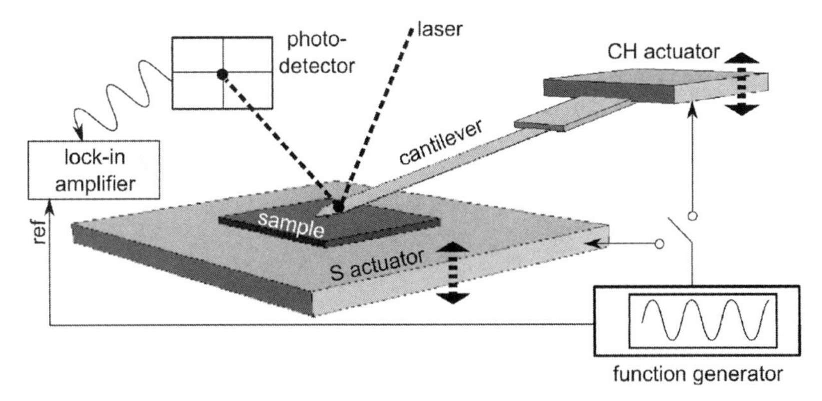

Figure 5.2 Instrumentation schematic for operation in FMM and CR modes. A function generator is used to sinusoidally drive a piezoelectric actuator to excite the cantilever. The actuator is located either beneath the sample (S actuator) or in the cantilever holder (CH actuator). The motion of the cantilever at the excitation frequency is detected with the AFM photodetector by lock-in techniques.

continuously in repulsive contact. The high frequencies of the dynamic excitation (kilohertz to megahertz) used in FMM and CR methods also allow for improved sensitivity to small changes in mechanical properties compared to quasi-static approaches, due to time averaging of the cantilever vibration response.

The instrumentation to accomplish dynamic contact AFM is shown schematically in Figure 5.2. Dynamic excitation is achieved through some type of actuation. Here we consider actuation by a piezoelectric element external to the cantilever, although other actuation approaches exist. Typically, piezoelectric actuation involves a single-frequency (sine-wave) excitation from a function generator. Because the tip is in contact and only relative motion matters, actuation can be accomplished either through the sample or from the clamped base of the cantilever. Sample actuation is accomplished by a piezoelectric element external to the commercial AFM located beneath the sample (S actuator in Figure 5.2). Actuation at the cantilever base involves a piezoelectric element in the cantilever holder (CH actuator in Figure 5.2) such as the one more commonly used for tapping mode operation.

As indicated by the arrows with thick dashed lines in Figure 5.2, the relative tip–sample motion occurs in an out-of-plane direction (direction normal to the sample plane). Although out-of-plane motion is typically used for FMM and CR methods, some other dynamic contact methods involve in-plane (parallel to the sample plane) motion. For instance, the primary difference between FMM and shear modulation force microscopy (SM-FM) is out-of-plane versus in-plane excitation motion. Methods with different actuation approaches are discussed in later sections.

Another thread that links various dynamic contact AFM methods is frequency-specific detection. Because the cantilever actuation occurs at a single frequency, it is necessary to measure the cantilever response at the same frequency. Frequency-specific or alternating current (AC) detection is typically performed with lock-in techniques that use the excitation sine wave for a reference, as indicated in Figure 5.2. A key measurement advantage to frequency-specific operation at relatively high

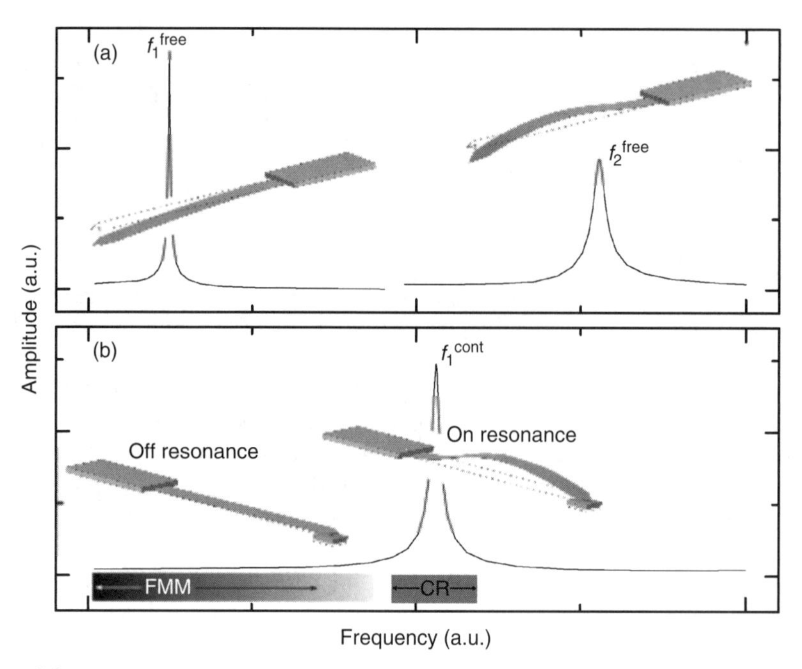

Figure 5.3 Amplitude spectra of cantilever response versus excitation frequency. (a) Tip out of contact (free space). Also shown are the cantilever mode shapes for excitation at frequencies f_1^{free} and f_2^{free} of the first (lowest) and second free flexural resonances, respectively. (b) Tip in contact with sample. The horizontal bands indicate the frequency range of operation for force modulation microscopy (FMM) and contact resonance (CR) methods. The cantilever mode shapes are shown for excitation at f_1^{cont}, the frequency of the first (lowest) flexural resonance ("on resonance") and at a frequency away from the contact resonance peak ("off resonance"). The faint outlines (wireframes) indicate the undeformed positions of the cantilever and sample. All results were calculated by finite-element analysis methods.

frequencies is that it minimizes contributions from extraneous DC or low-frequency signals (e.g., thermal drift, $1/f$ noise).

Another common theme in dynamic contact AFM methods is the cantilever response to high-frequency actuation. As is well known for modes such as intermittent contact, the cantilever motion exhibits a resonant response. The spectrum in Figure 5.3(a) was obtained from a finite-element analysis (FEA) model of the cantilever and shows the amplitude of the cantilever's flexural motion versus frequency for the tip in free space. It can be seen that the amplitude is much larger at certain preferred or natural frequencies. These frequencies, the so-called free resonances, can be determined from dynamic models for the cantilever beam with the assumption of clamped-free boundary conditions; their values depend on the cantilever geometry and material properties. The insets in Figure 5.3(a) indicate the cantilever mode shapes for the two lowest-order free resonances calculated by the FEA model. When performing dynamic contact AFM experiments, it is important to realize that the cantilever also exhibits resonant behavior when the tip is in contact with a surface. As shown in the amplitude spectrum in Figure 5.3(b) calculated by an FEA model for the tip in repulsive contact,

Table 5.1 Characteristics of Dynamic Contact Imaging Methods Including Force Modulation Microscopy (FMM), Shear Modulation Force Microscopy (SM-FM), and Contact Resonance Force Microscopy (CR-FM)*

		Excitation Frequency			
Method	Mode	Relative to Resonance	Range (kHz)	Excitation Orientation	Information Obtained
Force modulation	FMM	Off resonance	1–100	Out of plane	Quantitative images
	SM-FM		1–10	In plane	Qualitative images
Contact resonance	Spectroscopy	On (near) resonance	100–1000	Out of plane	Quantitative points
	Amplitude imaging				Qualitative images
	CR-FM				Quantitative images

*Typical literature values are given for the frequency range of excitation. Note that in-plane modulation is possible with CR methods but is not included here.

each "contact resonance (CR) frequency" is higher than that of the corresponding eigenmode for the tip in free space. In addition to the cantilever's material properties and geometry, the CR frequencies depend on contact parameters such as the applied force and sample properties such as elastic modulus.

Although all dynamic contact AFM methods utilize high-frequency excitation, they can be differentiated by whether the frequency is close to, or far from, a resonance peak. Off-resonance methods, which include FMM and SM-FM, operate in regions of the spectrum where there is little or no change in amplitude for small frequency shifts. CR methods deliberately operate at frequencies very close to a cantilever resonance ("on" resonance). As described below, on-resonance methods exploit the large shifts in amplitude with frequency to achieve increased sensitivity. Table 5.1 summarizes the similarities and differences between various dynamic contact methods.

5.1.2 Common Analysis Approach: Contact Mechanics

Finally, all dynamic contact AFM methods require data analysis to relate the experimentally measured quantities to the sample's mechanical properties. Analysis of both FMM and CR data relies on the theory of contact mechanics, which describes how the tip and the sample interact when in contact. For this discussion, the chief contact mechanics quantity of interest is the contact stiffness k that characterizes the elastic interaction between the tip and the sample. The contact stiffness k is defined by the differential force required to displace the tip–sample

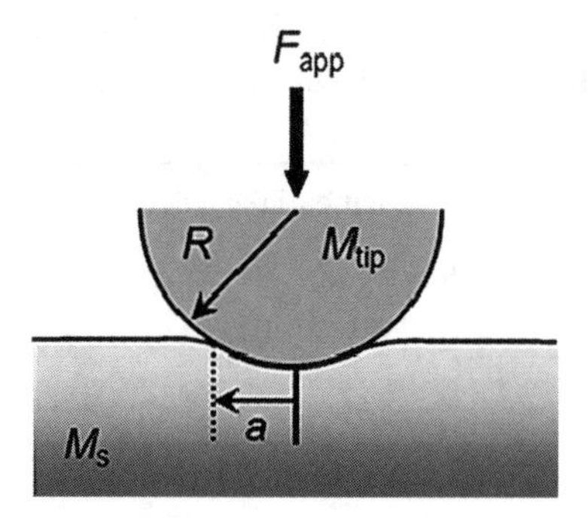

Figure 5.4 Hertzian contact parameters.

contact by a certain distance. The analysis approach used for each measurement method to determine k from the experimental data is discussed separately below. Once k has been determined, the same model for contact mechanics applies, regardless of the measurement technique.

Another chapter in this book contains a detailed discussion of contact mechanics concepts. Here, only the most pertinent elements are summarized. As shown in Figure 5.4, Hertzian contact describes the elastic interaction of a hemispherical tip with radius of curvature R and a flat surface (half plane). A static load F_{app} is applied to the tip normal to the sample. A purely elastic interaction between the tip and the sample is assumed, which can be experimentally approximated if F_{app} is sufficiently high compared to any adhesion force F_{adh}. For elastically isotropic materials, the tip and the sample form a circular contact region with radius a given by [1]

$$a = \sqrt[3]{\frac{3RF_{app}}{4E^R}} \tag{5.1}$$

where E^R is the reduced system modulus between the tip (subscript "tip") and the test sample (subscript "test") and is given by

$$E^R = \left(\frac{1}{M_{tip}} + \frac{1}{M_{test}}\right)^{-1} \tag{5.2}$$

Here, $M = E/(1-\nu^2)$ is the indentation or plane strain modulus for a material with Young's modulus E and Poisson's ratio ν. The indentation modulus is also a valid quantity for anisotropic materials. In this case, it is given by a combination of second-order elastic constants that depends on the material's symmetry [2]. The normal (vertical) contact stiffness k between the tip and sample is related to the quantities in Eqs. (5.1) and (5.2) by

$$k = 2aE^R \tag{5.3}$$

Despite its simplicity, the Hertzian contact model can be used effectively in many cases. However, a model that includes additional effects may be needed to accurately describe some experimental conditions. For instance, Derjaguin–Müller–Toporov (DMT) or Johnson–Kendall–Roberts (JKR) models [1] are often invoked to include adhesion forces in measurements of more compliant materials such as polymers.

In the rest of this chapter, we consider FMM and contact resonance methods in detail. We believe that CR methods represent the evolution of FMM into a state-of-the-art approach for highly sensitive, quantitatively accurate nanomechanical measurements. We first introduce the basic physical concepts of each approach and explain how they are implemented experimentally. Then we describe how the measured quantities can be related to the mechanical properties of the sample. Examples of experimental results are included to illustrate key features and further familiarize readers with the techniques. After describing FMM and CR techniques separately, we compare their relative features and merits. We also briefly discuss additional dynamic contact methods (e.g., SM-FM). We hope this chapter serves as a practical introduction to dynamic contact AFM methods and allows readers to assess the value of such methods for their own research.

5.2 FORCE MODULATION MICROSCOPY (FMM)

5.2.1 General Concepts

Force modulation microscopy was one of the earliest AFM techniques developed for imaging of nanomechanical contrast [3–5]. As indicated in Figure 5.3, FMM involves out-of-plane modulation of the tip–sample contact off resonance, that is, at a frequency away from the CR frequency. Working away from the resonance peaks ensures that they do not interfere with the amplitude response. Typically, a subresonant frequency is used; then the cantilever motion and the tip–sample contact motion generally correspond. For lower modulation frequencies (typically on the order of 1 kHz or less), excitation can often be achieved with the internal piezoelectric element used for imaging in the AFM instrument (the “Z piezo”). At higher frequencies, system resonances in the Z piezo preclude its use. Instead, either a high-frequency actuator in the cantilever holder (CH actuator in Figure 5.2) or beneath the sample (S actuator in Figure 5.2) is used.

For FMM data acquisition, the amplitude and phase of the cantilever motion at the excitation frequency are measured by the lock-in detector, as indicated in Figure 5.2. Thus, the instrumentation required for FMM measurements is relatively simple, consisting of a function generator and a lock-in detector. Most AFM instruments contain these capabilities internally. Several brands of commercial instruments provide FMM functionality as part of the operating platform. Other instruments allow sufficient access and control of the internal functions to enable the user to perform FMM with minimal effort. In AFM instruments that do not provide such access, an off-the-shelf function generator and lock-in detector can be used externally. Generally, this does not require additional software; the lock-in detector output amplitude and phase are connected to input channels of the AFM instrument. Then FMM imaging simply involves contact mode scanning at a constant excitation frequency by the function generator while the lock-in output signals are acquired.

5.2.2 Data Analysis and Interpretation

To understand how FMM measurements allow determination of elastic and viscoelastic properties, we refer to the parameters defined in the inset of Figure 5.1. The discussion here considers modulation of the sample (S actuator in Figure 5.2). For a constant sample modulation amplitude Δz, the resulting cantilever deflection amplitude is Δd, and the sample surface deforms by an amount $\delta = \Delta z - \Delta d$. The dynamic force ΔF acting on the material is found by $\Delta F = k\delta$, where k is the contact stiffness. As discussed above, the sample mechanical properties can be determined from k with use of contact mechanics. For a given cantilever spring constant k_L, it can be seen by considering the balance of forces that $\Delta F = k_L \Delta d$. Thus k can be determined (see Ref. [3]) by

$$k = \frac{\Delta F}{\Delta z - \Delta d} = \frac{k_L}{\Delta z / \Delta d - 1} \tag{5.4}$$

Determination of k with Eq. (5.4) first requires a calibration measurement to be performed. The amplitude of the cantilever response is measured on a sample that is considered to be infinitely stiff relative to the cantilever [3]. In this case there is no sample deformation, $\Delta z = \Delta d$, and a scale factor for the measured value of Δd is obtained. On the unknown material, this calibration factor and knowledge of k_L enable Eq. (5.4) to be solved for k. Finally, from the values of k (and other parameters such as the tip radius R), Eqs. (5.1) – (5.3) can be used to determine the sample's indentation modulus M.

It can be seen from Eq. (5.4) that obtaining accurate values of modulus with FMM depends on accurate measurements of both the cantilever spring constant k_L and the "lever sensitivity" calibration between cantilever amplitude and photodiode detector voltage. The time and effort involved in such measurements detract from the simplicity and ease of the original FMM experiment. Furthermore, the ability to discriminate elastically different sample regions (signal contrast) relies on the ability to measure differences in δ. For maximum sensitivity, the cantilever should be chosen so that the contact stiffness k is relatively similar to k_L. (Contact stiffness k can be estimated with use of approximate values for tip radius, force, and sample modulus.) Typically, the result is that FMM measurements on very stiff materials ($M > 50$ GPa) require very stiff cantilevers (up to $k_L \approx 1000$ N/m). However, $k_L \leq 50$ N/m for most commercial AFM cantilevers. Therefore, FMM is more readily applicable to compliant samples such as polymers and biological materials. For such materials, typical values of k_L range from approximately 0.1 to 3 N/m.

In addition, the phase signal gives information about the viscoelastic response. First, the phase angle is measured on a very stiff sample that is assumed to exhibit no dissipation. In theory, any additional phase lag measured on unknown materials can then be attributed to viscoelastic dissipation. More complex analyses than that discussed above have been developed to measure viscoelastic properties [4, 6, 7], but they lie outside the scope of this chapter.

An appealing feature of the FMM method is the simplicity of the mathematical model for data interpretation. If the amplitude response with frequency is assumed to be flat, the analysis approach described above is a straightforward path to quantitative modulus information. However, as FMM techniques evolved, it became apparent that the vibrational response of the cantilever was not perfectly flat with frequency away from the contact resonances [8, 9]. In calibration measurements made at identical drive amplitudes and frequencies to measurements on test or unknown samples, amplitude and phase spectra revealed that a variety of spurious and resonant vibrations could occur [9]. If these additional vibrations were affected by the stiffness of the contact, they could not be accommodated in the analysis of FMM amplitude and could lead to erroneous results. The following discussion on contact resonance techniques illustrates why performing FMM in the presence of unaccounted for vibrations is difficult, while also showing how amplitude analysis can be avoided if one measures the resonant frequency of the cantilever in contact.

5.2.3 Examples and Applications

Because FMM techniques were originally developed more than 20 years ago, they have been applied to a wide range of material systems. As discussed above, the range of k_L for commercially available cantilevers means that FMM is well suited to more compliant materials. Literature results include fiber-reinforced composites [3, 10], polymers [7, 11], organic monolayers [4–6, 12], and soft biological materials [13, 14]. In some cases (e.g., [3, 5, 13, 14]), analysis methods such as those discussed above were applied to obtain quantitative nanomechanical information.

Example results are shown in Figure 5.5. Figures 5.5(a)–5.5(c) show FMM images for a monomeric diamino-diethylene glycol-pentacosadiynoic acid (DPDA) polymer brush attached to a rigid silicon (Si) substrate [4]. The topographic height image in Figure 5.5(a) is equivalent to a contact-mode AFM image but was acquired in parallel with the images in Figure 5.5(b) and 5.5(c) for the FMM amplitude and phase, respectively. The height image in Figure 5.5(a) shows that the polymer brush regions extend above the rigid substrate. The amplitude image in Figure 5.5(b) shows lower FMM amplitudes on the brush regions than on the Si substrate. The reduced cantilever vibration amplitude corresponds to increased deformation of the polymer brush, indicating that it is more compliant. It can be seen from the phase image in Figure 5.5(c) that, whereas the FMM amplitude decreases on the brush, the corresponding phase angle increases. This behavior indicates greater viscous dissipation in the brush than in Si. The images in Figures 5.5(d) and 5.5(e) correspond to FMM measurements on a composite sample containing carbon fibers in an epoxy

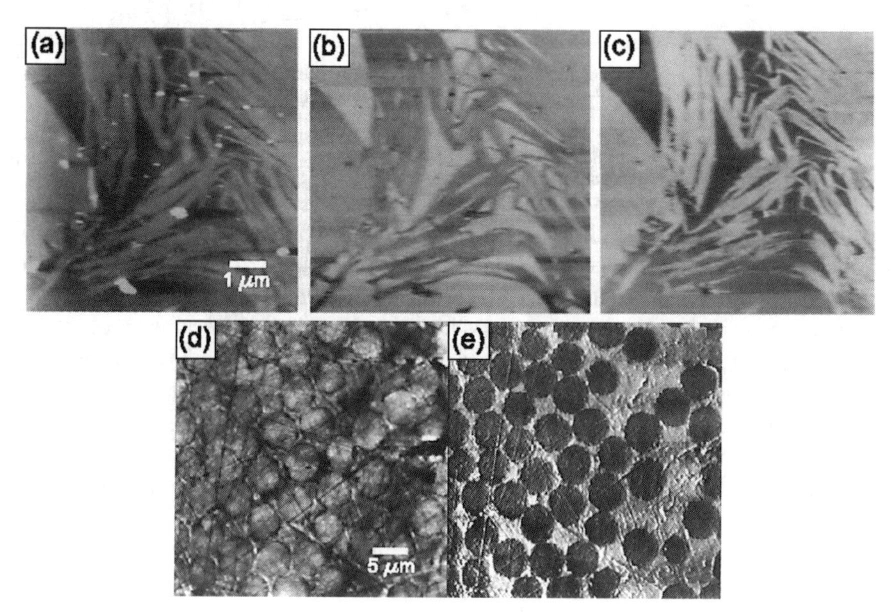

Figure 5.5 Example of FMM results. (a) Topography (z scale ~3 nm). (b) FMM amplitude and (c) FMM phase for a polymeric DPDA monolayer on Si. The sample modulation amplitude was 6 nm, and the modulation frequency was 15.6 kHz. (Reprinted from *Biophysical Journal*, Vol. 64, M. Radmacher, R. W. Tillmann, and H. E. Gaub, Imaging viscoelasticity by force modulation with the atomic force microscope, pp. 735–742, copyright 1993, with permission from Elsevier via RightsLink.) (d) Topography (z scale 150 nm) and (e) FMM amplitude (z scale 1 nm) for a carbon fiber/epoxy composite. (Republished with permission of Annual Reviews, Inc., from "Characterization of polymer surfaces with atomic force microscopy," S. N. Maganov and D. H. Reneker, vol. 27, pp. 175–222, 1997 via Copyright Clearance Center.)

matrix [10]. The amplitude image in Figure 5.5(e) acquired of the fibers in cross section shows strong contrast between the two very dissimilar elastic materials. Both of these examples illustrate the appeal of FMM imaging: With a relatively simple experimental apparatus, it provides clear qualitative contrast of elastic and viscoelastic properties with nanoscale spatial resolution.

Force modulation microscopy measurements are not limited to only a stiff, sharp AFM tip interacting with a continuous solid substrate. By attaching a polymer microsphere to the end of the cantilever, FMM has been used to probe the viscoelastic response of very compliant materials such as gels and cells [7, 13, 14]. This approach was also used to obtain quantitative modulus values by modifying the analysis model described above [13, 14]. The fact that the polymer microsphere is more compliant than a standard silicon or silicon nitride tip means that the depth of the stress field is reduced. This configuration allows thin compliant structures, such as the edge of a cell, to be characterized without substrate influence. Also, the easily varied tip diameter can be used to reduce contact pressure on fragile specimens.

Attached polymer microspheres were also recently used in FMM experiments to investigate the viscoelastic properties of suspended polyvinyl alcohol (PVA)

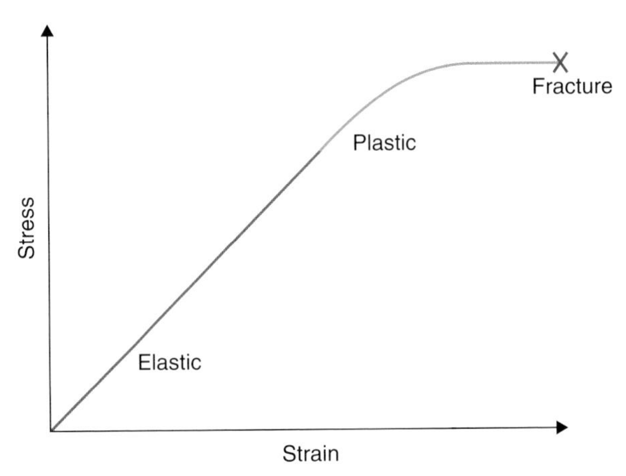

Figure 1.2 Basic stress vs. strain curve showing three mechanical regimes of elastic deformation, plastic deformation, and fracture.

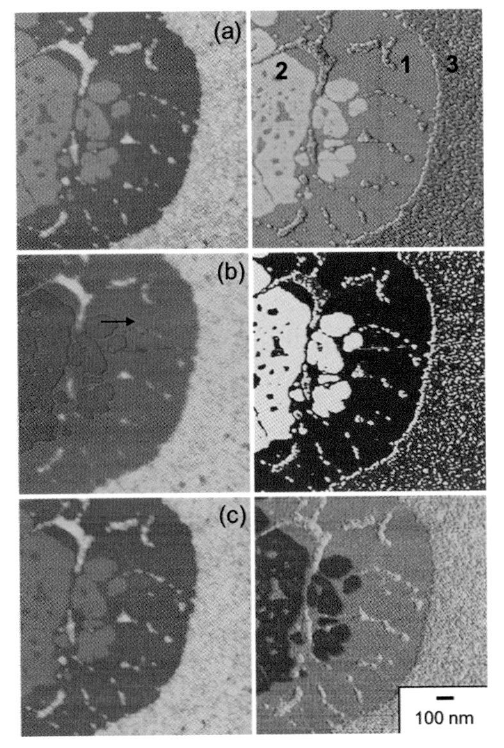

Figure 4.3 Height (left) and phase (right) images of a polyvinyl alcohol thin film containing three characteristic components. Three drive amplitudes produce free oscillation resonance amplitudes (A_0) of (a) 14, (b) 10, and (c) 4 nm corresponding to uniform repulsive, coexisting, and uniform attractive regimes, respectively. In height image, bright contrast refers to topographically high and dark contrast refers to topographically low. In phase image, bright contrast refers to high phase and dark contrast refers to low phase.

Scanning Probe Microscopy in Industrial Applications: Nanomechanical Characterization,
First Edition. Edited by Dalia G. Yablon.

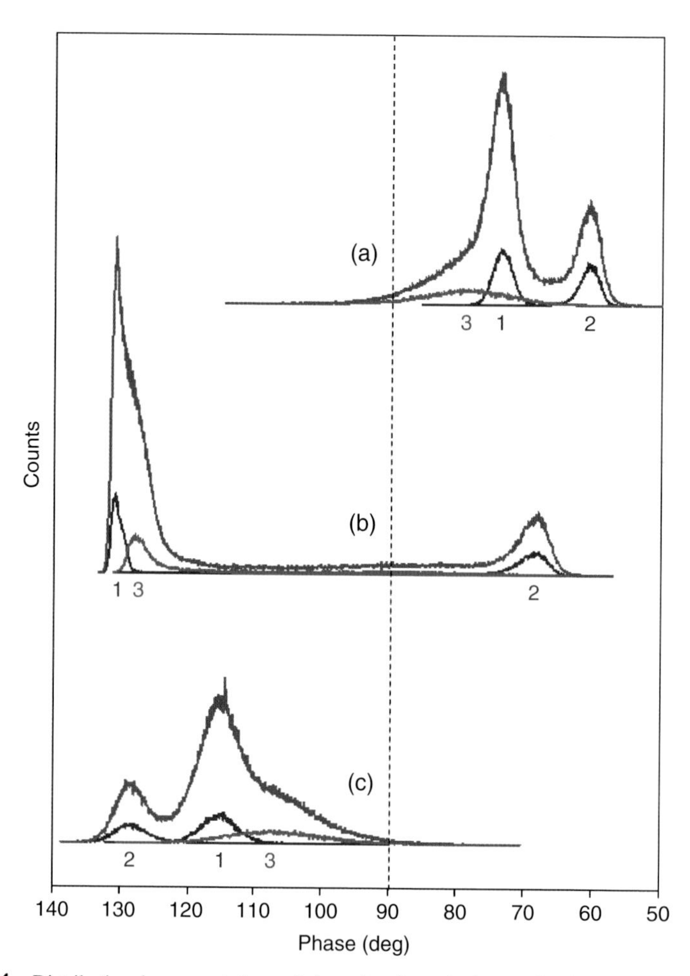

Figure 4.4 Distributional representations of phase data from the images in Figure 4.3 for film components 1–3. Multipeak histograms are from entire phase image; smaller single peaks derive from individual film components.

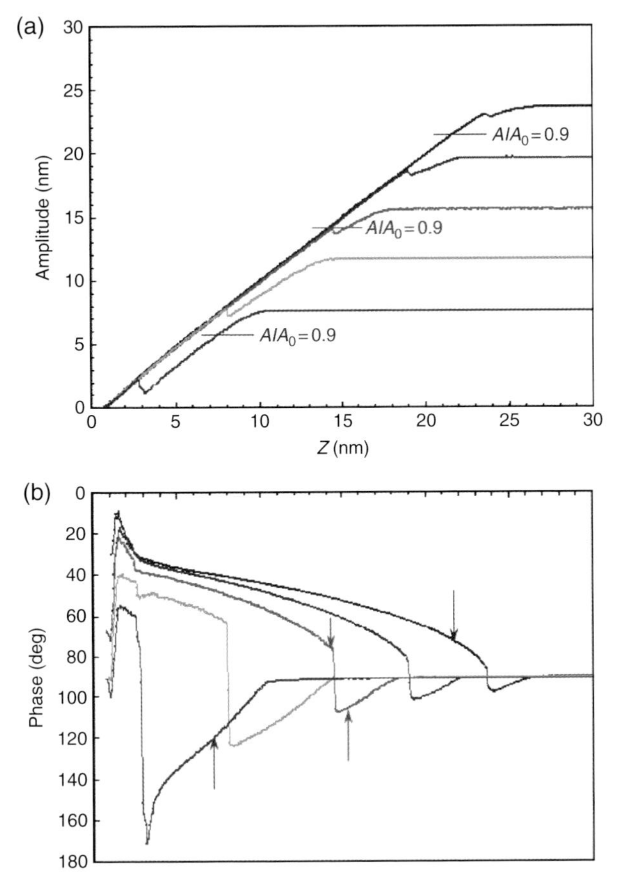

Figure 4.6 (a) Amplitude-Z and (b) phase-Z curves on a silicon wafer driving at resonance frequency and at five different drive amplitudes, producing free oscillation amplitudes A_0 ranging from 7.6 to 23.8 nm. Dashed horizontal lines denote typical setpoint amplitudes for repulsive-regime (top curve) and attractive-regime (bottom curve) imaging states at a setpoint amplitude ratio $A/A_0=0.9$.

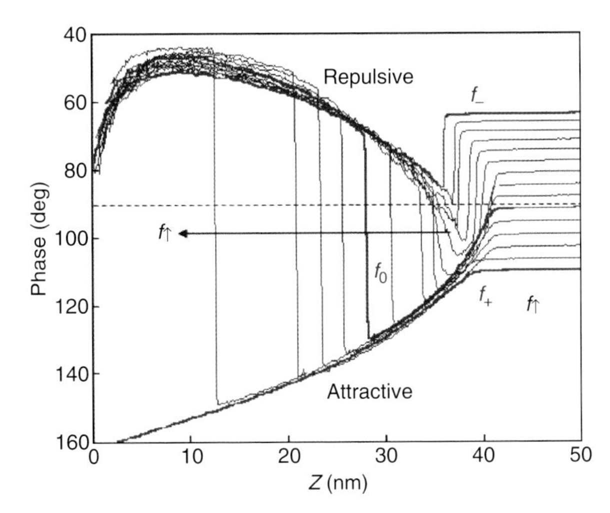

Figure 4.7 Phase-Z curves obtained at the 15 values of drive frequency. Arrows indicate the effect of increasing drive frequency on both the free oscillation phase value (right) and the discontinuous transition from attractive to repulsive regime (left).

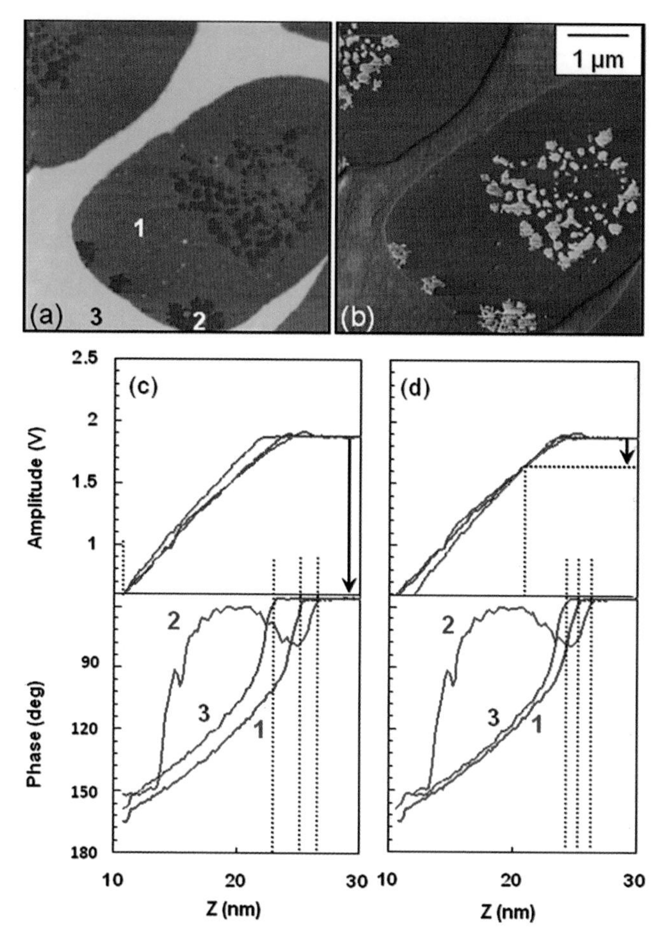

Figure 4.8 (a) Height and (b) phase images of a three-component PVA surface. (c) Raw amplitude-Z and phase-Z curves acquired at the three locations denoted in (a). (d) Curves from (c) Z shifted to achieve the same Z value at the amplitude value of 1.65 V.

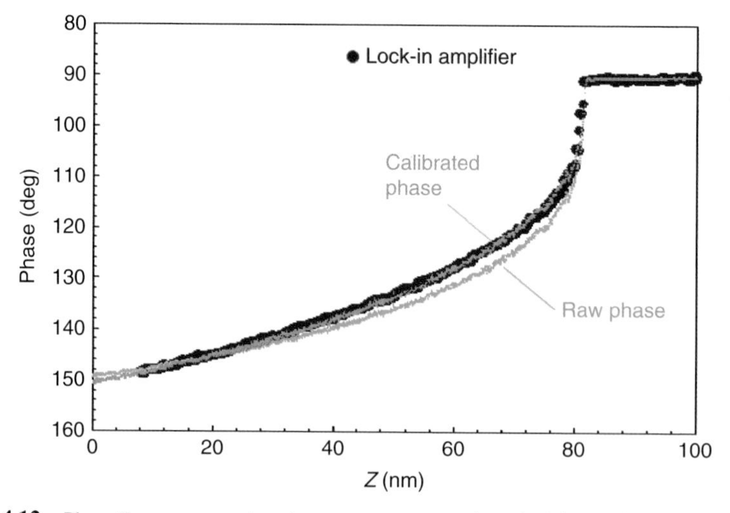

Figure 4.12 Phase-Z curve comparing phase measurements using a lock-in amplifier (circles) to raw and nonlinearly calibrated measurements (lines) using the Extender electronics.

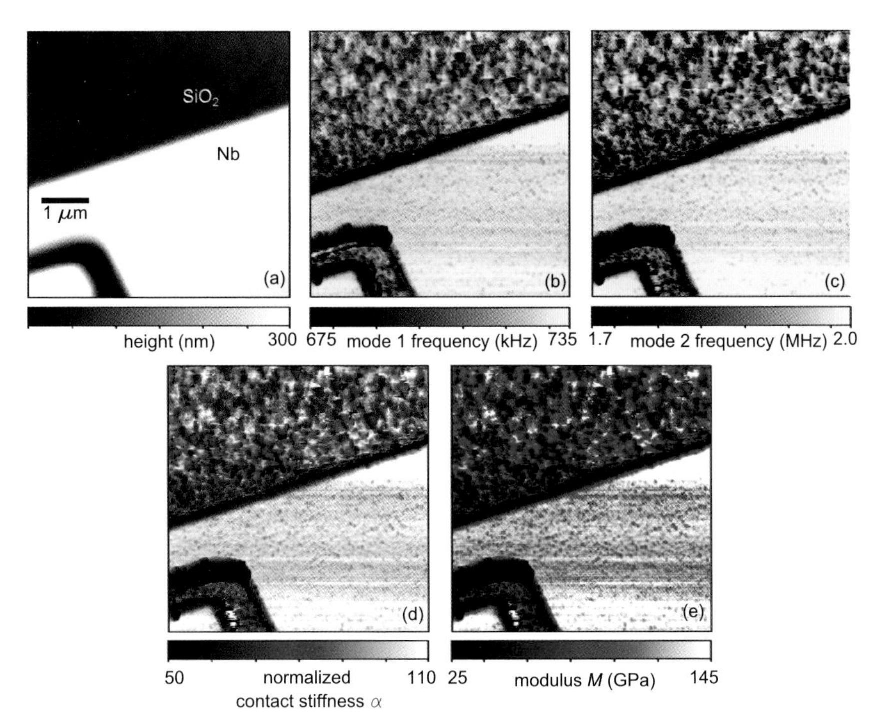

Figure 5.9 Example CR-FM results for a sample with a patterned Nb film on a SiO_2 blanket film. (a) Contact-mode topography. (b) and (c) CR frequency images for the first and second flexural eigenmodes, respectively. (d) Normalized contact stiffness $\alpha = k/k_L$ calculated from (b) and (c). (e) Map of indentation modulus M calculated from (d). The cantilever's nominal spring constant was $k_L = 45$ N/m. The resonance frequencies of the cantilever's first two flexural modes in free space were $f_1^{\text{free}} = 171.7$ kHz and $f_2^{\text{free}} = 1048.0$ kHz.

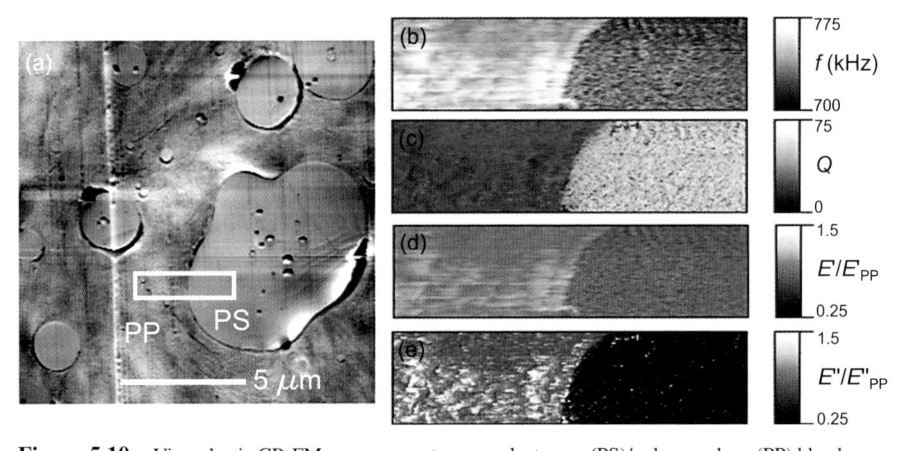

Figure 5.10 Viscoelastic CR-FM measurements on a polystyrene (PS)/polypropylene (PP) blend. (a) Topography image (z scale 100 nm) with PS domain and PP matrix indicated. (b) and (c) CR-FM maps of frequency and quality factor, respectively, for the second flexural eigenmode in the region indicated by the box in (a). (d) and (e) Calculated maps of storage modulus E' and loss modulus E'', respectively, normalized to their average values on polypropylene. The cantilever's nominal spring constant was $k_L = 3$ N/m, and the resonance frequency of the second flexural mode in free space was $f_2^{\text{free}} = 423.6$ kHz.

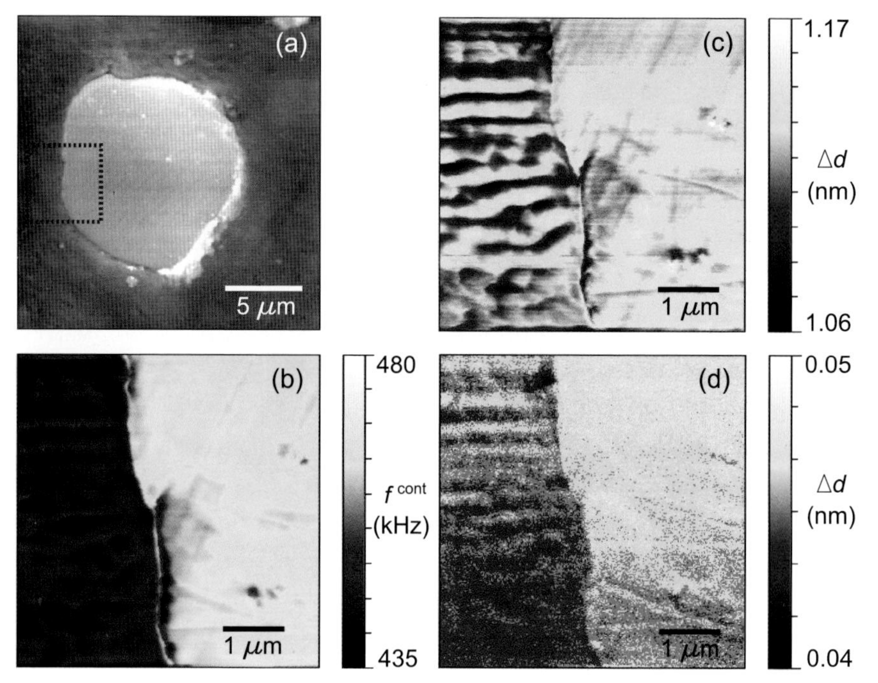

Figure 5.12 Comparison of dynamic contact AFM methods. The sample was a composite containing a cellulose fiber in a polypropylene matrix. (a) Tapping-mode topography image (z scale 400 nm). The box indicates the region used for the images in (b)–(d). (b) CR-FM frequency image. The cantilever's nominal spring constant was $k_L = 15$ N/m, and the free resonance frequency of the first flexural mode was $f_1^{free} = 106$ kHz. (c) and (d) FMM amplitude images for modulation at 1.5 and 100 kHz, respectively. The images in (b)–(d) were acquired with the same cantilever.

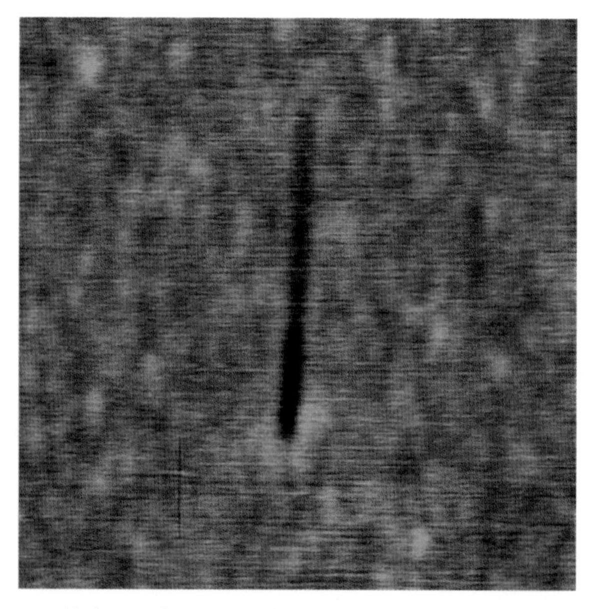

Figure 6.1 Topographic image of scratch produced during force-Z measurements due to lateral tip displacement as vertical force increases (top to bottom).

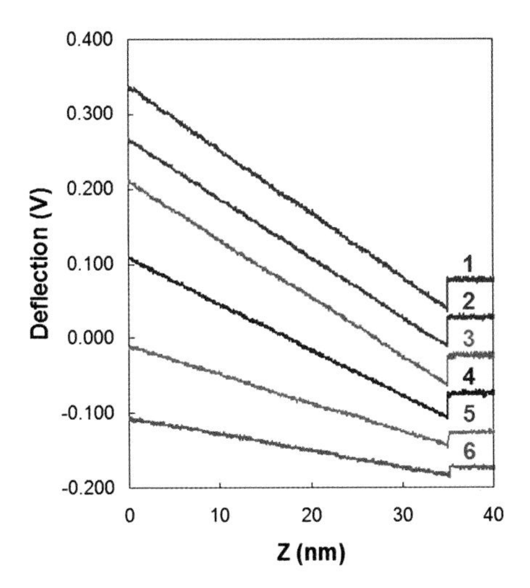

Figure 6.2 Approach force curves for different laser spot positions near evenly spaced along the length of the cantilever: 1 near tip end, 6 near base, and 3 near middle.

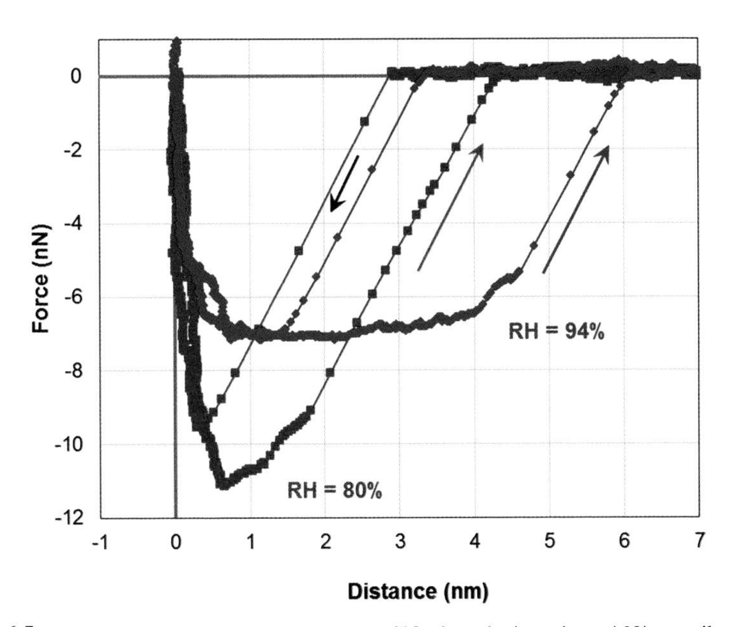

Figure 6.5 Distance dependences of force between SiO_2 tip and mica using a 4-N/m cantilever at two high humidities.

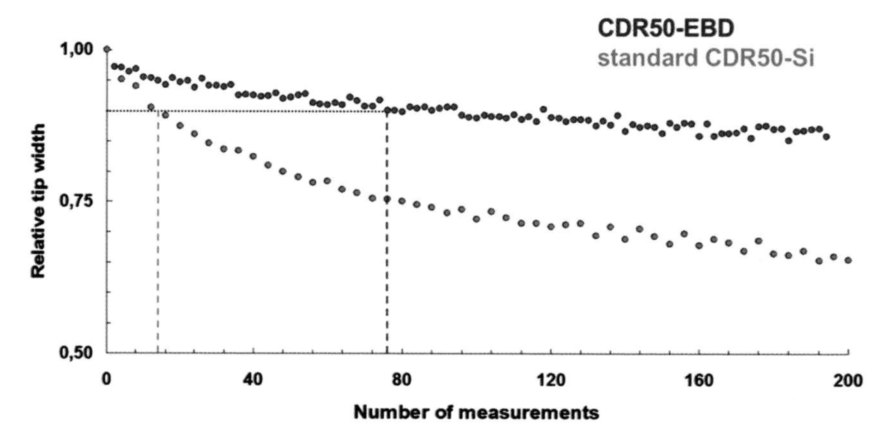

Figure 8.9 Relative tip width of an EBIP-manufactured CD-AFM tip (blue, CDR50-EBD) and a commercial state-of-the-art Si-based CD-AFM tip (orange, CDR50-Si) as a function of the number of measurements. The gray horizontal line represents the 10-percentage reduction of the tip width related to the initial tip width. The dashed vertical line points to a 10-percentage reduction after ~15 measurements for the standard CDR50-Si, while the CDR50-EBD exhibits the corresponding tip width wear after ~75 measurements (dashed blue vertical line).

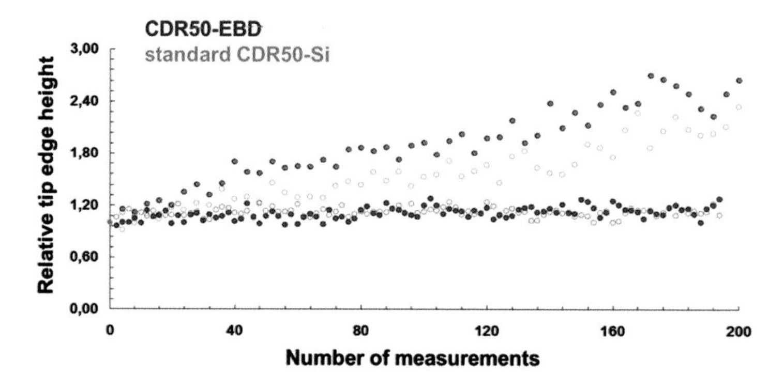

Figure 8.10 Relative tip edge height of a HDC/DLC CD-AFM tip (blue, CDR50-EBD) and a standard Si-based CD-AFM tip (orange, CDR50-Si) as a function of the number of measurements. Filled circles and open circles correspond to the right tip edge height and the left tip edge height, respectively.

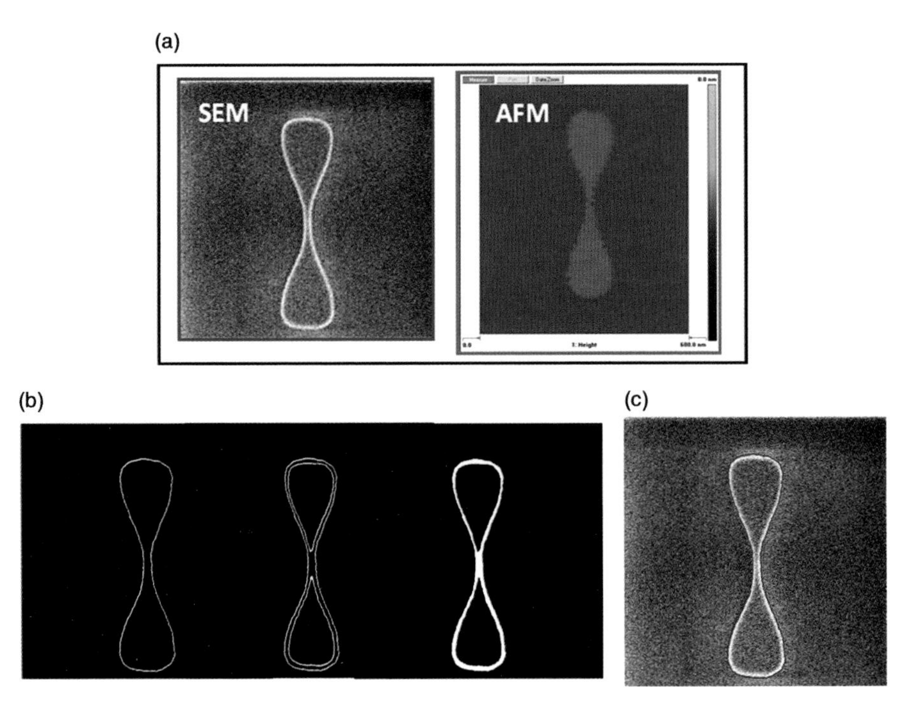

Figure 8.13 (a) Example of a single nanowire device measured by two techniques: CD-SEM and CD-AFM; (b) extraction of *X* andY contour information from CD-SEM and CD-AFM images; and (c) contour superimposition on an original CD-SEM image.

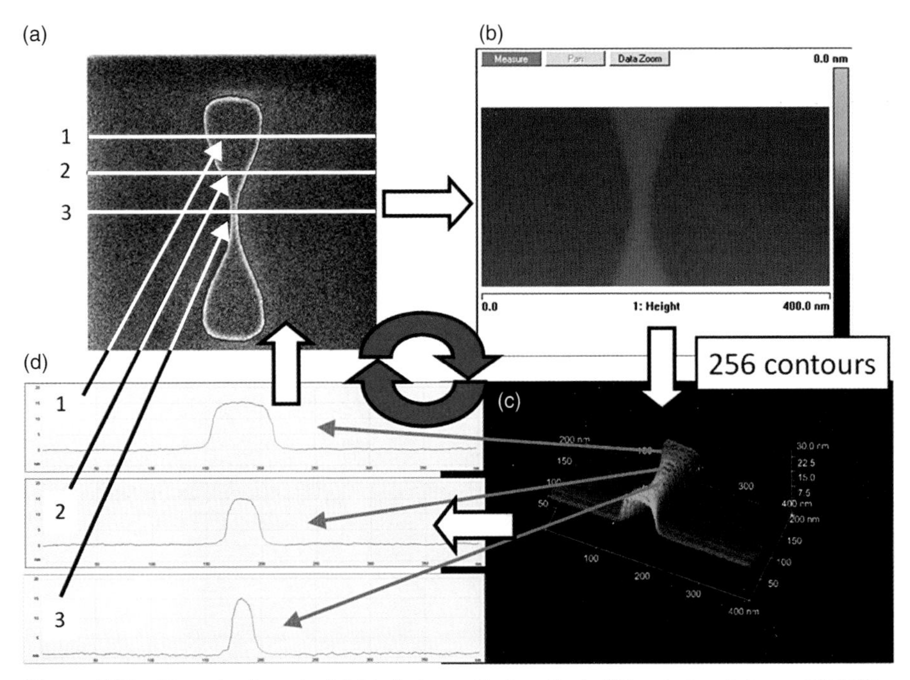

Figure 8.14 Example of a potential data fusion method applies to CD metrology between CD-SEM (a) and CD-AFM (b) and (c) through multiple contour extractions (d).

Figure 9.1 A 5-by-5-μm tapping phase image of a tire sidewall compound where the black phase is NR, the brown phase is EPDM, the yellow phase is CIIR, the individual round white particle aggregate is silica, and the large white flaky particle is TiO_2 (titanium dioxide).

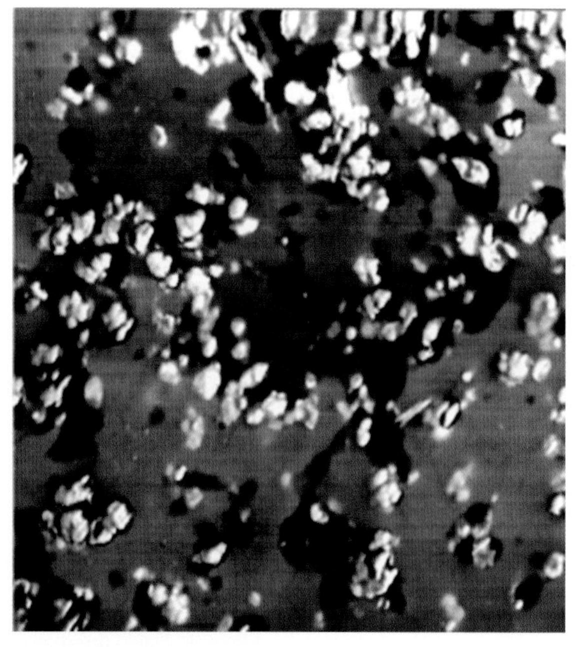

Figure 9.2 A 2.5-by-2.5-μm tapping phase image of an IIR (butyl rubber) and EPDM blend illustrating the preferential partition of the bright white CB particles into the dark EPDM phase. (White particles: CB; yellow matrix phase: IIR; dark areas: EPDM.)

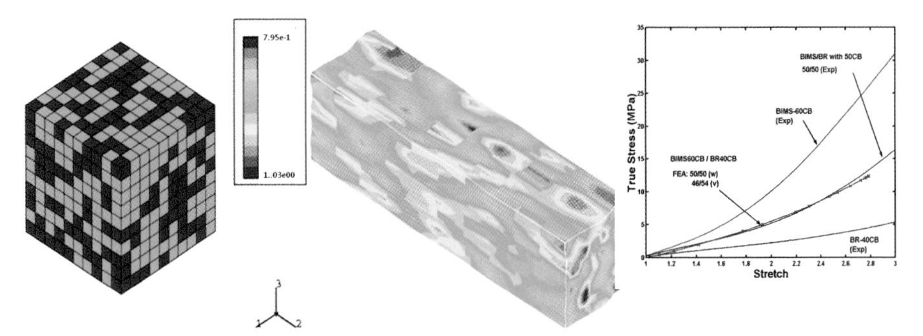

Figure 9.3 Finite-element simulation of a CB-filled blend of BIMS and BR (50/50 by weight). Image processing of tapping phase AFM micrographs quantified the phase partition of CB with 60% CB in BIMS and 40% CB in BR. Monte Carlo simulation was used to create the random blend morphology shown.

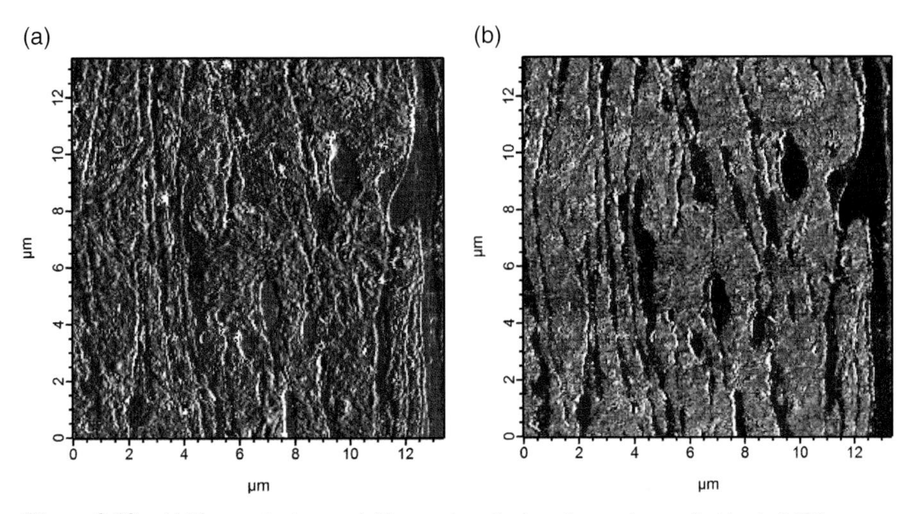

Figure 9.12 (a) First mode phase and (b) second mode phase image shown of a blend of 60% polyethylene (PE), 20% polypropylene (PP), and 20% polystyrene (PS).

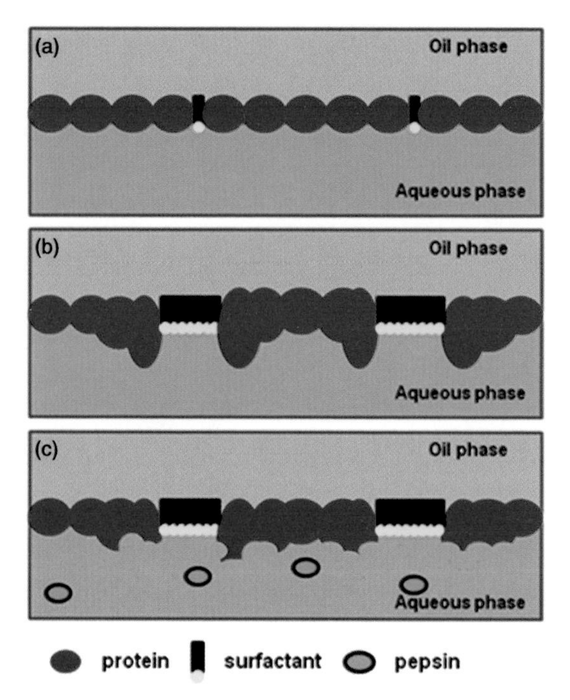

Figure 10.3 Schematic illustration of the synergy between surfactants and enzymes in the destruction of interfacial protein films during digestion. (a) Initial penetration of the protein film by surfactant leads to nucleation and domain growth (b) pushing out previously shielded regions of the proteins which become susceptible to pepsinolysis (c).

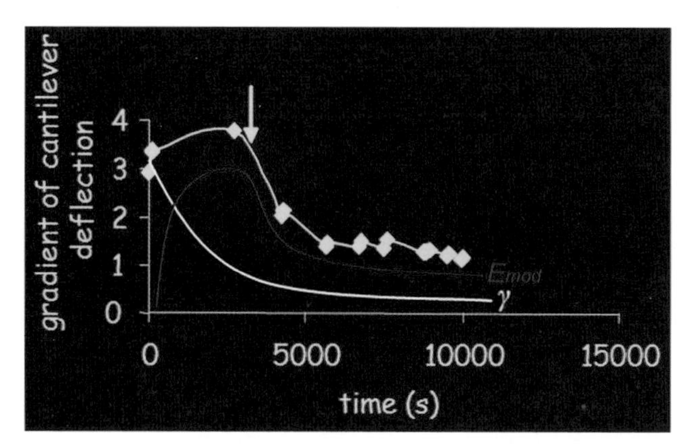

Figure 10.6 Effect of in-situ exchange of the interfacial film on a pair of emulsion droplets. The yellow line links the AFM data points (arrow marks the time point for addition of surfactant). Red and white lines illustrate the overall trends for the interfacial elastic modulus and interfacial tension, respectively, as protein is exchanged for surfactant. Note, the Y axis values are depicted in arbitrary units.

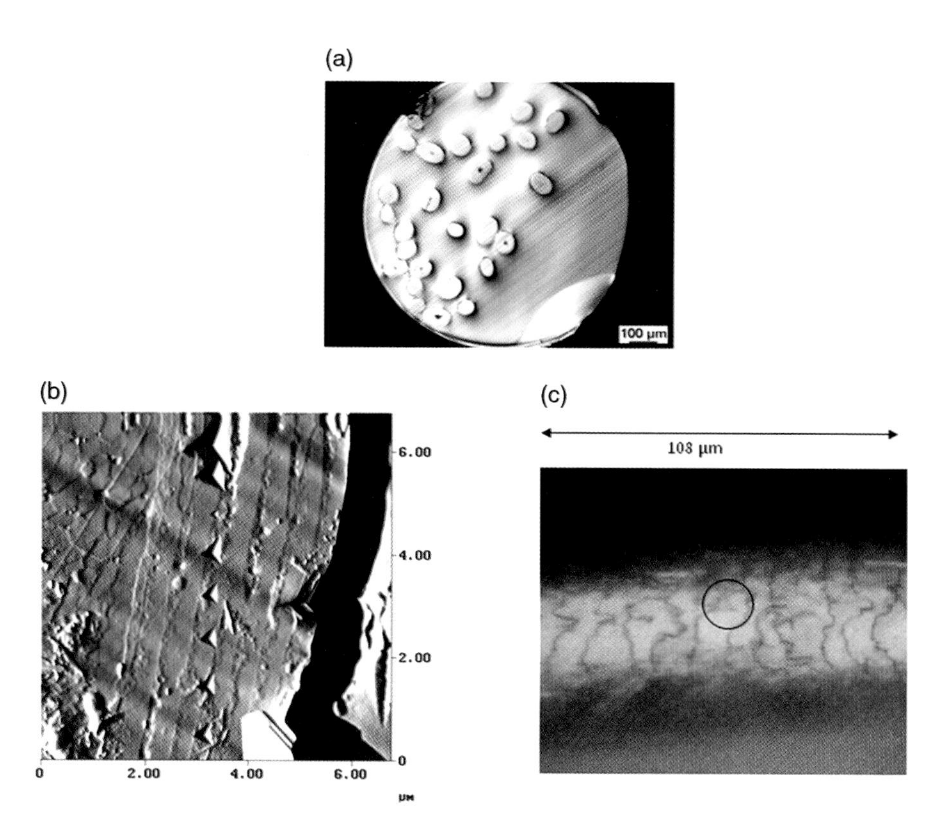

Figure 12.3 Examples of types of indentation of hairs. (a) Microtomed hair cuts embedded in a resin. (b) Example of precise indents on the cuticle of hair performed using an AFM-based indentation. (c) Examples of indents done at the surface of hair fibers using instrumented nanoindentation.

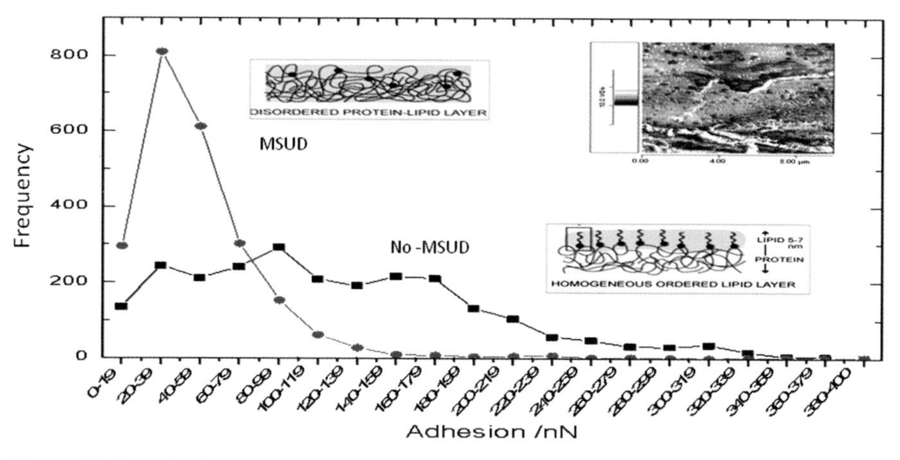

Figure 12.9 Adhesion between an AFM tip and hair of a patient lacking 18-MEA lipid or hair of a healthy individual.

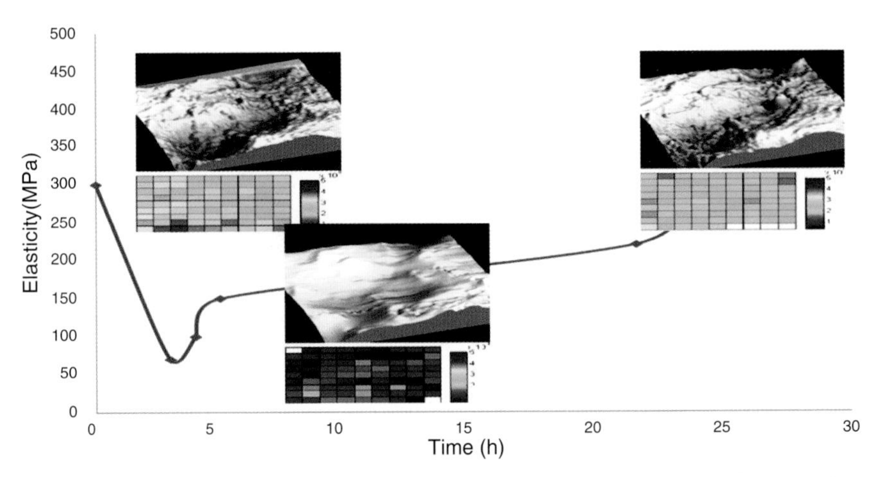

Figure 12.10 Changes in mechanical properties of the stratum corneum after applying a layer of glycerol onto its surface (AFM nanoindentation measurements).

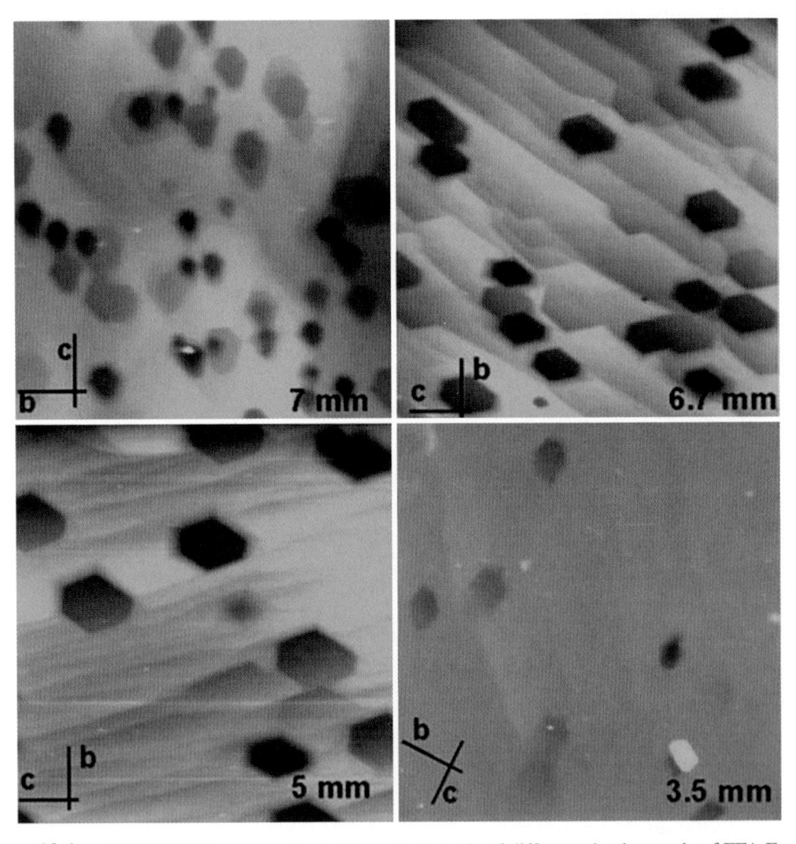

Figure 13.1 AFM height images collected in contact mode of different sized crystals of FFA Form I (1 0 0) face etched with *n*-pentane, with dark or red representing lower topography and bright or yellow representing higher topography. AFM scan size ~100 μm. Crystal size (lower right) and crystallographic orientation (lower left) are labeled on each image. Reprinted with permission from Ref. [7].

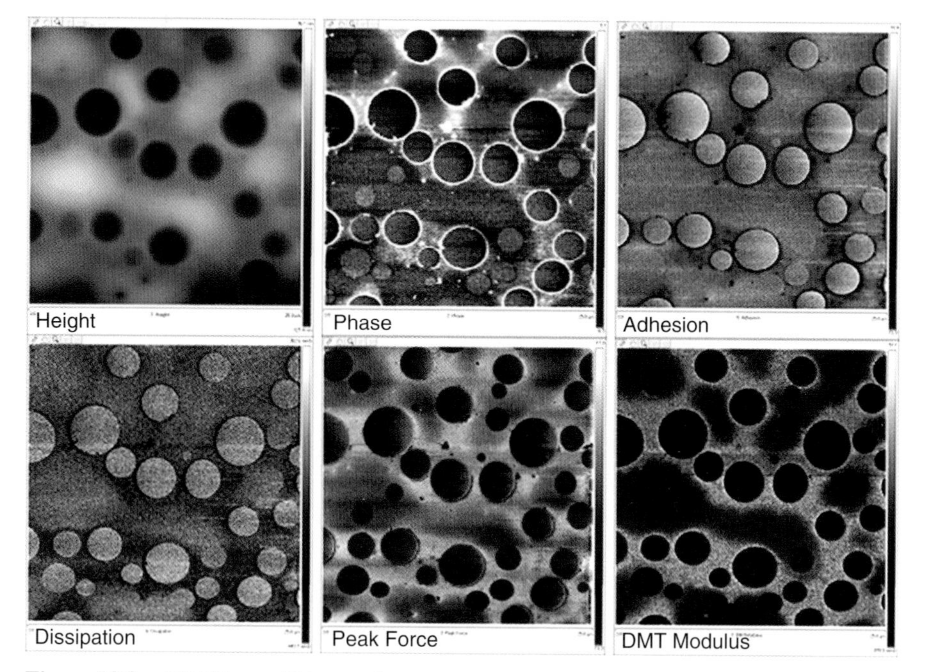

Figure 13.2 AFM HarmoniX image of humidity-induced phase separation of an amorphous solid dispersion. Scan size is 25 μm.

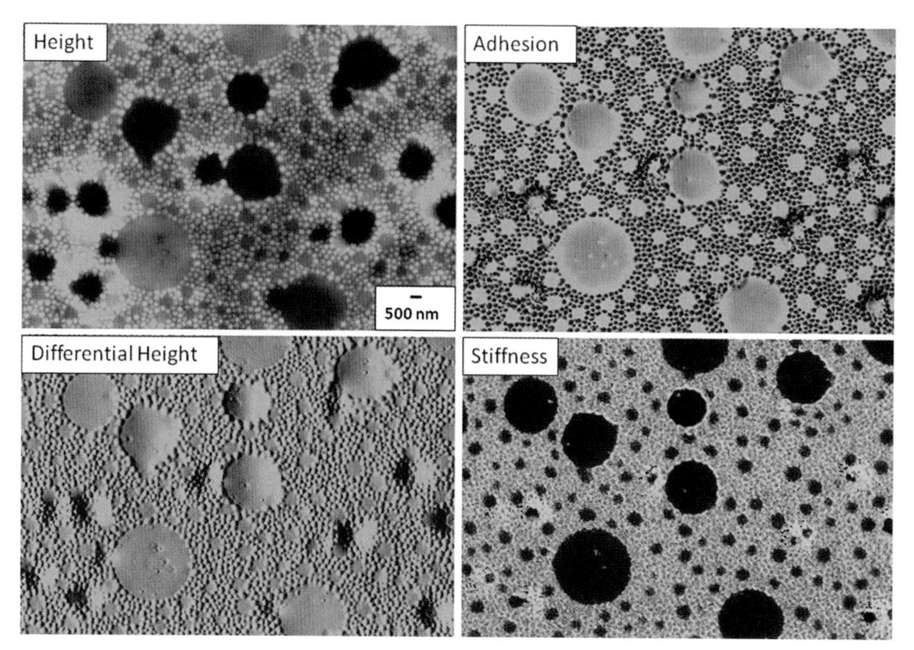

Figure 15.5 Multiple image types acquired in D-PFM in air on a PBMA–PLMA–dexamethasone coating.

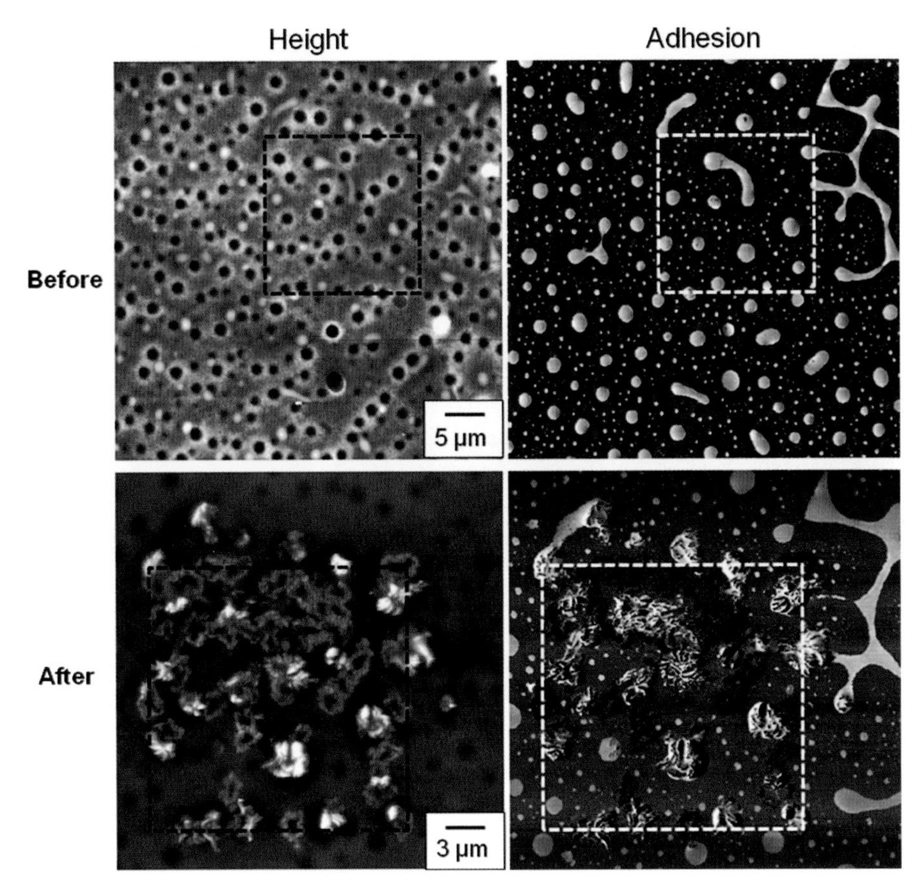

Figure 15.8 Height and adhesion image pairs of a PBMA–PLMA–dexamethasone coating acquired in D-PFM in air before (top) and after (bottom) a subregion was raster scanned at an elevated temperature of 80 °C (as designated with dashed squares).

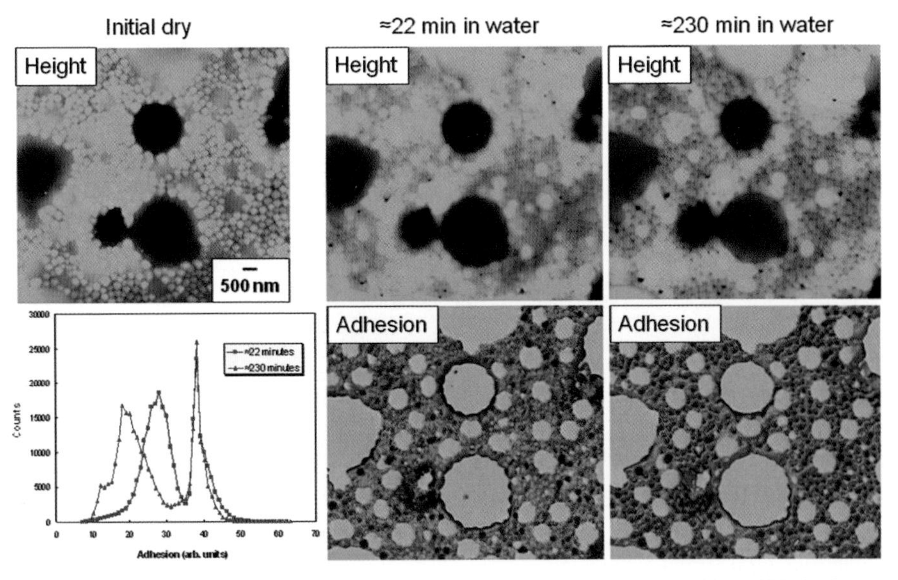

Figure 15.9 Height image of a PBMA–PLMA–dexamethasone coating acquired in D-PFM in air (top left) and height/adhesion image pairs of the same region while immersed in water for 22 minutes (middle) and 230 minutes (right). Histogram (bottom left) of the two adhesion images.

nanofibers [7]. Fibers have the advantage that on the suspended region, the cantilever vibration amplitude (and thus the contact stiffness) are affected primarily by the fiber's bending rigidity and not the contact mechanics response. Fitting the data to predictions across the entire length of the fiber allowed the quantitative viscoelastic response to be determined. Moreover, by performing experiments at several FMM excitation frequencies, the frequency dependence of the viscoelastic response was examined. These results highlight another strong point of FMM methods. As long as the narrow frequency bands where cantilever resonances occur are avoided, a wide range of frequencies with high-frequency resolution can be probed with FMM. In contrast, contact resonance methods are by definition limited to very specific frequency ranges. However, as discussed in the next section, CR methods possess other distinct advantages for quantitative characterization of mechanical properties.

5.3 CONTACT RESONANCE (CR) TECHNIQUES

5.3.1 Basic Concepts

Contact resonance techniques are the second broad class of dynamic contact AFM methods. As in FMM, a piezoelectric actuator is used for CR excitation. However, a key difference between CR and FMM methods is the relative frequency of operation. As indicated in Figure 5.3, FMM methods operate at frequencies far from the cantilever resonances in contact, while CR methods deliberately operate in the vicinity of a resonance. Near-resonance operation exploits the fact that the CR frequency and amplitude depends on the sample's elastic modulus due to tip–sample interaction forces. For the idealized case of constant-force scanning on a perfectly flat sample with no tip wear, changes in the CR frequency can be attributed to modulus variations between various sample components. Qualitative modulus information can be obtained by monitoring the relatively large shifts in signal amplitude that occur near resonance. This approach is analogous to FMM methods that produce contrast images from the phase or amplitude of the cantilever motion. In contrast, CR methods for quantitative modulus information (CR spectroscopy at a point or CR-FM imaging) rely on precise measurements of resonance frequency.

Standard CR experiments involve one or more flexural (bending) eigenmodes of the cantilever. Therefore, the excitation motion is normal (vertical) to the sample plane. The CR frequency of the first flexural mode can range from a few tens to several hundreds of kilohertz, depending on the cantilever stiffness. These frequencies are often beyond the reach of piezoelectric elements internal to a standard AFM instrument (those used for imaging or for tapping-mode tuning in the cantilever holder). Therefore, to date most CR work has involved homemade setups with an external actuator beneath the sample (S in Figure 5.2) or a custom piezoelectric element in the cantilever holder (CH in Figure 5.2) that was integrated into the commercial AFM. Such a setup can be achieved relatively easily with off-the-shelf components (see, e.g., [15, 16]). CR functionality is also currently provided by at least two commercial AFM instruments [17, 18]. The main requirements for integrating

CR methods in any AFM are that (a) the response of the AFM photodiode must be sufficiently fast to detect this high-frequency motion, and (b) external access to the photodiode's high-frequency signal is needed.

Several related methods can be designated by the term "contact resonance techniques." Historically, names of CR techniques developed by various researchers have often included the word "acoustic" or "ultrasonic" due to the frequency range of vibrations involved. For example, one of the first CR methods was called atomic force acoustic microscopy (AFAM) [19, 20] and involved excitation of the cantilever resonance by an actuator beneath the sample. A second method involving actuation by a custom piezoelectric element in the cantilever holder was named ultrasonic AFM (UAFM) [21, 22]. Mathematical analyses [20, 23, 24] show that if damping terms are small, the CR frequencies are equivalent for both excitation methods. However, the amplitude of the cantilever vibration and other details of the response spectrum differ. In spite of similar-sounding names, dynamic AFM methods such as ultrasonic force microscopy (UFM), heterodyne force microscopy, scanning near-field ultrasonic holography, and surface acoustic wave-assisted scanning probe microscopy are not included in our discussion for various reasons. For instance, in UFM the tip does not remain in contact; in fact, it relies on how the tip "jumps off" the sample. Further information about these methods and the nanomechanical information they give, as well as more on CR techniques, is available elsewhere [16].

It is worth noting that CR methods were originally developed in the mid- to late-1990s to determine elastic properties of very stiff materials (those with modulus greater than ~50 GPa). At the time, lack of commercial cantilevers with sufficiently high spring constants precluded sensitive measurements with F–D spectroscopy (static force curves) or FMM methods on stiff materials. Due to the cantilever stiffening effect for on-resonance operation, measurements with CR methods can achieve higher contrast on stiff materials than measurements performed with quasi-static methods and a similar stiffness cantilever. Improvements in cantilever microfabrication techniques in the past decade have somewhat altered this situation. However, CR techniques remain a valuable methodology for obtaining quantitative nanomechanical information on stiff materials.

5.3.2 Measurement Approaches

Quantitative Point Measurements: CR Spectroscopy

Contact resonance spectroscopy measurements are those performed at a fixed (stationary) position on the sample (i.e., scanning turned off). The excitation frequency of the actuator is incrementally increased, and the amplitude of the cantilever motion at each frequency is measured with a lock-in amplifier. In this way, a spectrum of the contact resonance amplitude (and phase, if desired) is acquired. Examples of CR amplitude spectra acquired on thin films of silica (SiO_2) and niobium (Nb) are shown in Figure 5.6. For reference, the modulus of Nb is greater than that of SiO_2; literature values for the indentation modulus of bulk SiO_2 and polycrystalline Nb are $M(SiO_2) \approx 75$ GPa and $M(Nb) \approx 125$ GPa. The experimental

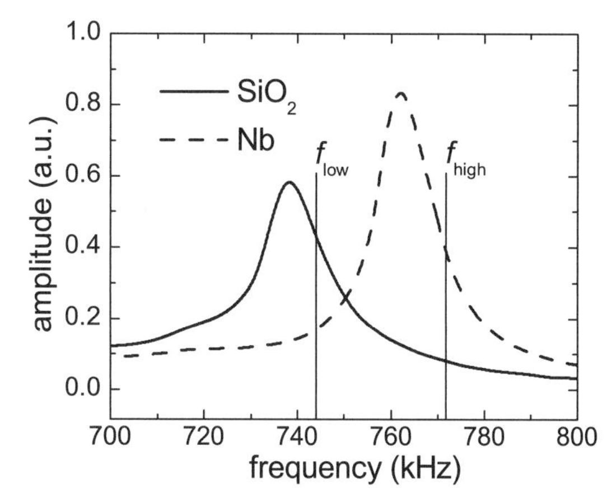

Figure 5.6 Contact resonance spectra obtained on SiO_2 and Nb films. For the same applied force, Nb has a higher CR frequency than SiO_2, consistent with the modulus trend for the bulk materials $M(\mathrm{Nb}) > M(\mathrm{SiO_2})$. Vertical lines indicate the frequencies f_{low} and f_{high} used to discuss how the relative amplitude detected in qualitative experiments (CR amplitude imaging) can change depending on the choice of excitation frequency.

response is similar to the theoretically predicted response in Figure 5.3(b), with a clearly identified resonance peak. The CR frequency can be chosen simply by determining the frequency of maximum amplitude or through more advanced methods (e.g., Lorentzian curve fitting). The data analysis method to determine the modulus from the measured CR frequency is described in a later section. However, even from the raw spectra, the general principle can be observed that a material with higher modulus (here, Nb) exhibits a higher resonance frequency.

The nanoscale spatial resolution afforded by AFM means that CR spectroscopy has been applied to a variety of micro- and nanostructures that cannot be evaluated with more conventional methods such as nanoindentation. Furthermore, the ability of CR methods to probe very stiff materials has proven valuable for characterization of new materials. For instance, the fundamental properties of diverse nanotubes, nanorods, nanowires, and nanocrystals have been investigated with CR spectroscopy. Size-dependent effects have been observed (e.g., [25]). Another class of materials well suited to CR spectroscopy is ultrathin films [26–28]. The relatively small depths probed by the AFM tip in contact enable CR spectroscopy methods to directly measure films on the order of tens of nanometers thick without substrate effects [27].

There are several practical issues involved in obtaining reliable experimental CR spectra. One such issue can be loosely called the transfer function of the piezoelectric actuator. This term includes effects such as (a) the spatial and frequency dependence of the actuator's response to electric excitation, (b) the efficiency of coupling between the actuator and the cantilever (through the tip or the cantilever holder, depending on the actuation scheme), and (c) the presence of additional system resonances besides the contact resonances. A related issue could be called the transfer function of the detector. This includes (a) the variation in deflection amplitude along the length of

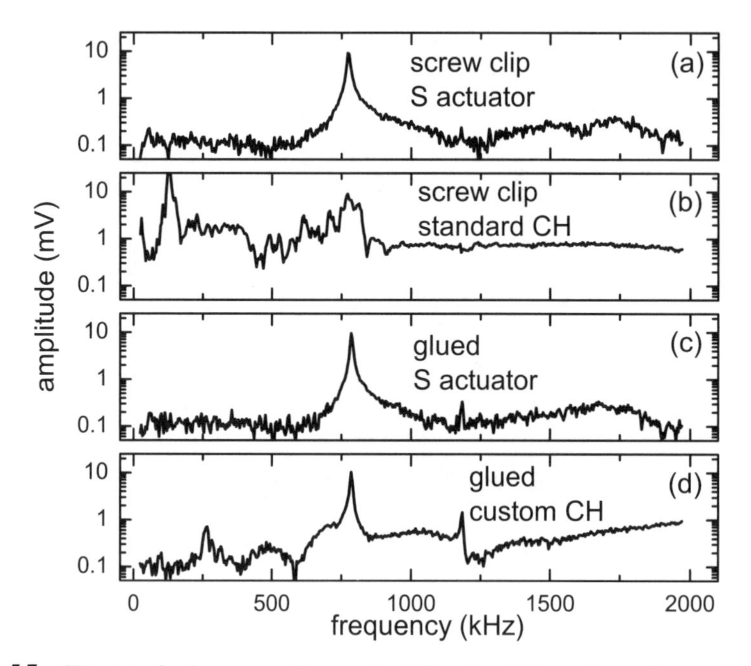

Figure 5.7 CR spectra for the same cantilever under different excitation and mounting conditions: (a) by an actuator beneath the sample and the cantilever attached to the holder with the standard screw clip, (b) by the AFM instrument's "standard" (tapping mode) piezoelectric element in the cantilever holder (CH) and attached with the standard screw clip, (c) by the sample actuator and the cantilever glued into the holder, and (d) by a specialized damped actuator in the cantilever holder and the cantilever glued into holder. The spectra are for the first flexural eigenmode of a cantilever with $f_1^{\text{free}} = 180$ kHz and nominal spring constant $k_L = 45$ N/m.

the cantilever for a given CR eigenmode and laser spot position [15] and (b) the spatial and frequency dependence of the photodiode detector response. Consideration of all these effects are important for CR data acquisition.

These effects are illustrated in Figure 5.7, which shows several CR spectra acquired with the same cantilever under different mounting and actuation conditions. The spectrum in Figure 5.7(a) was obtained by use of a highly damped, large actuator intended for ultrasonic inspection [29] mounted beneath the sample. In contrast, the CR spectrum indicated in Figure 5.7(b) was acquired with the standard cantilever holder (CH) piezoelectric element used for tapping-mode actuation. For both spectra, the cantilever was mounted in the holder with the standard screw-clip mounting. The spectrum obtained with the CH actuator contains background signals and extraneous peaks that hinder identification of the CR frequencies. These signals may be due to resonant modes of the small, undamped CH actuator. The cantilever mounting conditions can also affect CR measurements [30, 31] by causing small shifts in the CR frequencies or amplitudes and creating spurious resonances. To minimize such effects, current best practices require an improved cantilever mounting method such as gluing the cantilever into the holder with fast-setting epoxy, as in Figure 5.7(c).

A custom CH actuator can also be implemented to excite the cantilever resonances [21]. As shown in Figure 5.7(d), a combination of a custom, strongly damped CH actuator and glue mounting of the cantilever can produce a much purer spectrum than that from a typical CH actuator and a screw-clip mount. However, it may still be difficult to obtain the flat baseline provided by sample actuation. Because results depend strongly on the specific AFM instrument used, general conclusions about the best mounting and actuation methods should not be drawn from these results.

Another consideration for contact resonance measurements is the choice of eigenmode. Use of the second, third, or higher flexural eigenmode instead of the first (lowest-order) mode can optimize measurement sensitivity and ensure accurate data analysis with established models. The choice depends on the experimental range of $\alpha = k/k_L$, the contact stiffness normalized by the cantilever spring constant. As a general rule, increasingly higher-order eigenmodes are needed as α increases (i.e., increasing sample modulus for a given cantilever stiffness). These concepts have been theoretically understood for some time [15, 32], but only recently have they been fully utilized experimentally [33, 34].

Qualitative Imaging: CR Amplitude Imaging

Although CR spectroscopy (point) measurements are often useful, many other applications require images of the spatial distribution in mechanical properties. Imaging can be performed straightforwardly with the apparatus for CR spectroscopy or FMM measurements [15, 35]. The resulting images provide qualitative information only, but they are easy to acquire and are valuable in many applications. In this approach, the actuator frequency is set near to the CR frequency. As in FMM, a contact-mode scan is performed with the excitation frequency held constant, and an image of the lock-in amplitude output is acquired.

In the resulting "CR amplitude image," the signal intensity corresponds to the amplitude of the cantilever vibration at the excitation frequency. In turn, the variation in cantilever vibration amplitude throughout the sample depends on the relative elastic stiffness or modulus of various sample components [36]. Image contrast can thus be enhanced or suppressed by the choice of excitation frequency. Additionally, a reversal or "inversion" in image contrast can be observed by changing the excitation frequency. This can be understood by further examination of Figure 5.6. If the excitation frequency is relatively low (e.g., f_{low} in Figure 5.6), the cantilever vibration amplitude will be higher for regions with lower elastic modulus. More compliant regions (SiO_2 in this example) will thus appear brighter in the image. Operation at an excitation frequency closer to the CR frequency of stiffer sample components (e.g., f_{high} in Figure 5.6) decreases the vibration amplitude on more compliant regions. In this case, stiffer regions (Nb in this example) will appear brighter. (Indeed, contrast inversion will also occur at an excitation frequency lower than, but still close to, the CR frequency of stiffer component.) Although the exact behavior of the inversion depends on the excitation frequency used and the combination of materials involved, the general concept of inversion broadly applies to CR amplitude images.

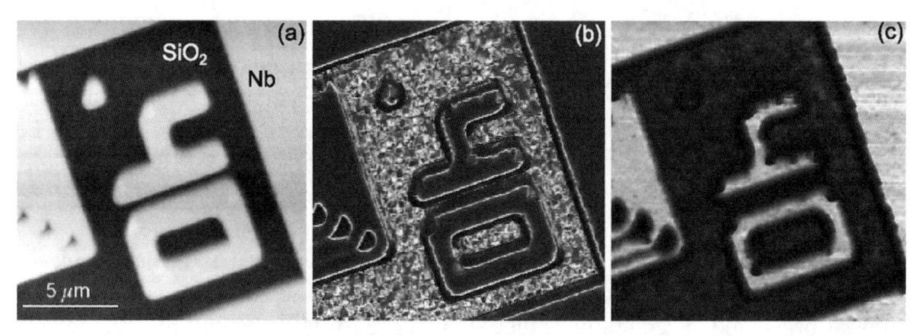

Figure 5.8 Example of CR amplitude imaging for a sample with a patterned Nb film on a blanket film of SiO_2. In each image, black represents the minimum value, and white corresponds to the maximum value. (a) Topography image (z range 360 nm) acquired in contact mode at the same time as the image in (b). (b) Amplitude image of the cantilever vibration with excitation frequency f = 751 kHz. (c) Amplitude image at f = 786 kHz. The cantilever's nominal spring constant was k_L = 45 N/m. The resonance frequencies of the cantilever's first two flexural modes in free space were f_1^{free} = 180.0 kHz and f_2^{free} = 1101.0 kHz.

Examples of amplitude imaging are shown in Figure 5.8. The sample was a silicon wafer with a blanket film of SiO_2 several hundred nanometers thick. A patterned Nb film was sputtered on top of the silica layer. (It may be useful to refer to Figure 5.6 in the following discussion. Although the exact values of the excitation and CR frequencies differ slightly between Figures. 5.6 and 5.8, the general situation is identical.) The contact-mode topography image in Figure 5.8(a) reveals that the Nb film is ~350 nm thick. Figures 5.8(b) and 5.8(c) are amplitude images that show the relative amplitude of cantilever vibration at a given excitation frequency. In each image, lighter shades indicate higher amplitude values, while darker shades correspond to lower amplitude values. The two images were acquired at two different excitation frequencies in the vicinity of the CR frequency for the cantilever's first flexural mode. The image in Figure 5.8(b) was acquired at 751 kHz, similar to f_{low} in Figure 5.6. It can be seen that the SiO_2 regions of the image are brighter than the Nb regions. The image in Figure 5.8(c) was acquired at a higher excitation frequency (786 kHz, similar to f_{high} in Figure 5.6), and here the Nb regions are brighter. This contrast inversion implies that the CR frequency of the Nb film is higher than that of the SiO_2 film. Because higher contact resonance frequencies imply greater elastic modulus, we conclude that the modulus of the Nb film is greater than that of the SiO_2 film.

The images in Figures 5.8(b) and 5.8(c) also reveal more scatter in the signal for the SiO_2 film. Such scatter is caused by small shifts in the tip–sample contact area and indicates that the surface roughness of the SiO_2 film is greater than that of the Nb film. Narrow, bright lines and wider, dark bands are additionally noticed in Figure 5.8(b) at the interfaces between the SiO_2 and Nb films. These features indicate topography-induced frequency shifts caused by transient changes in the contact area as the tip moves up or down the relatively large height step between materials. The information in these narrow regions is not deemed reliable.

Figures 5.8(b) and 5.8(c) also illustrate the issues that could arise if FMM analysis were applied to measurements of amplitude in the vicinity of the contact resonance frequency. With FMM analysis, one would conclude that the modulus of SiO_2 is higher than that of Nb in Figure 5.8(b) because $d(SiO_2) > d(Nb)$. However, the same logic applied to Figure 5.8(c) would lead to the conclusion that the modulus of SiO_2 is lower than that of Nb. It is known from bulk measurements and complementary AFM measurements that only the contrast in Figure 5.8(c) correctly represents the modulus trend. The amplitude contrast when operating near resonance is a convolution of cantilever dynamics and surface mechanics and thus cannot be directly correlated to mechanical properties.

Quantitative Imaging: Contact Resonance Force Microscopy (CR-FM)

The contrast in CR amplitude images allows regions of different elastic modulus to be easily identified. However, further interpretation of CR amplitude images is challenging, for example, to determine the stiffness ratios of various components. For detailed analysis of nanoscale elastic properties, quantitative imaging—*mapping*—is required. To obtain quantitative images, the contact resonance frequency must be detected at each position as the tip is scanned. A single such frequency image could provide more information than an entire series of amplitude images.

We suggest the term "contact resonance force microscopy" (CR-FM) to denote those techniques for quantitative contact resonance imaging, namely measuring the CR frequency or other resonance properties at each image pixel. CR-FM differs from CR spectroscopy, which involves the spectrum at a fixed sample position, and CR amplitude imaging, which yields qualitative images of amplitude contrast. Several methods for CR-FM have been implemented with different names, primarily related to their actuation and excitation approach. For instance, CR-FM could be performed with AFAM (referring to an actuator beneath the sample) or band excitation (referring to the nature of the electrical signal sent to the actuator). By using the term CR-FM to encompass all these methods, we hope to highlight their common themes without losing their individual features, as well as to simplify terminology.

In CR-FM, the key measurement challenge is that the CR frequency f^{cont} can vary significantly throughout an image, depending on the contact stiffness of different sample components. A straightforward approach is to perform a sweep over the same frequency range at each image pixel [37–40]. The frequency range of the sweep must be wide enough to include all possible peaks. However, depending on the desired frequency resolution, the acquisition speed of this approach is not practical. If acquisition rates similar to those of point measurement techniques are assumed, a single image could take a few days to obtain.

Achieving CR-FM at practical imaging rates has been accomplished in several ways. Most methods involve use of feedback over a limited frequency interval in order to "track" the CR frequency as it changes with position. Methods include phase-locked loop approaches [41–44], use of chirped frequency band(s) [45–47], and simultaneous excitation of all frequencies in a band [48]. Some commercial

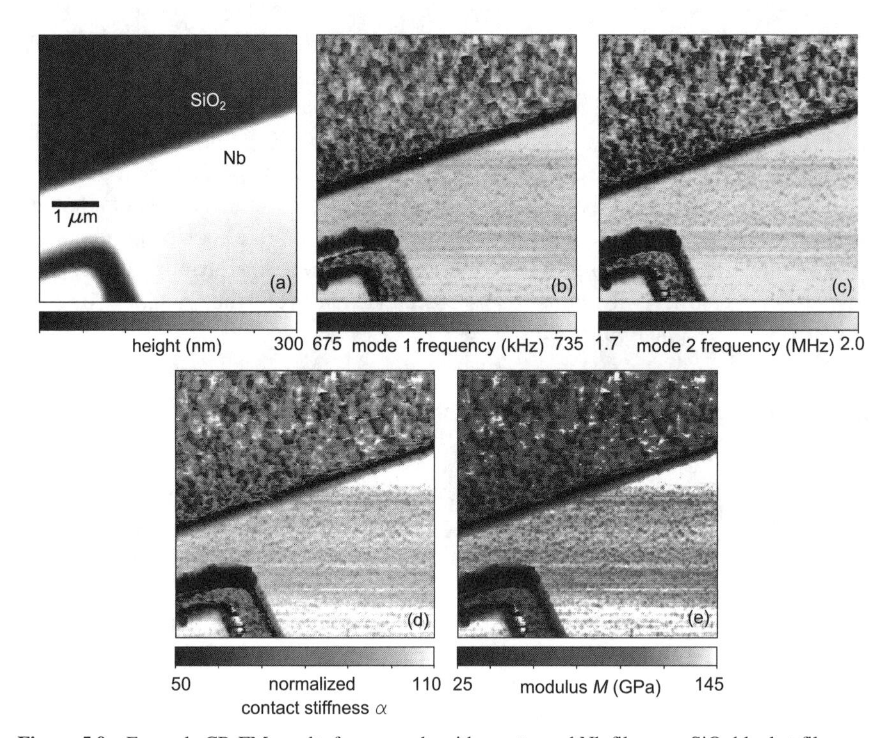

Figure 5.9 Example CR-FM results for a sample with a patterned Nb film on a SiO_2 blanket film. (a) Contact-mode topography. (b) and (c) CR frequency images for the first and second flexural eigenmodes, respectively. (d) Normalized contact stiffness $\alpha = k/k_L$ calculated from (b) and (c). (e) Map of indentation modulus *M* calculated from (d). The cantilever's nominal spring constant was $k_L = 45$ N/m. The resonance frequencies of the cantilever's first two flexural modes in free space were $f_1^{\text{free}} = 171.7$ kHz and $f_2^{\text{free}} = 1048.0$ kHz. For color details, please see color plate section.

AFM instruments offer CR-FM functionality in various ways, including excitation over a fixed frequency range with no feedback [17, 49] and adaptive feedback based on two excitation frequencies [18, 50].

An example of quantitative CR-FM imaging is shown in Figure 5.9. The images correspond to another region of the SiO_2/Nb sample shown in Figure 5.8. The topography image in Figure 5.9(a) and the two CR frequency images in Figures 5.9(b) and 5.9(c) were acquired simultaneously with a modified version of our custom CR-FM electronics [46]. The frequency images in Figures 5.9(b) and 5.9(c) for the two lowest flexural eigenmodes of the cantilever reveal directly that the CR frequency in the Nb regions is higher than that in the SiO_2 regions. Similar to the amplitude images above, the edges of the film show large frequency shifts due to topography effects. The rest of the figure is discussed later in the chapter.

Numerous material systems with micro- to nanoscale features have been evaluated with contact resonance imaging techniques. Of the many reported studies, examples include carbon-fiber-reinforced polymer composites [22, 51], teeth [52], and piezoelectric ceramics [36, 37, 52]. Applications studied with quantitative imaging

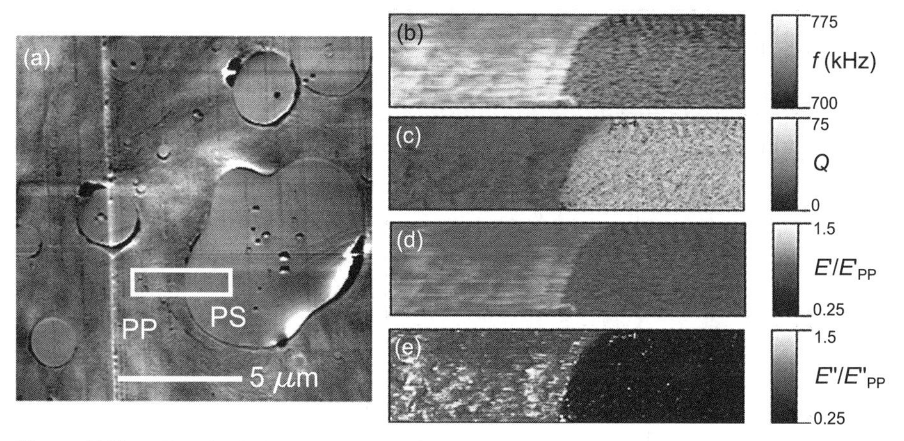

Figure 5.10 Viscoelastic CR-FM measurements on a polystyrene (PS)/polypropylene (PP) blend. (a) Topography image (z scale 100 nm) with PS domain and PP matrix indicated. (b) and (c) CR-FM maps of frequency and quality factor, respectively, for the second flexural eigenmode in the region indicated by the box in (a). (d) and (e) Calculated maps of storage modulus E' and loss modulus E'', respectively, normalized to their average values on polypropylene. The cantilever's nominal spring constant was k_L = 3 N/m, and the resonance frequency of the second flexural mode in free space was f_2^{free} = 423.6 kHz. For color details, please see color plate section.

(CR-FM) include carbon-fiber-reinforced polymer composites [41], epoxy-silica nanocomposites [53], piezoelectric domains in $BaTiO_3$ [37], and diamond-like carbon films [54]. CR methods have also been exploited to investigate the nanoscale fiber–matrix interphase region in composites both qualitatively and quantitatively [33, 51] and to study anisotropic properties of nanocrystalline nickel thin films [55].

Contact resonance force microscopy methods have also been extended to achieve mapping of viscoelastic properties. Viscoelastic information can be obtained by considering the relative damping in the resonance in addition to its frequency f^{cont}. Several approaches have been presented [24, 39, 44, 56]; our approach [56–58] makes use of the CR quality factor Q^{cont} in a commercial AFM platform [18, 59]. This viscoelastic CR-FM approach ultimately yields two experimental images for the values of f^{cont} and Q^{cont} at each image pixel.

Figure 5.10 shows viscoelastic CR-FM results from a polystyrene (PS)/polypropylene (PP) blend that was cryomicrotomed to minimize surface roughness. As seen in the topography image in Figure 5.9(a), the blend exhibits a continuous PP phase and discrete PS domains that range from approximately 1 to 20 μm in diameter. The experimental images of f^{cont} and Q^{cont} in Figures 5.9(b) and 5.9(c) reveal that the PS/PP blend exhibits little frequency contrast but large quality factor contrast. The small contrast in f^{cont} and large contrast in Q^{cont} suggest that qualitatively, the two phases have similar elastic stiffnesses but different damping characteristics. However, because the true damping is a function of both f^{cont} and Q^{cont}, a quantitative evaluation of the stiffness and damping requires further analysis. This analysis is discussed at the end of the next section.

5.3.3 Data Analysis and Interpretation

In CR techniques, the sample's elastic properties are deduced from the measured frequencies f^{free} and f^{cont}. In the data analysis, these frequencies are first related to the tip–sample interaction force by means of a model for the dynamic motion of the cantilever. The analytical model is described in detail elsewhere (e.g., [15, 16, 20]); finite-element analysis methods have also been used [26, 34, 60, 61]. Appendix A contains a step-by-step "recipe" for the entire data analysis procedure; here, we describe the general concepts involved.

Two models for cantilever dynamics are shown in Figure 5.11. The discussion here is restricted to the main concepts of the analytical model. One aspect worth noting is that the cantilever is modeled as a distributed mass with spring constant k_L. Unlike the case in some other AFM modes, the point mass (harmonic oscillator) approximation does not always accurately describe the contact vibration [9, 62, 63]. Note that highly rectangular, uncoated, single-crystal silicon cantilevers are often used in experiments to simulate the idealized models in Figure 5.11 as closely as possible.

In the simplest model shown in Figure 5.11(a), the tip–sample interaction is considered to be entirely elastic and to act in a direction normal (vertical) to the sample surface. The interaction is represented by a spring of stiffness k that denotes the contact stiffness. Finally, as depicted in Figure 5.11, the tip is not located at the free end of the cantilever but at a relative position $\gamma = L_1/L$ (with $0 \le \gamma \le 1$) from the clamped end. Consideration of this system with Euler–Bernoulli beam theory leads to a characteristic equation that gives the ratio $\alpha = k/k_L$ in terms of the relative tip position γ and the wavenumber x of the resonance [62]. The measured values of f^{free} and f^{cont} are used to determine x, which allows the characteristic equation to be solved for the normalized contact stiffness α as a function of γ. The final value of $\alpha(\gamma)$ is chosen either for a given value of γ or with a "mode crossing" approach [64], in which the value of γ is chosen that gives equal values of α for two different eigenmodes.

Once the values of normalized contact stiffness $\alpha = k/k_L$ have been calculated, the elastic modulus M_{test} of the test or unknown sample can be determined through a second model for the tip–sample contact mechanics. A logical approach is to apply the basic equations for Hertzian contact. In principle, Eqs. (5.1) and (5.3) and the calculated values of α could be combined to determine the reduced modulus E^{R} [and

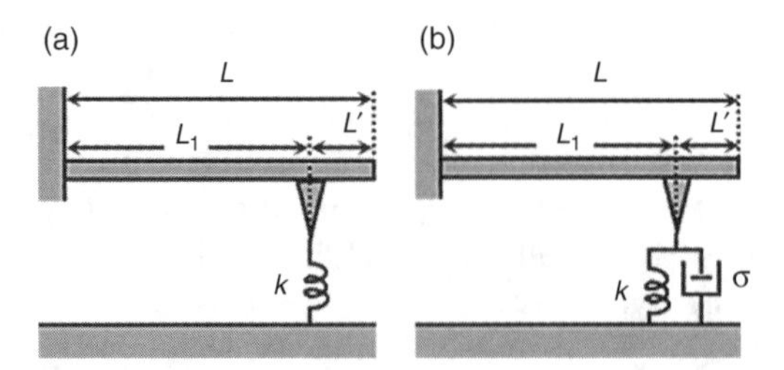

Figure 5.11 Models for cantilever dynamics. (a) Tip in contact, normal (vertical) elastic interaction forces. (b) Tip in contact, normal elastic and dissipative (damping) forces.

subsequently M_{test} with Eq. (5.2)]. However, this approach requires precise knowledge of micro- and nanoscale quantities that are difficult to measure accurately: the tip radius R, the applied force F_{app}, and the cantilever stiffness k_L. A more practical measurement technique that has been shown to yield accurate results is a relative or comparative approach [26, 36]. Measurements are performed at constant force in alternation on the unknown sample (subscript "test") and a calibration sample (subscript "cal"). The mechanical properties of the calibration sample are considered known, either through independent measurements with another technique or from literature values. Then [36]

$$E_{test}^{R} = E_{cal}^{R}\left(\frac{\alpha_{test}}{\alpha_{cal}}\right)^{3/2} = E_{cal}^{R}\left(\frac{k_{test}}{k_{cal}}\right)^{3/2} \tag{5.5}$$

where the exponent $m = 3/2$ denotes Hertzian contact. (The value $m = 1$ can be used for a flat punch tip geometry.) The indentation modulus M_{test} of the test specimen can be determined from the experimental values of α from Eqs. (5.2) and (5.5) with knowledge of M_{tip}. ($M_{tip} = 165$ GPa for <100> silicon is usually assumed, although other approaches have been taken [38].) The calibration approach avoids difficult measurements of absolute quantities such as k_L, F_{app}, R, and a. It can also minimize the influence of tip wear on the measured values. For heterogeneous samples, an "internal" calibration approach can be performed in which one component or region acts as the calibration sample [33, 58].

Figure 5.9(d) shows a map of the normalized contact stiffness α calculated from the frequency images in Figures 5.9(b) and 5.9(c) for the Nb/SiO_2 sample. The image was calculated on a pixel-by-pixel basis with the mode-crossing analysis approach described above. Although just a contact stiffness map may be sufficient in some cases, often a map of the indentation modulus M_{test} is needed. To obtain a modulus map, we assumed that the mean value of E^R in the Nb regions corresponded to the literature value $M = 125$ GPa for polycrystalline Nb. The average value of α in the Nb region of the image was used as the calibration value in Eq. (5.5). Figure 5.9(e) shows the resulting modulus map. The average value of M in the SiO_2 film region is 73 GPa, consistent with literature values for the modulus of bulk SiO_2 ($M \approx 75$ GPa). This simple example shows the capabilities of CR-FM for quantitative modulus mapping.

For data analysis to obtain viscoelastic properties, the model in Figure 5.11(b) is used to describe the cantilever dynamics [56, 63, 65]. The effect on the tip–sample interaction of the specimen's combined elastic and viscoelastic properties is modeled by including a viscous damper (dashpot) in parallel with the elastic spring (Kelvin–Voigt mechanical equivalent). Application of Euler–Bernoulli beam theory to this model leads to a new characteristic equation in terms of the flexural mode's complex wavenumber and the relative tip position γ. With this equation, the normalized contact stiffness α and damping coefficient β can be determined from the experimental values of frequency (f^{free} and f^{cont}) and quality factor (Q^{free} and Q^{cont}). Contact mechanics are then used to determine the viscoelastic properties from the values of α, β, and f^{cont}. Assuming sphere-plane Hertzian contact, the reduced storage modulus E'^{R} and reduced loss modulus E''^{R} are obtained from

$$E'^{\text{R}}_{\text{test}} = E'^{\text{R}}_{\text{cal}} \left(\frac{\alpha_{\text{test}}}{\alpha_{\text{cal}}} \right)^{3/2} \tag{5.6}$$

and

$$E''^{\text{R}}_{\text{test}} = E''^{\text{R}}_{\text{cal}} \left(\frac{f^{\text{cont}}_{\text{test}} \beta_{\text{test}}}{f^{\text{cont}}_{\text{cal}} \beta_{\text{cal}}} \right)^{3/2} \tag{5.7}$$

where $f^{\text{cont}}_{\text{cal}}$, E'_{cal}, and E''_{cal} are the CR frequency, storage modulus, and loss modulus, respectively, of the calibration material. Assuming that Poisson's ratio ν is similar for the unknown and calibration materials, the storage modulus E'_{test} and loss modulus E''_{test} can be determined from the reduced complex modulus $E^{*\text{R}}_{\text{test}}$ given by

$$E^{*\text{R}}_{\text{test}} = E'^{\text{R}}_{\text{test}} + iE''^{\text{R}}_{\text{test}} = \left[\frac{1-\nu^2_{\text{test}}}{E'_{\text{test}} + iE''_{\text{test}}} + \frac{1}{M_{\text{tip}}} \right]^{-1} \tag{5.8}$$

Note that damping in the silicon tip has been neglected (i.e., $E''_{\text{tip}} = 0$). Recent work involves an expression for the loss tangent $\tan \delta = E''/E'$ that does not require calibration data [66].

This viscoelastic CR-FM analysis approach was used on the data in Figures 5.10(b) and 5.10(c) for the PP/PS blend to determine images of normalized contact stiffness α and damping β. The images of α and β and the frequency image in Figure 5.10(b) were used to calculate the maps in Figures. 5.10(d) and 5.10(e) of the storage modulus E'/E'_{PP} and loss modulus E''/E''_{PP} relative to the PP values, respectively. For a relative calibration of the images, the viscoelastic properties of the polypropylene phase were assumed to be those of bulk, homogeneous polypropylene. Values for α_{cal}, β_{cal}, and $f^{\text{cont}}_{\text{cal}}$ were determined from masked averages of the PP phase in the α, β, and f^{cont} maps. The maps in Figures 5.10(d) and 5.10(e) show very good agreement with results from bulk dynamic mechanical analysis measurements that have been shifted to the frequency range of CR-FM [58]. These results are especially promising for characterizing materials in which viscoelastic behavior contributes strongly to the overall mechanical response.

5.4 COMPARISON OF FMM AND CR-FM

As a comparison of FMM and CR-FM methods, images were acquired with both techniques on the same sample with the same cantilever. The sample was a composite of cellulose fibers in a polypropylene matrix. An ultramicrotome was

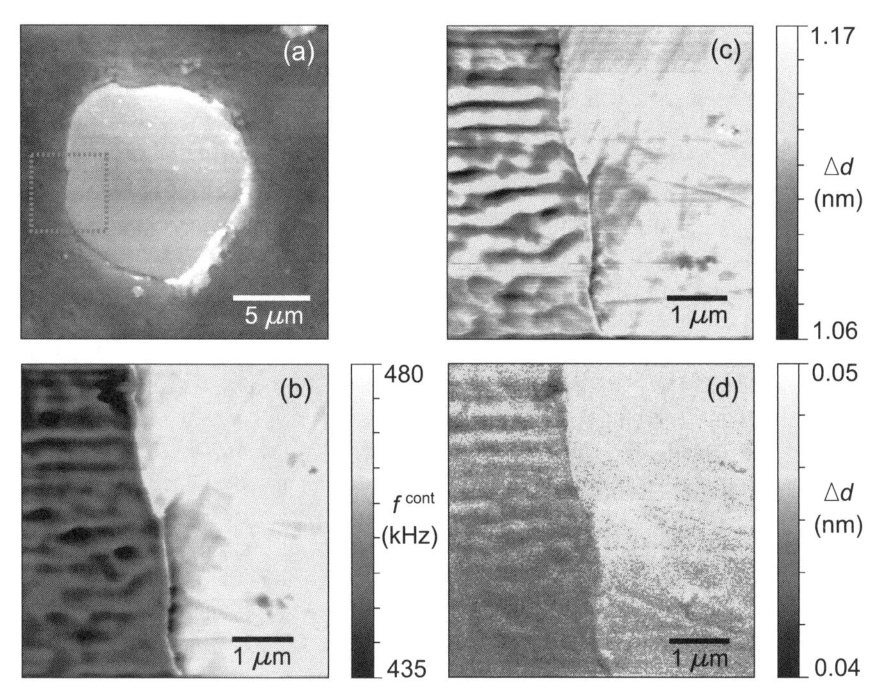

Figure 5.12 Comparison of dynamic contact AFM methods. The sample was a composite containing a cellulose fiber in a polypropylene matrix. (a) Tapping-mode topography image (z scale 400 nm). The box indicates the region used for the images in (b)–(d). (b) CR-FM frequency image. The cantilever's nominal spring constant was k_L = 15 N/m, and the free resonance frequency of the first flexural mode was f_1^{free} = 106 kHz. (c) and (d) FMM amplitude images for modulation at 1.5 and 100 kHz, respectively. The images in (b)–(d) were acquired with the same cantilever. For color details, please see color plate section.

used to prepare a smooth surface containing the fibers in cross section. Values of the indentation modulus $M_{\text{matrix}} = (3.7 \pm 0.9)$ GPa and $M_{\text{fiber}} = (15.1 \pm 1.1)$ GPa for the matrix and the fiber, respectively, were determined by nanoindentation. For a given sample, the most suitable cantilever for CR-FM generally has a lower spring constant k_L than one that provides optimal contrast in FMM. For instance, previous CR-FM experiments on similar composite samples used cantilevers with nominal $k_L = 3$ N/m [33], while a cantilever with $k_L > 40$ N/m is more typical for FMM. As a compromise, a cantilever with nominal $k_L = 15$ N/m was used here.

Figure 5.12 shows raw experimental data with both FMM and CR-FM on the composite sample. The tapping-mode topography image in Figure 5.12(a) indicates that in spite of the ultramicrotoming surface preparation, there was a relatively large difference in height between the fiber and the matrix. The CR-FM frequency image in Figure 5.12(b) and the FMM amplitude images in Figures 5.12(c) and 5.12(d) were acquired in the region indicated by the box in Figure 5.12(a), where the height difference was relatively low. The FMM amplitude image in Figure 5.12(c), which was acquired at 1.5 kHz, shows a larger amplitude response than the image in Figure 5.12(d), which was acquired at 100 kHz. Although the lower-frequency image exhibits overall larger amplitudes, the contrast between the stiffer fiber and more compliant matrix appears less. In addition, the relative stiffness of the fiber and matrix cannot be conclusively evaluated in the lower-frequency FMM image in Figure 5.12(c). However, the higher-frequency FMM image clearly shows larger amplitudes on the fiber, consistent with higher contact stiffness [see Eq. (5.4)] and hence higher modulus. The CR-FM frequency map in Figure 5.12(b) provides even better material contrast. It displays a clear bimodal distribution in contrast, with the CR frequency approximately 30 kHz higher on the fiber than on the matrix.

However, a clear advantage of FMM is the flexibility of excitation frequency. The choice of FMM excitation frequency is virtually limitless compared to CR methods, which are constrained to a finite number of resonance peaks. Most techniques for mechanical property measurements on macroscale (bulk) samples operate at relatively low frequencies, in the range from approximately 1 to 100 Hz. In theory, FMM can easily probe in this range, allowing direct comparison and validation with established methods. However, as mentioned in Section 5.2.2, in practice it can be difficult to identify a region where the response is flat with frequency and does not contain extraneous signals. On the other hand, the lower limit of CR methods is typically in the tens of kilohertz; more often, measurements are made at hundreds of kilohertz. Thus, validation and comparison of CR measurements on materials with strongly frequency-dependent properties pose significant challenges.

A final point of comparison concerns the analysis methods used to obtain quantitative values of contact stiffness and, ultimately, modulus. FMM benefits from a simple analysis procedure with a limited number of parameters, while CR methods rely on more complex dynamic beam models to determine contact stiffness. Appropriate cantilevers must be chosen for these beam models to apply, and the limits of a given model must be understood. However, when applied correctly, the contact resonance–beam model approach has the advantage that quantitative analyses are possible even when the contact stiffness k is much greater than the cantilever spring constant k_L (i.e., the normalized contact stiffness $\alpha = k/k_L >> 1$). It is worth noting that the low-α regime where FMM is usually applied can also be probed and analyzed with CR methods. Furthermore, much simpler (point-mass-type) models for data analysis provide sufficient accuracy in this regime.

5.5 OTHER DYNAMIC CONTACT APPROACHES

As mentioned in the introduction, dynamic contact AFM methods are not limited solely to out-of-plane modulation to determine modulus. A variety of dynamic modulation schemes can be used to probe a number of contact-sensitive properties. This section presents a brief overview of other dynamic contact methods and results and provides references to more in-depth discussion.

5.5.1 In-Plane Modulation Schemes

A number of researchers have utilized in-plane or shear modulation of a tip in contact with a surface. As with out-of-plane modulation, both off-resonance and on-resonance operation may be performed. Likewise, analysis may be either qualitative or quantitative.

Off-resonance excitation in the lateral or in-plane direction is used in shear modulation force microscopy (SM-FM), the in-plane analog to FMM. The cantilever is modulated either by the imaging *X* or *Y* piezoelectric actuators transverse to the long axis of the cantilever or by a sample actuator with a shear piezoelectric element oriented in the desired direction. Instead of detecting the vertical signal of the photodiode, SM-FM uses lock-in amplification of the horizontal (left–right) signal, which is more commonly used for nanoscale friction measurements. As in FMM, the amplitude and phase of the lock-in output correspond to the resultant cantilever modulation. A key advantage of SM-FM over FMM is that the normal force is effectively constant throughout the measurement, ensuring a constant stress field beneath the tip. Like FMM, the SM-FM signal is proportional to the tip–sample contact stiffness; however, here the lateral contact stiffness is probed. Because the lateral spring constant of a cantilever is orders of magnitude stiffer than its normal (flexural) stiffness, SM-FM measurements with a compliant cantilever can be much more sensitive to stiffness variations than FMM measurements with the same cantilever.

By carefully calibrating the torsional sensitivity of the cantilever, SM-FM has been used for quantitative measurements of lateral contact stiffness [67]. More commonly, SM-FM has been used as a qualitative measurement of isobaric contact stiffness. An example is the determination of glass transition temperature T_g in thin polymer films [68, 69]. Thin films undergo a significant softening during the glass transition, which tends to simultaneously decrease modulus and increase contact area. By observing the SM-FM response during a ramped or ramp-and-hold heating cycle, a qualitative kink in amplitude at T_g was observed. A second application where SM-FM has shown considerable benefit is the study of the formation of viscoelastic contacts [70]. By superimposing the SM-FM modulation on a standard F–D experiment, hysteresis in the contact stiffness due to creep of the tip into the polymer was observed.

On-resonance measurements of the cantilever's torsional motion also provide valuable information. CR amplitude images of the tip's lateral motion at frequencies near that of the torsional resonance showed elastic and damping contrast [71]. Analysis of the CR frequencies of torsional modes yielded data on in-plane surface properties [72]. If the frequencies of both flexural and torsional contact resonances are measured, shear elastic properties such as Poisson's ratio or shear modulus can be determined separately from Young's modulus [73, 74]. If the amplitude of the in-plane excitation is gradually increased, slip is induced in the torsional CR experiments, and high-frequency frictional behavior can be examined [43, 75]. These results and others suggest that with use of different dynamic contact methods, quantitative data may be achieved on a much wider range of mechanical properties than is currently available.

5.5.2 Beyond Modulus: Measurements of Other Properties

The nature of the tip–sample contact mechanics and the dynamic excitation schemes involved in dynamic contact methods mean that they are a sensitive probe of other properties besides elasticity or viscoelasticity. Because the stress field extends beneath the sample surface, contact techniques can be sensitive to buried structures and interfaces with stiffness or load transfer heterogeneity. CR imaging has been used to identify or characterize subsurface dislocations in graphite [76], adhesion and delamination of thin films [77, 78], and buried (subsurface) voids and nanoparticles [79, 80]. Because contact stiffness depends on tip properties such as geometry, CR methods are also suitable for monitoring tip health and wear [81, 82] Finally, through nonlinear (large excitation amplitude) CR spectroscopy, detailed information about the tip–sample interaction forces in both the repulsive and attractive regime has been obtained [83]. This sampling of results hints at the potential of dynamic contact methods for further application to a wide range of new areas.

5.6 SUMMARY AND CONCLUSIONS

In this chapter, we have described AFM methods for nanomechanical characterization that involve operating in continuous tip–sample contact and monitoring the dynamic response of the cantilever. The use of frequency-specific excitation and detection minimizes extraneous signals present in quasi-static methods, leading to increased measurement sensitivity. Force modulation microscopy and related methods such as shear modulation force microscopy operate at relatively low excitation frequencies away from cantilever resonances. Differences in the amplitude signal at a given excitation frequency provide contrast in contact stiffness or elastic modulus, while phase differences yield damping or dissipation contrast. In principle, FMM can be performed over a wide range of frequencies, offering measurement flexibility as well as frequency-dependent property characterization. Moreover, the model for quantitative analysis of FMM data is appealingly simple and does not depend on excitation frequency. Despite these advantages, quantitative FMM measurements are often hampered in practice by the nonideal nature of the cantilever response with spurious signals and extraneous peaks.

Contact resonance techniques are an emerging class of dynamic contact AFM methods that advance the state of the art in quantitative nanomechanical measurements. In contrast to FMM, CR methods operate at relatively high excitation frequencies deliberately near the cantilever resonance peaks. Qualitative modulus contrast is readily obtained by monitoring amplitude changes at a single excitation frequency, similar to FMM. Improved sensitivity or contrast is gained by large shifts in amplitude near the resonance peaks. However, the real strength of CR methods lies in the ability to provide accurate mechanical property data. Analysis of CR frequencies with analytical models for the cantilever dynamics yields quantitative modulus information. Due to the dynamic stiffening of the cantilever, CR methods are applicable to much stiffer materials than can be accessed by FMM or F–D spectroscopy (static force curves). Improved measurement accuracy on very compliant polymers and biomaterials will require modifications to current approaches, such as contact mechanics models that include adhesive forces. CR methods are now sufficiently mature for more widespread use by the materials community. However, the learning curve for new users to become familiar with the experimental and analysis methods is still somewhat steep. Measurement innovations may be possible to overcome this hurdle and enable more rapid adoption. In addition, continued commercialization of these methods will enable wider access to the greater community.

Overall, dynamic contact AFM methods have much potential for characterization of nanomechanical properties. We hope that the information presented here provides readers with confidence in these methods and encourages them to apply the methods in their own research. By providing nanomechanical information on a range of material systems, dynamic contact AFM tools will contribute to further development of advanced materials and will play a significant role in future materials industries.

ACKNOWLEDGMENTS

We gratefully acknowledge the contributions of current and former NIST co-workers including S. Campbell, R. Geiss, M. Kopycinska-Müller, A. Kos, and P. Rice, as well as interactions with NIST colleagues including R. Cook, J. Kelly, J. Pratt, D. Smith, C. Stafford, and G. Stan. We also value collaborations with other researchers including W. Arnold, U. Rabe, and co-workers (Fraunhofer Institute for Nondestructive Testing IZFP, Saarbrücken, Germany); Y. Ding and students (University of Colorado-Boulder); C. Prater (Anasys Instruments); R. Proksch, A. Gannepalli, and co-workers (Asylum Research, Santa Barbara, CA); J. Turner and P. Yuya (University of Nebraska-Lincoln); S. Wang and S. Nair (University of Tennessee-Knoxville); and D. Yablon and A. Tsou (ExxonMobil Research and Engineering). The Nb/SiO_2 sample was provided by G. Hilton (NIST), and the fiber–composite sample was provided by S. Nair (University of Tennessee-Knoxville). Commercial equipment and instruments are identified in order to adequately specify experimental procedures. Such identification does not imply recommendation or endorsement by the NIST, nor that the items identified are necessarily the best available for the purpose. Contribution of NIST; not subject to copyright in the United States. All NIST publications are available at http://www.nist.gov/publication-portal.cfm.

APPENDIX 5A DATA ANALYSIS PROCEDURE FOR CONTACT RESONANCE SPECTROSCOPY MEASUREMENTS

Perform Contact Resonance Spectroscopy Experiments

In the measurement, a spectrum of the cantilever vibration amplitude for a particular flexural eigenmode versus frequency is acquired at a single sample position. All measurements performed in contact must be acquired at the same applied force. The CR

Table 5A.1 Example Results for Free Resonance Data with Flexural Modes $n=1,2$ of a Cantilever with Nominal Stiffness $k_L=40$ N/m

f_1^{free} (kHz)	A_1 (kHz$^{-1/2}$)	f_2^{free} (kHz)	A_2 (kHz$^{-1/2}$)
154.67	0.1508	960.92	0.1514

Table 5A.2 Example Results for Contact Resonance Data*

Material	f_1^{cont} (kHz)	y_1L	f_2^{cont} (kHz)	y_2L	α	γ	E^R (GPa)	M (GPa)
Si (cal)	651.35	3.848	1535.0	5.933	52.4	0.9711	82.6	165.1
Ti (test)	627.50	3.777	1492.0	5.849	46.6	0.9837	69.2	119.1

*The values of E^R_{cal} and M_{cal} were based on literature values. The values of E^R_{test} and M_{test} were obtained with $m=3/2$ in Eq. (5A.7).

frequency is determined simply by choosing the frequency of maximum amplitude or by a more sophisticated method such as curve fitting. At a minimum, the following measured values are needed:

1. The "free" frequency f_n^{free} for two flexural modes n and $n+1$ (tip out of contact)
2. The CR frequencies f_n^{cont} for the same two modes on the test (unknown) material
3. The CR frequencies f_n^{cont} for the same two modes on a calibration material

To illustrate the discussion, example results are shown in Tables 5A.1 and 5A.2. These results can also be used to check the reader's personal calculations. The test sample was a titanium (Ti) film (thickness ~200 nm) deposited on a <001> silicon (Si) wafer. A region of bare Si adjacent to the Ti film served as the calibration sample. A cantilever with nominal stiffness $k_L=40$ N/m was used. For such cantilevers, the lowest two flexural modes ($n=1$, 2) have been shown to yield accurate results on materials of comparable modulus.

Determine the Parameter A_n for Each Flexural Mode n from the Free Frequency f_n^{free}

The parameter A_n that characterizes the cantilever is defined by

$$A_n^2 = \frac{(x_n L)^2}{f_n^{free}} \tag{5A.1}$$

The wavenumber $x_n L$ for the flexural mode n is found from the following characteristic equation for cantilever vibration in free space:

$$1 + \cos x_n L \cosh x_n L = 0 \tag{5A.2}$$

The first two roots of Eq. (5A.2) are $[x_1L, x_2L] = [1.8751, 4.6941]$. Thus, A_n is calculated by

$$A_n = \frac{1.8751 \text{ or } 4.6941}{\sqrt{f_n^{\text{free}}}} \tag{5A.3}$$

Determine the Value of y_nL from the CR Frequency f_n^{cont} and the Parameter A_n for Each Mode n and for Both the Test and Calibration Materials

$$y_nL = A_n\sqrt{f_n^{\text{cont}}} \tag{5A.4}$$

The wavenumber y_nL corresponds to the characteristic equation for cantilever vibration with the tip in contact [Eq. (5A.5)].

Calculate the Normalized Contact Stiffness $\alpha = k/k_L$ versus Relative Tip Position $\gamma = L_1/L$ from y_nL for Each Measurement

$$\alpha = \frac{k}{k_L} = \frac{2}{3}(y_nL\gamma)^3 \frac{1+\cos\ y_nL \cosh\ y_nL}{D} \tag{5A.5}$$

where

$$\begin{aligned} D = &\left[\sin y_nL(1-\gamma)\cosh y_nL(1-\gamma) - \cos y_nL(1-\gamma)\sinh y_nL(1-\gamma)\right] \\ &\left[1-\cos y_nL\gamma\cosh y_nL\gamma\right] \\ &-\left[\sin y_nL\gamma\cosh y_nL\gamma - \cos y_nL\gamma\sinh y_nL\gamma\right]\left[1+\cos y_nL(1-\gamma)\cosh y_nL(1-\gamma)\right] \end{aligned}$$

The cantilever parameters L and L_1 are defined in Figure 5.11.

Determine the "Mode-Crossing" Values of α and γ for the Mode Pair (n, n+1) on Both the Test and Calibration Materials

Plot α versus γ for both modes in a given measurement and identify the value of γ where the two modes "cross," that is, have the same value of α. This value of α is considered to be the experimental result. Typically, γ is in the range $0.9 \leq \gamma \leq 1.0$. The values of γ may differ slightly from the physical tip position measured with optical or electron microscopy. Figure 5A.1 shows this mode-crossing plot for the CR data in Table 5A.2 for the Si calibration sample.

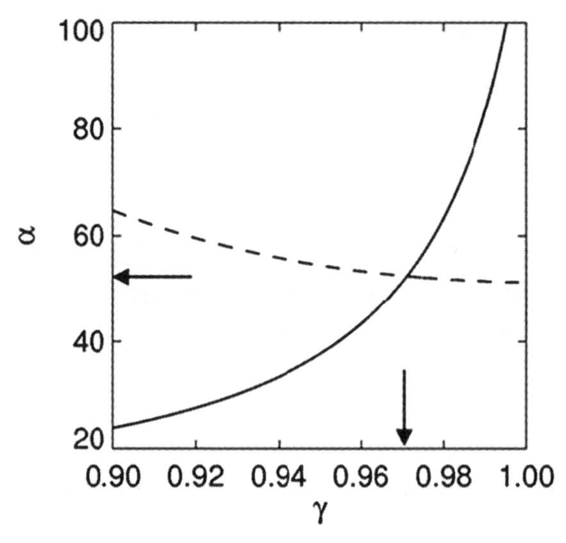

Figure 5A.1 Example of "mode-crossing" plot with normalized contact stiffness α as a function of relative tip position γ for the two lowest flexural modes ($n=1$, solid line; $n=2$, dashed line). Arrows indicate the experimental values of α and γ determined from the plot.

Determine the Reduced Modulus E^{R}_{cal} for the Calibration Sample

For this case, the indentation modulus of the silicon wafer was assumed to be $M_{cal}=165.1$ GPa. The value of M_{cal} can also be determined experimentally with another technique [e.g., instrumented (nano-) indentation]. Usually, $M_{tip}=165.1$ GPa for <001> silicon is assumed.

$$E^{R}_{cal}=\left(\frac{1}{M_{tip}}+\frac{1}{M_{cal}}\right)^{-1} \tag{5A.6}$$

Calculate the Reduced Modulus E^{R}_{test} from the Values of $\alpha_{test}=k_{test}/k_L$ on the Unknown Sample and $\alpha_{cal}=k_{cal}/k_L$ on the Calibration Sample

$$E^{R}_{test}=E^{R}_{cal}\left(\frac{k_{test}}{k_{cal}}\right)^{m}=E^{R}_{cal}\left(\frac{k_{test}/k_L}{k_{cal}/k_L}\right)^{m}=E^{R}_{cal}\left(\frac{\alpha_{test}}{\alpha_{cal}}\right)^{m} \tag{5A.7}$$

The exponent m depends on the tip shape. Usually, values are calculated for $m = 1$ (flat punch tip) and $m = 3/2$ (hemispherical tip) and cited as upper and lower bounds. If calibration measurements are performed before and after each set of unknown measurements, each set of measurements on the unknown sample will provide two values of $E_{\text{test}}^{\text{R}}$.

Calculate the Indentation Modulus M_{test} of the Test Sample from the Value of $E_{\text{test}}^{\text{R}}$

$$E_{\text{test}}^{R} = \left(\frac{1}{M_{\text{tip}}} + \frac{1}{M_{\text{test}}} \right)^{-1} \tag{5A.8}$$

This is the final result. Typically, multiple measurements are made, and an average and standard deviation are reported. For comparison, literature values of M for bulk Ti are in the range from approximately 110 to 125 GPa.

REFERENCES

1. JOHNSON, K. L. *Contact Mechanics*, Cambridge University Press: Cambridge, UK, 1985.
2. VLASSAK, J. J., and NIX, W. D. Measuring the Elastic Properties of Anisotropic Materials by Means of Nanoindentation Experiments, *J Mech Phys Solids* **42** (1994): 1223–1245.
3. MAIVALD, P., BUTT, H. J., GOULD, S. A. C., PRATER, C. B., DRAKE, B., GURLEY, J. A., ELINGS, V. B., and HANSMA, P. K. Using Force Modulation to Image Surface Elasticities with the Atomic Force Microscope, *Nanotechnology* **2** (1991): 103–106.
4. RADMACHER, M., TILLMANN, R. W., and GAUB, H. E. Imaging Viscoelasticity by Force Modulation with the Atomic Force Microscope, *Biophys J* **64** (1993): 735–742.
5. OVERNEY, R. M., MEYER, E., FROMMER, J., GÜNTHERODT, H.-J., FUJIHIRA, M., TAKANO, H., and GOTOH, Y. Force Microscopy Study of Friction and Elastic Compliance of Phase-separated Organic Thin Films, *Langmuir* **10** (1994): 1281–1286.
6. JOURDAN, J. S., CRUCHON-DUPEYRAT, S. J., HUAN, Y., KUO, P. K., and LIU, G. Y. Imaging Nanoscopic Elasticity of Thin Film Materials by Atomic Force Microscopy: Effects of Force Modulation Frequency and Amplitude, *Langmuir* **15** (1999): 6495–6504.
7. YANG, N., WONG, K. K. H., de BRUYN, J. R., and HUTTER, J. L. Frequency-dependent Viscoelasticity Measurement by Atomic Force Microscopy, *Meas Sci Technol* **20** (2009): 025703.
8. YAMANAKA, K., OGISO, H., and KOLOSOV, O. Ultrasonic Force Microscopy for Nanometer Resolution Subsurface Imaging, *Appl Phys Lett* **64** (1994): 178–180.
9. MAZERAN, P. E., and LOUBET, J. L. Force Modulation with a Scanning Force Microscope: an Analysis, *Trib Lett* **3** (1997): 125–132.
10. MAGANOV, S. N., and RENEKER, D. H. Characterization of Polymer Surfaces with Atomic Force Microscopy, *Annu Rev Mater Sci* **27** (1997): 175–222.
11. GALUSKA, A. A., POULTER, R. R., and MCELRATH, K. O. Force Modulation AFM of Elastomer Blends: Morphology, Fillers and Cross-linking, *Surf Interface Anal* **25** (1997): 418–429.
12. BAR, G., RUBIN, S., PARIKH, A. N., SWANSON, B. I., ZAWODZINSKI, T. A., and WHANGBO, M.-H. Scanning Force Microscopy Study of Patterned Monolayers of Alkanethiols on Gold. Importance of Tip–Sample Contact Area in Interpreting Force Modulation and Friction Force Microscopy Images, *Langmuir* **13** (1997): 373–377.

13. Mahaffy, R. E., Shih, C. K., MacKintosh, F. C., and Käs, J. Scanning Probe-based Frequency-dependent Microrheology of Polymer Gels and Biological Cells, *Phys Rev Lett* **85** (2000): 880–883.
14. Mahaffy, R. E., Park, S., Gerde, E., Käs, J., and Shih, C. K. Quantitative Analysis of the Viscoelastic Properties of Thin Regions of Fibroblasts Using Atomic Force Microscopy, *Biophys J* **86** (2004): 1777–1793.
15. Hurley, D. C. Contact Resonance Force Microscopy Techniques for Nanomechanical Measurements, in *Applied Scanning Probe Methods Vol. XI*, Bhushan, B., and Fuchs, H., Eds., Springer: Berlin, 2009, pp. 97–138.
16. Marinello, F., Passeri, D., and Savio E., Eds. *Scanning Probe Acoustic Techniques*, Springer: Berlin, 2012.
17. NT-MDT. AFAM. http://www.ntmdt.com/page/afam (accessed 4 May 2012).
18. Asylum Research. Bimodal Dual AC Imaging. http://www.asylumresearch.com/Applications/BimodalDualAC/BimodalDualAC.shtml (accessed 4 May 2012).
19. Rabe, U., and Arnold, W. Acoustic Microscopy by Atomic Force Microscopy, *Appl Phys Lett* **64** (1994): 1493–1495.
20. Rabe, U. Atomic Force Acoustic Microscopy, in *Applied Scanning Probe Methods, Vol. II*, Bhushan, B., and Fuchs, H., Eds., Springer: Berlin, 2006, pp. 37–90.
21. Yamanaka, K., and Nakano, S. Ultrasonic Atomic Force Microscope with Overtone Excitation of Cantilever, *Jpn J Appl Phys* **35** (1996): 3787–3792.
22. Yamanaka, K., Kobari, K., and Tsuji, T. Evaluation of Functional Materials and Devices Using Atomic Force Microscopy with Ultrasonic Measurements, *Jpn J Appl Phys* **47** (2008): 6070–6076.
23. Burnham, N. A., Gremaud, G., Kulik, A. J., Gallo, P.-J., and Oulevey, F. Materials' Properties Measurements: Choosing the Optimal Scanning Probe Microscope Configuration, *J Vac Sci Technol B* **14** (1996): 1308–1312.
24. Kareem, A. U., and Solares, S. D. Characterization of Surface Stiffness and Probe-Sample Dissipation Using the Band Excitation Method of Atomic Force Microscopy: a Numerical Analysis, *Nanotechnology* **23** (2012): 015706.
25. Stan, G., Ciobanu, C. V., Parthangal, P. M., and Cook, R. F. Diameter-dependent Radial and Tangential Elastic Moduli of ZnO Nanowires, *Nano Lett* **7** (2007): 3691–3697.
26. Hurley, D. C., K. Shen, K., Jennett, N. M., and Turner, J. A. Atomic Force Acoustic Microscopy Methods to Determine Thin-film Elastic Properties, *J Appl Phys* **94** (2003): 2347–2354.
27. Kopycinska-Müller, M., Geiss, R. H., and Hurley, D. C. Contact Mechanics and Tip Shape in AFM-based Nanomechanical Measurements, *Ultramicroscopy* **106** (2006): 466–474.
28. Kopycinska-Müller, M., Striegler, A., Schlegel, R., Köhler, B., and Wolter, K.-J. Mechanical Characterization of Thin Films by Use of Atomic Force Acoustic Microscopy, *Adv Eng Mater* **13** (2011): 312–318.
29. Olympus NDT. Contact transducers (V106 or V103), http://www.olympus-ims.com/en/ultrasonic-transducers/contact-transducers/ (accessed 4 May 2012).
30. Rabe, U., Hirsekorn, S., Reinstädtler, M., Sulzbach, T., Lehrer, C., and Arnold, W. Influence of the Cantilever Holder on the Vibrations of AFM Cantilevers, *Nanotechnology* **18** (2007): 044008.
31. Tsuji, T., Kobari, K., Ide, S., and Yamanaka, K. Suppression of Spurious Vibration of Cantilever in Atomic Force Microscopy by Enhancement of Bending Rigidity of Cantilever Chip Substrate, *Rev Sci Instr* **78** (2007): 103703.
32. Turner, J. A., and Wiehn, J. S. Sensitivity of Flexural and Torsional Vibration Modes of Atomic Force Microscope Cantilevers to Surface Stiffness Variations, *Nanotechnology* **12** (2001): 322–330.
33. Nair, S. S., Wang, S., and Hurley, D. C. Nanoscale Characterization of Natural Fibers and Their Composites Using Contact Resonance Force Microscopy, *Compos Pt A—Appl Sci Manuf* **41** (2010): 624–631.
34. Killgore, J. P., and Hurley, D. C. Low-Force AFM Nanomechanics with Higher-Eigenmode Contact Resonance Spectroscopy, *Nanotechnology* **23** (2012): 055702.
35. Rabe, U., Scherer, V., Hirsekorn, S., and Arnold, W. Nanomechanical Surface Characterization by Atomic Force Acoustic Microscopy, *J Vac Sci Technol B* **15** (1997): 1506–1511.
36. Rabe, U., Amelio, S., Kopycinska, M., Hirsekorn, S., Kempf, M., Göken, M., and Arnold, W. Imaging and Measurement of Local Mechanical Material Properties by Atomic Force Acoustic Microscopy, *Surf Interf Anal* **33** (2002): 65–70.

37. Rabe, U., Kopycinska, M., Hirsekorn, S., Muñoz Saldaña, J., Schneider, G. A., and Arnold, W. High-Resolution Characterization of Piezoelectric Ceramics by Ultrasonic Scanning Force Microscopy Techniques, *J Phys D Appl Phys* **35** (2002): 2621–2635.
38. Stan, G., and Price, W. Quantitative Measurements of Indentation Moduli by Atomic Force Acoustic Microscopy Using a Dual Reference Method, *Rev Sci Instr* **77** (2006): 103707.
39. Arinéro, R., Lévêque, G., Girard, P., and Ferrandis, J. Y. Image Processing for Resonance Frequency Mapping in Atomic Force Modulation Microscopy, *Rev Sci Instr* **78** (2007): 023703.
40. Huey, B. D. AFM and Acoustics: Fast, Quantitative Nanomechanical Mapping, *Annu Rev Mater Res* **37** (2007): 351–385.
41. Yamanaka, K., Maruyama, Y., Tsuji, T., and Nakamoto, K. Resonance Frequency and *Q* Factor Mapping by Ultrasonic Atomic Force Microscopy, *Appl Phys Lett* **78** (2001): 1939–1941.
42. Kobayashi, K., Yamada, H., and Matsushige, K. Resonance Tracking Ultrasonic Atomic Force Microscopy, *Surf Interface Anal* **33** (2002): 89–91.
43. Steiner, P., Roth, R., Gnecco, E., Glatzel, T., Baratoff, A., and Meyer, E. Modulation of Contact Resonance Frequency Accompanying Atomic-scale Stick–slip in Friction Force Microscopy, *Nanotechnology* **20** (2009): 495701.
44. Stan, G., King, S. W., and Cook, R. F. Nanoscale Mapping of Contact Stiffness and Damping by Contact-Resonance AFM, *Nanotechnology* **23** (2012): 215703.
45. Hurley, D. C., Kos, A. B., and Rice, P. In *Scanning-Probe and Other Novel Microscopies of Local Phenomena in Nanostructured Materials*, Kalinin, S. V., Goldberg, B., Eng, L. M., and Huey, B. D., Eds., *MRS Symposium Series 838E*, Materials Research Society: Warrendale, PA, 2005, pp. O8.2.1–O8.2.6.
46. Kos, A. B., and Hurley, D. C. Nanomechanical Mapping with Resonance Tracking Scanned Probe Microscope, *Meas Sci Technol* **19** (2008): 015504.
47. Kopycinska-Müller, M., Striegler, A., Schlegel, R., Kuzeyeva, N., Köhler, B., and Wolter, K.-J. Dual Resonance Excitation System for the Contact Mode of Atomic Force Microscopy, *Rev Sci Instr* **83** (2012): 043703.
48. Jesse, S., Kalinin, S. V., Proksch, R., Baddorf, A. P., and Rodriguez, B. J. The Band Excitation Method in Scanning Probe Microscopy for Rapid Mapping of Energy Dissipation on the Nanoscale, *Nanotechnology* **18** (2007): 435503.
49. Asylum Research. Band Excitation Scanning Probe Microscopies: Traveling through the Fourier Space http://www.asylumresearch.com/Applications/BandExcitation/BandExcitation.shtml (accessed 4 May 2012).
50. Rodriguez, B. J., Callahan, C., Kalinin, S. V., and Proksch, R. Dual-Frequency Resonance-Tracking Atomic Force Microscopy, *Nanotechnology* **18** (2007): 475504.
51. Zhao, W., Singh, R. P., and Korach, C. S. Effects of Environmental Degradation on Near-fiber Nanomechanical Properties of Carbon Fiber Epoxy Composites, *Compos Pt A—Appl Sci Manuf* **40** (2009): 675–678.
52. Shin, J., Rodriguez, B. J., Baddorf, A. P., Thundat, T., Karapetian, E., Kachano, M., Gruverman, A., and Kalinin, S. V. Simultaneous Elastic and Electromechanical Imaging by Scanning Probe Microscopy: Theory and Applications to Ferroelectric and Biological Materials, *J Vac Sci Technol B* **23** (2005): 2102–2108.
53. Preghenella, M., Pegoretti, A., and Migliaresi, C. Atomic Force Acoustic Microscopy Analysis of Epoxy-Silica Nanocomposites, *Polymer Testing* **25** (2006): 443–451.
54. Passeri, D., Bettucci, A., Germano, M., Rossi, M., Alippi, A., Sessa, V., Fiori, A., Tamburri, E., and Terranova, M. L. Local Indentation Modulus Characterization of Diamondlike Carbon Films by Atomic Force Acoustic Microscopy Two Contact Resonance Frequencies Imaging Technique, *Appl Phy Lett* **88** (2006): 121910.
55. Kopycinska-Müller, M., Caron, A., Hirsekorn, S., Rabe, U., Natter, H.; Hempelmann, R., Birringer, R., and Arnold, W. Quantitative Evaluation of Elastic Properties of Nano-crystalline Nickel Using Atomic Force Acoustic Microscopy, *Z Phys Chem* **222** (2008): 471–498.
56. Yuya, P. A., Hurley, D. C., and Turner, J. A. Contact Resonance Atomic Force Microscopy for Viscoelasticity, *J Appl Phys* **104** (2008): 074916.
57. Yuya, P. A., Hurley, D. C., and Turner, J. A. Relationship between Q-factor and Sample Damping for Contact Resonance AFM Measurement of Viscoelastic Properties, *J Appl Phys* **109** (2011): 113528.

58. Killgore, J. P., Yablon, D. G., Tsou, A. H., Gannepalli, A., Yuya, P. A., Turner, J. A., Proksch, R., and Hurley, D. C. Viscoelastic Property Mapping with Contact Resonance Force Microscopy, *Langmuir* **27** (2011): 13983–13987.
59. Gannepalli, A., Yablon, D. G., Tsou, A. H., and Proksch, R. Mapping Nanoscale Elasticity and Dissipation Using Dual Frequency Contact Resonance AFM, *Nanotechnology* **22** (2011) 355705.
60. Arinéro R., and Lévêque, G. Vibration of the Cantilever in Force Modulation Microscopy Analysis by a Finite Element Model, *Rev Sci Instr* **74** (2003): 104–111.
61. Espinoza Beltrán, F. J., Scholz, T., Schneider, G. A., Muñoz-Saldaña, J., Rabe, U., and Arnold, W. In *Proceedings of ICN&T 2006*, Meyer, E., Hegner, M., Gerber, C., and Güntherodt, H.-J., Eds., *J. Phys. Conference Series 61*, IOP Publishing: Bristol, UK, 2007, pp. 293–297.
62. Rabe, U., Janser, K., and Arnold, W. Vibrations of Free and Surface-Coupled Atomic Force Microscope Cantilevers: Theory and Experiment, *Rev Sci Instr* **67** (1996): 3281–3293.
63. Turner, J. A., Hirsekorn, S., Rabe, U., and Arnold, W. High-Frequency Response of Atomic-Force Microscope Cantilevers, *J Appl Phys* **82** (1997): 966–979.
64. Rabe, U., Amelio, S., Kester, E., Scherer, V., Hirsekorn, S., and Arnold, W. Quantitative Determination of Contact Stiffness Using Atomic Force Acoustic Microscopy, *Ultrasonics* **38** (2000): 430–437.
65. Rabe, U., Turner, J., and Arnold, W. Analysis of the High-Frequency Response of Atomic Force Microscope Cantilevers, *Appl Phys A* **66** (1998): S277–S282.
66. Campbell, S. E., Ferguson, V. L., and Hurley, D. C. Nanomechanical Mapping of the Osteochondral Interface with Contact Resonance Force Microscopy and Nanoindentation, *Acta Biomater* **8** (2012): 4389–4396.
67. Carpick, R. W., Ogletree, D. F., and Salmeron, M. Lateral Stiffness: A New Nanomechanical Measurement for the Determination of Shear Strengths with Friction Force Microscopy, *Appl Phys Lett* **70** (1997): 1548–1550.
68. Ge, S., Pu, Y., Zhang, W., Rafailovich, M., Sokolov, J., Buenviaje, C., Buckmaster, R., and Overney, R. M. Shear Modulation Force Microscopy Study of Near Surface Glass Transition Temperatures, *Phys Rev Lett* **85** (2000): 2340–2343.
69. Overney, R., Buenviaje, C., Luginbühl, R., and Dinelli, F. Glass and Structural Transitions Measured at Polymer Surfaces on the Nanoscale, *J Therm Anal Calorim* **59** (2000): 205–225.
70. Wahl, K., and Stepnowski, S. Viscoelastic Effects in Nanometer-Scale Contacts under Shear, *Tribol Lett* **5** (1998): 2–5.
71. Kawagishi, T., Kato, A., Hoshi, Y., and Kawakatsu, H. Mapping of Lateral Vibration of the Tip in Atomic Force Microscopy at the Torsional Resonance of the Cantilever, *Ultramicroscopy* **91** (2002): 37–48.
72. Song, Y., and Bhushan, B. Quantitative Extraction of In-plane Surface Properties Using Torsional Resonance Mode of Atomic Force Microscopy, *J Appl Phys* **97** (2005): 083533.
73. Yamanaka, K., and Nakano, S. Quantitative Elasticity Evaluation by Contact Resonance in an Atomic Force Microscope, *Appl Phys A* **66** (1998): S313–S317.
74. Hurley, D. C., and Turner, J. A. Measurement of Poisson's Ratio with Contact-Resonance Atomic Force Microscopy, *J Appl Phys* **102** (2007): 033509.
75. Reinstädtler, M., Rabe, U., Scherer, V., U. Hartmann, U., Goldade A., Bhushan, B., and Arnold, W. On the Nanoscale Measurement of Friction Using Atomic-Force Microscope Cantilever Torsional Resonances, *Appl Phys Lett* **82** (2003): 2604–2606.
76. Tsuji, T., and Yamanaka, K. Observation by Ultrasonic Atomic Force Microscopy of Reversible Displacement of Subsurface Dislocations in Highly Oriented Pyrolytic Graphite, *Nanotechnology* **12** (2001): 301–307.
77. Hurley, D. C., Kopycinska-Müller, M., Langlois, E. D., Kos, A. B., and Barbosa, N. Mapping Substrate/Film Adhesion with Contact-Resonance-Frequency Atomic Force Microscopy, *Appl Phys Lett* **89** (2006): 021911.
78. He, C., Zhang, G., Wu, B., and Wu, Z. Subsurface Defect of the SiO_x Film Imaged by Atomic Force Acoustic Microscopy, *Opt Laser Eng* **48** (2010): 1108–1112.
79. Parlak Z., and Degertekin, F. L. Contact Stiffness of Finite Size Subsurface Defects for Atomic Force Microscopy: Three-dimensional Finite Element Modeling and Experimental Verification, *J Appl Phys* **103** (2008): 114910.

80. Killgore, J. P., Kelly, J. Y., Stafford, C. M., Fasolka, M. J., and Hurley, D. C. Quantitative Subsurface Contact Resonance Force Microscopy of Model Polymer Nanocomposites, *Nanotechnology* **22** (2011): 175706.
81. Kopycinska-Müller, M., Geiss, R., Müller, J., and Hurley, D. C. Elastic-Property Measurements of Ultrathin Films Using Atomic Force Acoustic Microscopy. *Nanotechnology* **16** (2005): 703–709.
82. Killgore, J. P., Geiss, R. H., and Hurley, D. C. Continuous Measurement of Atomic Force Microscope Tip Wear by Contact Resonance Force Microscopy, *Small* **7** (2011): 1–5.
83. Rupp, D., Rabe, U., Hirsekorn, S., and Arnold, W. Nonlinear Contact Resonance Spectroscopy in Atomic Force Microscopy, *J Phys D: Appl Phys* **40** (2007): 7136–7145.

Chapter 6

Guide to Best Practices for AFM Users

Greg Haugstad

University of Minnesota, Minneapolis, MN

The aim of this chapter is a practical guide to help the atomic force microscopy (AFM) user effectively implement some of the methodologies described in this book and to help understand the various artifacts and associated challenges encountered in practical imaging. These artifacts arise from a number of nonidealities, calibration issues, and other difficulties. The problems tend to be more glaring when one seeks quantitative, rather than qualitative, information, or absolute rather than relative numbers. But having even a qualitative sense of what the AFM is actually measuring—rather than some idealization of this measurement—will help the user to better understand the technique and in some cases enable more meaningful experimentation [1].

Common artifacts associated with all the methods described previously will be discussed including those involving force curve measurements (Chapter 3) and phase imaging (Chapter 4) and will draw upon the contact mechanics terminology explained in Chapter 2.

We begin our treatment in Section 6.1 with issues of vertical (quasi-static) force measurements versus Z displacement, derived from (a) the significant angle of inclination of the cantilever relative to sample surface, (b) nonlinearity in the split-photodiode measurement of laser spot displacement, (c) geometric optics and interferometric effects related to the laser beam trajectory, and (d) thermal drift in the baseline (zero) of force. In each case we note that *instrument design* plays an integral role in nonideal behavior. In Section 6.2 we focus on complications that are instead due to *physical interactions* at the tip–sample interface. These phenomena derive from capillarity and long-range material extension between tip and nominal sample surface; the latter is particularly problematic on soft (and thus mobile) materials but also can

Scanning Probe Microscopy in Industrial Applications: Nanomechanical Characterization, First Edition. Edited by Dalia G. Yablon.

result from adventitious sample surface contamination on what are ostensibly clean, rigid materials. These physical interactions affect most types of AFM use including single-point force curve measurements or imaging methods such as topography and phase. (Each of the above topics is treated in greater detail in Ref. [1].)

6.1 FORCE–DISTANCE MEASUREMENTS—INSTRUMENTAL SOURCES OF NONIDEALITY

6.1.1 Geometric Considerations for Tip–Cantilever

Vertical and Lateral Motion Crosstalk

It is first important to examine the geometric constraints of a tip–cantilever system. In a typical force curve acquired with a low to intermediate stiffness (say 0.1–3 N/m) cantilever on a glassy polymer, for example, one can easily execute several tens of nanometers of upward tip movement to produce only a few nanometers of vertical tip displacement into the sample. The cantilever cannot bend vertically (*Z*) *without also moving the tip horizontally* (*X*). For example, for $\Delta Z = 50$ nm (and treating the cantilever angle of inclination to be a constant, typically 10–15°), one can easily produce a horizontal displacement of ≈ 10 nm, which can be much greater than the vertical indentation of tip into sample, meaning a more trenchlike excursion of tip into sample. But there is also a physical issue: whether the tip actually moves 10 nm along *X*, or whether static friction instead keeps it pinned at a point. Given that the tip–sample contact diameter, for a typical ~10-nm-radius tip, is at most a few nanometers on a glassy polymer, a pinned tip is unlikely (pinning being more likely for, say, a 5-nm rather than 50-nm *Z* displacement). Thus the overall force curve, from the first touch during approach to the highest force achieved at turnaround to the final tip–sample separation during retraction, may involve stick-to-slide and slide-to-stick transitions at points *other than* the initial touch and final break of contact. Even within the "stick" regime there is necessarily a lateral component of force, due to cantilever geometry, in addition to the normal component (i.e., there are both normal and shear stresses in the contact). Knowing that one ultimately wishes to quantify mechanical properties such as elastic modulus, one should recognize that the lateral stress must play a role in the most precise, quantitative measurements of contact stiffness. This issue has been discussed in the context of force modulation microscopy by Mazeran and Loubet [2] as well as others.

Lateral tip movement suggests that one can confirm material *yield* during force curves by subsequently imaging the scratch, exemplified in Figure 6.1. Here a topographic trench approximately 300 by 20 nm in lateral dimension was produced in an ultrathin film of the glassy polymer polyvinyl alcohol (PVA, on mica) using a (Si_3N_4) tip a few tens of nanometers in nominal radius (cantilever stiffness $k \approx 0.6$ N/m) [3]. The trench was produced while ramping *Z* over a distance of microns to reach contact forces of order ~1 μN. This caused the tip to penetrate the PVA to the point of reaching the mica substrate. The height data from the image in Figure 6.1 reveals that the tip penetrated deeper into the material as *Z* increased

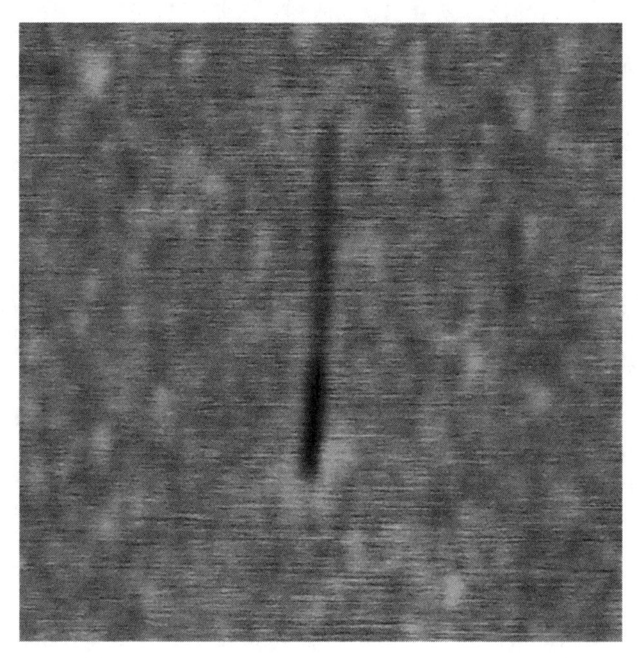

Figure 6.1 Topographic image of scratch produced during force-Z measurements due to lateral tip displacement as vertical force increases (top to bottom). For color details, please see color plate section.

(and thus the cantilever deflection, i.e., force); and an abrupt penetration from 0.27 to 0.76 nm depth is observed, a yield point where the contrast in the scratch goes from dark red to a much deeper dark red. The width of the trench also provides an estimate of the contact size during the plastic deformation process, an upper bound on the contact diameter under elastic conditions just prior to the onset of yield. This value could be cross-checked with the force curve data, at forces just below yield onset, as analyzed within a model of elastic contact mechanics.

Laser Spot Positioning

Another question related to cantilever inclination—specifically cantilever bending under force—is the importance of laser spot positioning. A flexible cantilever is the force sensor and so bending can occur at all points along the length of the cantilever. One benefits from understanding the mathematical shape of this bent cantilever, $z(x)$, and thus how the angle of inclination varies along its length. Deformable body mechanics can provide the expression $z(x)$ for a simple rectangular cantilever (i.e., ignoring the tip and complications of a slightly nonrectangular shape or reflective coatings), resulting in

$$z(x) = \frac{2Fx^2}{Ewt^3}(3L - x) \tag{6.1}$$

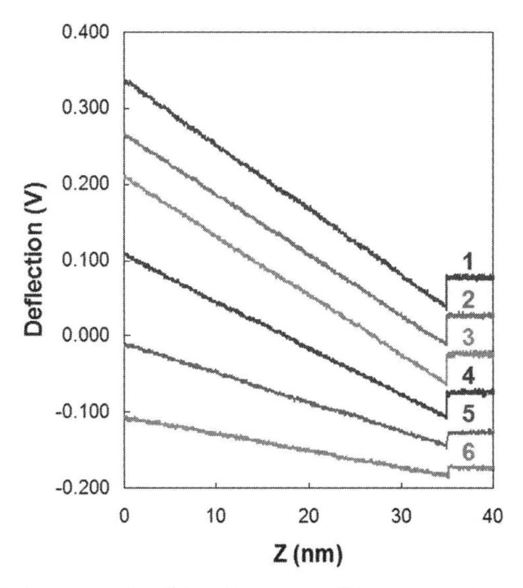

Figure 6.2 Approach force curves for different laser spot positions near evenly spaced along the length of the cantilever: 1 near tip end, 6 near base, and 3 near middle. For color details, please see color plate section.

where F is force, L is cantilever length, E is Young's modulus of cantilever, and w and t are the width and thickness of the cantilever cross section, respectively. For a rectangular silicon beam with $L=225\,\mu m$, $w=35\,\mu m$, $t=2.0\,\mu m$, $E=179\,GPa$, and $F=10$ nN localized at $x=L$, Eq. (6.1) produces a vertical displacement of 9 nm at $x=L$. One can plot Eq. (6.1) in this case to see that almost all of the bending involves the interval $0<x<125\,\mu m$, whereas the inclination is almost constant for $125<x<225\,\mu m$. This suggests that the position of the reflected laser spot at the photodiode should be nearly independent of spot position *as long as it lies along the outer half of the cantilever.*

The AFM user can approximately measure the amount of variation in cantilever bending by measuring the sensitivity S from the slope of a force curve in contact for different laser positions on the cantilever. Figure 6.2 contains six force curves acquired during approach (to silicon, in air) corresponding to six laser positions on a ~400-μm-long rectangular cantilever. For locations 1–3 where the laser spot is positioned on the outer half of the length of the cantilever (1 near end, 3 near middle, 2 in between), the slopes of deflection versus Z, or sensitivities S, are within 5% of 8.0 mV/nm. For positions 4–6 where the laser spot was located along the inner half of the length of the cantilever (6 near base, 4 at 40% of the distance to the end of the cantilever, 5 in between), the slopes substantially decrease to 5.8, 3.6, and 2.0 mV/nm, respectively. Clearly, accurate measurements of cantilever deflection, and thus force, require proper calibration to sensitivity S.

6.1.2 Nonlinearity in Split-Photodiode Detector

Another issue is the *linearity* of the force measurement scheme. The question is the photodiode/laser spot system: as already described in Chapter 1, the photodiode does not measure spot position per se [as more so the case with a charge

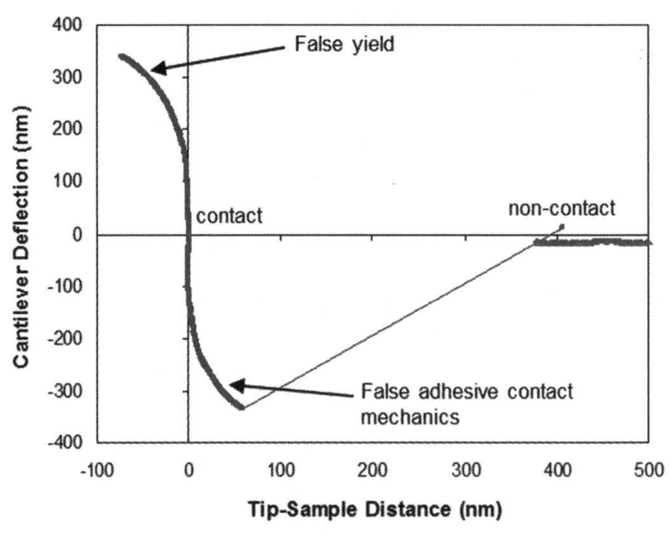

Figure 6.3 Retraction force curve spanning the entire range of measurable cantilever deflection, following conversion to distance abscissa.

coupled device (CCD) camera]; rather it measures the integrated spot intensity on either side of the split photodiode. Importantly the *spot intensity profile is not uniform*; rather, it has some variable cross-sectional intensity probably akin to Gaussian [4]. Thus as the laser spot displaces across the photodiode junction, the split photodiode's differential output changes linearly only if the movement across the detector is very small. If the laser spot movement at the split photodiode is large, meaning *several volts* of signal change (i.e., a large fraction of its measurement range), nonlinearity can produce a cubic response that comprises as much as ~10% of the overall signal (adding to or subtracting from the linear trend) [4]. The magnitude of this nonlinearity becomes glaring upon conversion of the abscissa to tip–sample distance, as exemplified in Figure 6.3. In this example the actual indentation is very small (on a metal), so as to produce a vertical data trend in contact on the scale shown. Such a vertical trend is indeed observed for cantilever deflections measured within approximately ±100 nm of zero output from the differential photodiode. (That is, the operator might limit ultimate loading to preserve linear response.) But upon retraction, a large tip–sample adhesion, such as shown in Figure 6.3, will force nonlinear response. Upon analyzing these data with adhesive contact mechanics models as discussed in Chapter 2, *inconsistencies between data and model are inevitable* at large values of negative deflection. Similarly, if raising the ultimate load to high positive forces in order to cause material yield, *an inherently nonlinear force–indentation signature of yield may be difficult to quantify* given the nonlinear measurement. The obvious solution is to use a much stiffer cantilever, whereby forces can reach yield points at cantilever deflection values that are within the linear measurement regime.

6.1.3 Geometric Optics and Interferometry

Imaging in Liquids: Effect of Refractive Index

One also should recognize that if using liquid immersion, the angle of the laser beam as it heads toward the split photodiode is dependent on the *index of refraction* of the immersion liquid, per Snell's law,

$$n_1 \sin\theta_1 = n_2 \sin\theta_2 \tag{6.2}$$

where n_1 and n_2 are the indices of refraction of the media (e.g., air and water) and θ_1 and θ_2 are the angles of the laser beam relative to the normal of the interface on either side of the interface, for example, air and water [1]. (Some liquid cells are closed, whereby laser passes from liquid to glass to air as it exits the cell. The liquid/glass and glass/air interfaces "cancel out," however, as described by Snell's law for the two-interface system). Importantly, for a given vertical tip displacement, the laser spot displacement at the photodiode will be *greater for the immersed case* than the in-air measurement. Obviously, this must be considered during force calibration (i.e., the determination of the sensitivity S is specific to the immersion medium, greater for liquids than air); but one also should recognize that the *force sensitivity* is greater, and the *force range* lesser, in liquid. Among other things this means that the onset of nonlinearity in the photodiode-based system will be reached at smaller cantilever deflections (forces) in liquid than in air. Thus it is best to characterize this onset in the medium of interest.

Optical Interference Effects between Laser and Cantilever

There are other nonidealities—again in the measurement of force—that derive from more esoteric *optical interferometric* effects, especially in first-generation instrument designs (e.g., that do not use low-coherence lasers as found in some later generations). Interference results, first, from the fact that the incident laser light is not completely reflected from the cantilever: laser intensity does not drop abruptly to zero at the "edge" of the laser spot. Instead finite laser intensity extends beyond the cantilever and reflects off the sample surface. Second, the cantilever can be so thin (~1 μm) that a significant fraction of the laser light intensity reaches the sample surface. The laser light reflected from the sample interferes with the laser light that is incident upon and reflected from the cantilever. As a result, *the distribution of laser intensity at the photodiode can vary as the distance between cantilever and sample varies, though unrelated to forces*. Most commonly this results in a wavy and tilted baseline in force curves in the noncontact regime, easily seen for Z ranges of many hundreds of nanometers.

For ordinary contact-mode imaging, the potential variation of contact force due to this artifact is usually small if using a soft cantilever. But if analyzing long-range forces in force curves, as in Chapter 3, the wavy baseline can be an issue. One strategy to reduce this effect is to coat cantilevers with reflective metals such as gold or aluminum, which are commonly available commercially.

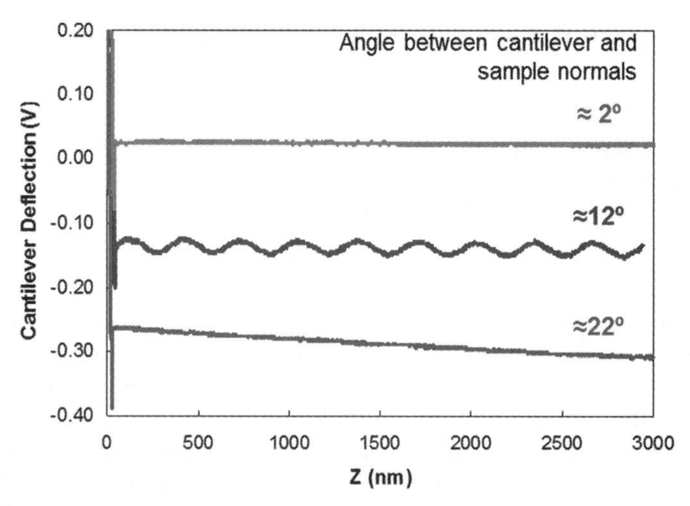

Figure 6.4 Effect on deflection baseline of variable cantilever and laser orientation with respect to sample surface.

Another strategy is using a low-coherence laser. The effect will also be reduced on nonreflective samples compared to highly reflective ones.

There are also strategies one can use to lessen the optical artifact. It is easy to verify that both the choice of laser position on the cantilever and the relative tilt between incident laser beam and sample normal affect the amplitude of interference fringes and the tilt of the baseline. Figure 6.4 compares force curve baselines (approach data) for different angles between the incident laser beam and the sample normal. In addition to the incident laser beam being parallel to the sample normal parallel, we also examine angles ranging of approximately ±10° from the 12° which is the fixed angle between the incident laser beam and cantilever normal (by instrument design). The corresponding angles between cantilever and sample thus examined were approximately 2°, 12°, and 22°, as indicated in Figure 6.4. This comparison was performed by tilting the head of the AFM via two independently adjustable threaded legs of a tri-leg assembly on which the head rests. We see that the wavy optical interference artifact *essentially vanishes* for a significant tilt between incident laser beam and sample normal, here 10° in either of two tilt directions. Second, we note that tilting in the direction that *reduces* the angle between cantilever and sample (angle of 2°) results in a quite horizontal baseline, whereas tilting in the opposite direction, such that the cantilever tilt relative to sample is *increased* (angle of 22°), results in a severely tilted baseline. Thus the former is the better strategy to eliminate the wavy artifact yet leave a nearly horizontal baseline.

6.1.4 Changing of Applied Force Due to Thermal Drift

A final note of caution regards the baseline (i.e., zero) of force curves: the possibility of substantial vertical drift due to thermal warm-up. A common observation with many AFM systems is a gradual downward drift of force baseline with passing time, significant in the first 1–2 hours since a cold start (at least tens of nanometers equivalent, in the measured cantilever deflection). In force curves this baseline is captured for each measurement cycle, thus thermal drift effects can be removed during data processing. The larger concern is *load-sensitive* contact-mode imaging—especially on moderately soft materials such as glassy polymers or organic crystals (whereas dynamic mode is more likely to be used on very soft materials). The total magnitude of drift can easily translate to several tens of nanonewtons of force, and hence substantial differences in indentation (and thereby apparent topography) and stress (normal and frictional, possibly producing wear). Clearly the operator must keep a close watch on this kind of thermal drift and repeatedly offset the effects of baseline drift. (Some commercial designs are less susceptible to thermal drift than others.)

An alternative approach is to use high-speed force curve mapping modes (commercially known as Witec's Digital Pulsed Force Mode or Bruker's PeakForce Tapping) where the setpoint deflection is continuously referenced to the baseline of force, given a touch and release at each image location. Though the baseline of force may be slowly drifting up or down, the feedback tracking of topography is continuously accounting for this drift. Of course, in this mode shear forces are largely removed and thus not available for materials analysis (i.e., friction force microscopy is not possible unless custom implemented with shear modulation methods). Another alternative is to switch to dynamic (a.k.a. tapping mode) AFM, where an alternating current (AC) signal is controlling the feedback circuit. Here again, sliding friction is largely precluded. But more significantly the method is not quasi-static, and thus a force cannot be experimentally measured, such that nanomechanical modeling cannot be employed except in the context of computer simulation.

6.2 FORCE–DISTANCE MEASUREMENTS—PHYSICAL SOURCES OF NONIDEALITY

6.2.1 Force Curve Hysteresis: Jumps to and from Contact, Role of Capillarity, and Force Gradient

In the force curve plots of the preceding section as well as in Chapter 3, one notes two additional "nonidealities" that result from real, meaningful physics: jumps to and from contact between the tip and the sample. The underlying physics behind these jumps is the attractive force between tip and sample, at minimum van der Waals (dipole–dipole) in character, but potentially involving hydrogen bonding (if a capillary water meniscus has formed) and/or longer range electrostatic forces. What is not ideal is "missing data": Because of vertical jumps in measured force, a given position of the Z scanner may produce uncontrolled variation in force (and thereby vertical tip position). In other words, we are not controlling the tip–sample distance as the *independent variable* as would be ideal. In an ideal measurement we have complete control of tip–sample distance, and simply measure force as a function of this distance. In the absence of attractive forces and provided only elastic (reversible) mechanical deformation of sample (and tip), force curves would contain a simple

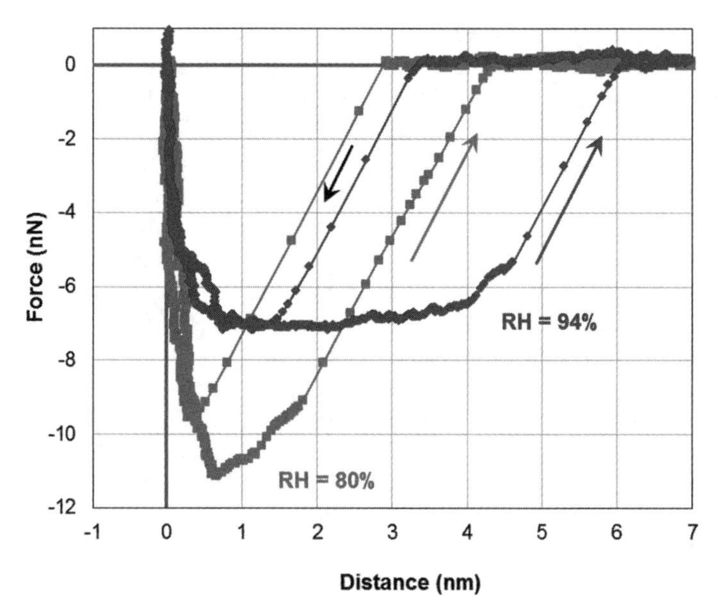

Figure 6.5 Distance dependences of force between SiO_2 tip and mica using a 4-N/m cantilever at two high humidities. For color details, please see color plate section.

onset of an increasingly repulsive force during approach, reversing to a decreasingly repulsive force during retraction, with a single value of force at each value of *Z*, irrespective of whether one is approaching or retracting in *Z*. In other words there would be *no hysteresis* in a plot of force versus *Z*.

In the real world, irreversible changes during such a *Z* cycle indeed may produce hysteresis (as seen in Chapter 3). Hysteresis usually means that there is something different about the interface during retraction compared to approach. The most common case, for measurements in air, is indeed the formation of a capillary meniscus at the tip–sample interface. Generally, such a meniscus or bridge will pull a tip to the surface, overcoming the Hookean spring force of resistance; then grow in strength while the tip is touching, such that more force is needed to break the meniscus during retraction. These common artifacts in air measurements of "snap into contact" and pull-out adhesion dip due to capillary meniscus are described in greater detail in Chapter 3. In addition, the (solid–solid) tip–sample interface itself can be hysteretic: atomic and/or molecular-scale structures can change during contact (especially on mobile surfaces such as polymers and organics), achieving a state of lower free energy and hence requiring more force to escape such a physical "bond."

To exemplify approach–retract hysteresis and other effects that are due to a water capillary meniscus, Figure 6.5 compares force–distance data acquired on cleaved mica at two high humidities, relative humidity (RH) = 80 and 94%, using a silicon tip (with native oxide) of nominal radius 10 nm, attached to a rectangular silicon cantilever of calibrated spring constant $k = 4.0$ N/m. Jumps to and from interaction are seen in both force–distance cycles: data segments of fixed slope (arrows) that seem to be "missing" data points in some intervals of distance. The black arrow indicates the jump to contact for curves collected at both humidities. The presence of a stable capillary bridge is also seen: flat or decreasing attractive force with increasing distance during retraction [1]. There are two noticeable differences in the two cases: (i) a larger magnitude of adhesion force at lower humidity (−11 nN compared to −7 nN) and (ii) a longer range adhesion force at higher humidity (roughly 5 nm compared to 2 nm at the jump out), indicating a longer capillary bridge. Observation (ii) indicates greater water presence in the interfacial system, likely a preexisting layer. With this in mind, observation (i) may be considered the beginning of a transformation to much lower adhesion in changing from air to liquid immersion. In the case of full immersion, capillary forces would vanish, of course (in the absence of bubbles or two liquid types); but also the *Hamaker constant often drops by an order of magnitude* such that attractive forces become very small. Moreover the onset of disjoining forces [e.g., hydration, Derjaguin-Landau-Verwey-Overbeek (DLVO)] can contribute to a reduction of adhesion within a capillary meniscus at high humidity [1]. We also note a departure from smooth force–distance trends, both in the steep negative-slope portion below the 1-nm distance (prior to break of "rigid contact") and in the positive-slope portion beyond the force minimum in each curve. (More than 100 repeatedly acquired force curves exhibited similar structures.) In particular, the force trend in rigid contact (0- to 0.5-nm distance) contains nearly vertical (high stiffness) segments separated by 0.2–0.3 nm, of the order of a single water molecule, suggestive of "layered" structural forces [1]. At RH=94% and distances >2 nm one also discerns little peak structures in the force curve that appear to be stick–slip in nature, possibly suggesting a degree of solidlike character in the water meniscus, as has been reported [5].

It is worth noting that one can use an order-of-magnitude stiffer cantilever to avoid a jump from contact and "missing" data (the case in Figure 6.5), but at the expense of signal–noise ratio. Force curve averaging (e.g., 10–20 data sets) can be employed to dramatically improve signal-to-noise ratio. The ability to discern weak force changes as noted in Figure 6.5 may be lost, however, because of the somewhat random location from one data set to the next (i.e., these variations average away).

Even in the absence of capillary forces, hysteresis may still occur when imaging in contact mode with soft cantilevers because of instabilities derived from other physical interactions such as electrostatic forces. This is covered in detail in Chapter 3 of Ref. [1].

6.2.2 Force Curve Hysteresis: Soft Surfaces

Another physical source of force curve hysteresis is material mobility at the tip–sample interface that results in either tip contamination or artifacts in molecular pulling experiments in the AFM. These effects are particularly egregious for long-chain molecular systems, whereby (usually) attractive interactions can range many tens or hundreds of nanometers of tip–sample distance. These molecules may be intrinsic to the sample itself or may result from tip contamination where the tip has previously encountered long-chain molecules: whether from previously analyzed polymeric samples or from contamination at some sites of otherwise clean surfaces (e.g., residual photoresist or near-ubiquitous silicone contamination). For example, it is possible for a "nanofibril" to

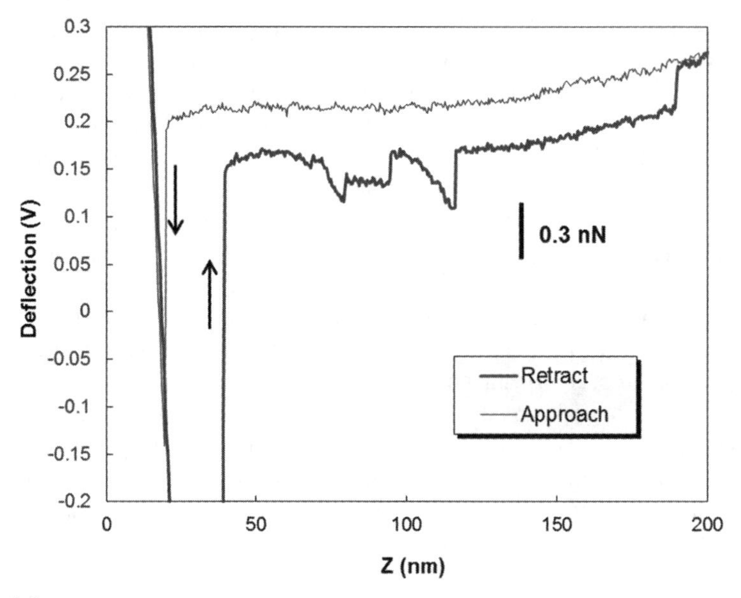

Figure 6.6 Raw force curve with a gelatin fibril bridging between tip and graphite surface in air.

become attached to the tip and have even softer mechanical response than a soft cantilever itself. Such a fibril thus can be the element whose "spring constant" is exceeded by the attractive force gradient. One may find that immediately after a particularly large jump to contact, the force curve exhibits a near-horizontal data trend with further approach, indicating a property of material attached to the tip, as opposed to the properties of the tip itself. For example, in the case of a fibril attached to the tip, the near-horizontal data could indicate *minimal resistance to the compression/collapse of a contaminant fibril.* Upon retraction, one often observes that the jump from contact in fact does not return to the zero force baseline; rather a weak attractive force remains, such that vanishing force is not achieved until a much larger retraction distance.

An example force curve cycle, between a gelatin fiber (attached to a tip) and a clean graphite surface where the effects of long-range forces can be observed, is presented in Figure 6.6 in ambient air. The approach data beyond contact displays a variable baseline from Z=30–200 nm due to the interferometry artifact discussed in Section 6.1, the equivalent of 0.3 nN of force "variation." The retraction data contains additional signatures of weak, but real, physical forces due to a long-chain fibril spanning between tip and sample. These can include both "plateau" and "sawtooth" kinds of force extension signatures. These behaviors are treated in detail in Chapters 2 and 3 of Ref. [1]. The more one uses AFM on soft and/or mobile materials, the more variations of these kinds of observations one may catalog. "Textbook" behavior becomes the exception rather than the rule.

It is worth pointing out here that most quantitative force curve analyses utilize Z ramp sizes that are very small compared to the full range of the Z scanner. Exceptions can include high-molecular-weight polymers that collectively extend into a long fibril

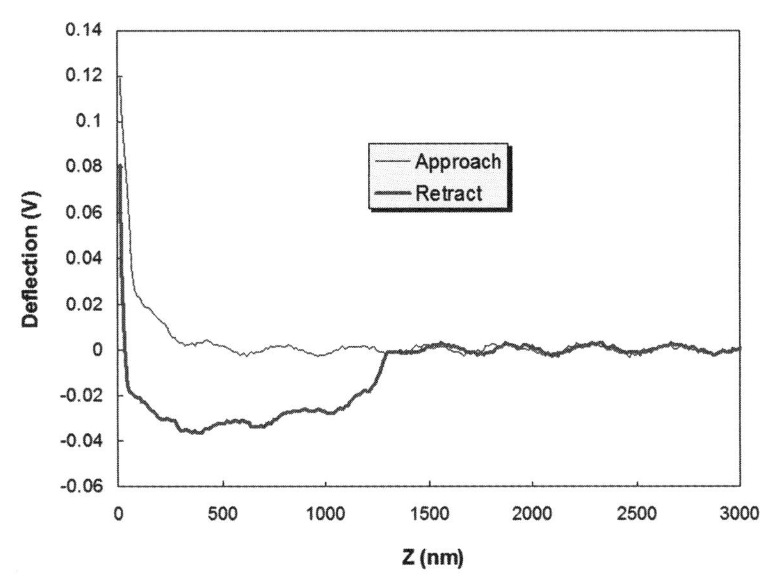

Figure 6.7 Raw force curve obtained on poly lauryl methacrylate illustrating the significance of the wavy optical baseline in both approach and retract data and further its dependence on *Z* scanner hysteresis.

between tip and sample (i.e., much longer than individual chains). In such a case, the *magnitude* of force may be small but its *range* very large. An example of a tip pulling on a long molecular, poly lauryl methacrylate where a large *Z* ramp size is required is shown in Figure 6.7; this force curve is further complicated by *Z* scanner hysteresis and laser interferometry, artifacts described above in Sections 6.2.1 and 6.1.3, respectively. The wavy baseline signal is apparent in both the approach data and the lengthy adhesive bridge during retraction, where polymer is clinging to the tip. One strategy is to acquire a "dummy" or "virtual" force curve, with the tip not quite reaching the sample but with the same size *Z* cycle: One can then mathematically fit this "baseline" with a multisinusoidal function of variable wavelength (as a function of *Z*) to account for nonlinearity and (with appropriate *Z* offsets) subtract this function from the data of interest to remove the wavy artifact.

REFERENCES

1. Haugstad, G. *Atomic Force Microscopy: Exploring Basic Modes and Advanced Applications*, Wiley: Hoboken, NJ, 2012.
2. Mazeran, P.-E., and Loubet, J. L. Force Modulation with a Scanning Force Microscope: an Analysis, *Trib Lett* **3** (1997): 125–132.
3. Haugstad, G. Contrasting Static-to-Kinetic Friction Transitions on Layers of an Autophobically Dewetted Polymer Film Using Fourier-Analyzed Shear Modulation Force Microscopy, *Trib Lett* **19** (2005): 49–58.
4. Haugstad, G., and Gladfelter, W. L. Force-Displacement Measurements in a Beam-Reflection Scanning Force Microscope: Calibration Issues, *Ultramicroscopy* **54** (1994): 31–40.
5. Choe, H., Hong, M.-H., Seo, Y., Lee, K., Kim, G., Cho, Y., Ihm, J., and Jhe, W. Formation, Manipulation, and Elasticity Measurement of a Nanometric Column of Water Molecules, *Phys Rev Lett* **95** (2005): 187801.

Chapter 7

Nanoindentation Measurements of Mechanical Properties of Very Thin Films and Nanostructured Materials at High Spatial Resolution

Steve J. Bull

School of Chemical Engineering and Advanced Materials, University of Newcastle, Newcastle upon Tyne, United Kingdom

7.1 INTRODUCTION

With the current drive to miniaturization of components there is a growing need to assess the mechanical properties of very small volumes of material in their as-processed state. Traditional mechanical testing approaches are not suitable in most cases, and alternative methods based on indentation testing have become increasingly popular. Continuously recording indentation techniques, often termed nanoindentation techniques, have been specifically developed for the assessment of mechanical response at high spatial resolution over the past 30 years [1, 2]. In these tests a continuous record of load and indenter displacement (as well as other properties such as contact stiffness) are made during a single test and the load–displacement curve produced is further analyzed to determine the mechanical properties of the material. For instance, it is usual to use the method of Oliver and Pharr [3] to determine the hardness and Young's modulus from the slope of the unloading curve. Interpreting the data from nanoindentation has been widely discussed and many reviews are available [e.g., 4, 5]. The technique can be used to assess bulk materials, coatings, and interfaces, but in many cases a composite response is produced

Scanning Probe Microscopy in Industrial Applications: Nanomechanical Characterization, First Edition. Edited by Dalia G. Yablon.

that is influenced by the surrounding material. It is, therefore, essential to determine the correct test protocol to use if mechanical data is to be generated that is free of such influences. This is discussed in this chapter.

7.2 BULK MATERIALS

In indentation testing of bulk materials there are three areas where care is necessary when designing an appropriate nanoindentation test. First, it is necessary to compare the scale of the deformation introduced by the test with the microstructure of the material since measurements at high spatial resolution will be considerably influenced by microstructure. Second, for very small structures, the dimensions of the device itself may be critical as measurements cannot be successfully carried out too close to an edge or on too thin a layer. Third, the validity of the measurements is affected by the mechanics of the indentation process itself, which depends on the properties of the materials assessed.

The traditional analysis methods based on the approach of Oliver and Pharr [3] do not account for the pileup of material around the indenter and direct measurement of the area of contact [e.g., by atomic force microscopy (AFM)] is preferred. Also there can be issues related to time-dependent deformation modes (creep or viscoelasticity), which means that the measured response is affected by the way it has been measured. These are discussed in more detail in the following sections.

7.2.1 Effect of Microstructure

Plastic Properties

When testing bulk polycrystalline materials, it is conventional to use a test load so that a large number of grains are sampled (>100) by the plastic zone associated with the indentation. The plastic zone associated with a hardness impression is approximately hemispherical and has a much larger diameter than the size of the impression, but its size is critically dependent on the material, specifically the material's ratio of Young's modulus, E, to hardness, H. For a soft metal, this ratio is higher (~100) than for a hard ceramic (~10) so the size of the plastic zone tends to be much bigger for metals than ceramics for the same indentation size. For hardness indentations with a Vickers pyramidal indenter (square-based pyramid), Lawn and co-workers showed that [6]

$$R = c\frac{d}{2}\left(\frac{E}{H}\right)^{1/2} \cot^{1/3}\theta \tag{7.1}$$

where d is the indentation diagonal, c is a constant ($c=1$ in the absence of significant pileup or sink-in), and θ is the indenter semiangle. For a Berkovich indenter generally used in nanoindentation studies, the diagonal is replaced by the indentation depth, δ, the constant is then $c/2=3.64$ [7] and $\theta=70.3°$ [8]. The radius of this plastic zone may be bigger than that of the impression. Finite-element studies have shown that this equation overestimates the plastic zone size and an alternative expression for plastic zone size has been developed. The relationship between radius of plastic zone, R_p, and residual depth of indent, δ_r, is given by [9]

$$\frac{R_p}{\delta_r} = \sqrt{0.3\pi}\sqrt{\frac{E_r}{Y}}\left[\left(\cot\theta\sqrt{\frac{E_r}{H}} + \frac{\varepsilon\pi}{2}\sqrt{\frac{H}{E_r}}\right) - 0.62\varepsilon\pi\sqrt{\frac{H}{E_r}}\right]^{-1} \tag{7.2}$$

In terms of the maximum indenter displacement, δ_{max}, Eq. (7.2) maybe approximated by an overall equation to determine the plastic zone radius, R_p, for a given maximum displacement:

$$\frac{R_p}{\delta_{max}} = -12.907\frac{H}{E_r} + 4.5451 \tag{7.3}$$

Equation (7.3) can be used to assess whether a substrate contribution to the hardness is measured for a coated system. We apply Eq. (7.3) to the example of an 800-nm layer of copper on silicon. The plastic zone radius R_p is set to 800 nm and, using the measured *H*/*E* ratio for copper determined at low contact depths, the maximum displacement, δ_{max}, determined from Eq. (7.3) is the largest indenter penetration where no substrate contribution to hardness is expected. The equation can also be used to estimate the critical maximum indenter displacement when the plastic deformation around an indentation interacts with a neighboring grain boundary if R_p is replaced by the grain radius in Eq. (7.3). For large grained materials there is often a transition from single-crystal behavior to polycrystalline behavior where grain size hardening (Hall–Petch behavior) is observed at this critical displacement.

For instance, Figure 7.1 compares the indentation creep and hardness data for aluminum in three different forms, a single crystal, a polycrystal with grain size of 20 μm, and a very thick polycrystalline coating with a grain size of 0.5 μm, which can be treated as a bulk material. Figure 7.1(a) shows the indent profile accompanied by the indentation data in Figure 7.1(b). As shown in Figure 7.1(a), the indentation test includes a hold at peak load (1 mN) in which the creep behavior is assessed—the creep strain is the fractional increase in indentation depth at a given time during the holding period. The residual depth of the indentation at the end of the test is about 150 nm in all cases. The hardness and creep behavior of the single crystal and 20-μm grain size material is almost identical, and it can easily be shown that the deforming volume is much smaller than a single grain using Eq. (7.3)—the radius of the plastic zone is about 4.5 times the maximum displacement at the start of the hold segment. However, the higher hardness and creep rate of the 0.5-μm grain size material is related to the fact that the deforming volume is now bigger than a single grain, and the effects of grain boundaries cannot be ignored.

A similar approach can be used to determine if hardness values for distributed phases can be assessed by nanoindentation methods. For instance, the plastic zone size must be smaller than the size of a precipitate and an appropriate test load should be chosen to achieve this. Usually some test indentations are made to

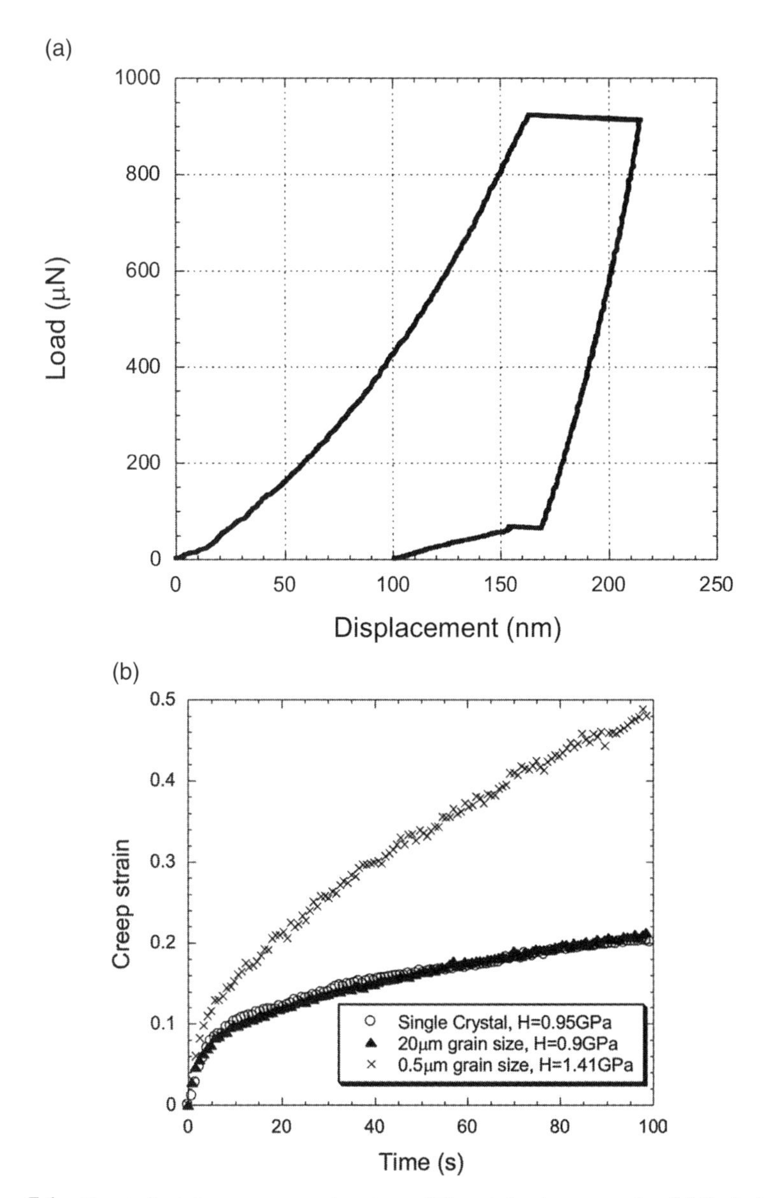

Figure 7.1 Nanoindentation creep experiments on different aluminum samples. (a) Typical load–displacement curve and (b) creep strain development during the hold period.

determine estimates for E and H and the typical relationship between indent depth and load, and then Eq. (7.3) is used to ensure that the data generated is not influenced by the surrounding material. The ability of modern nanoindentation systems to accurately position indentations is key for this. For instance, using a Hysitron triboindenter it is possible to determine the surface topography of the sample by AFM using the tip that in addition to making the indentation, will also choose the

Table 7.1 Hardness of Phases in Pearlite in 0.4% Carbon Steel Determined by Nanoindentation at High Spatial Resolution (250-μN peak load)*

Phase	Hardness (GPa)
Ferrite	2.4±0.1
Fe_3C	13.0±0.5
Pearlite average (microhardness, 0.25-N load)	3.4±0.7

* New sharp Berkovich tip, end radius 50 nm.

appropriate site to make an impression with an accuracy of 10 nm. This can be used to assess the hardness in regions where the microstructure has changed (e.g., regions of Martensite decay in gear steels where phase changes in the hardened steel surface (from Martensite to ferrite and iron carbide) are produced by contact stress cycles in service [10] or assess the hardness of the individual constituents of the Pearlite regions in a medium carbon steel, that is, individual regions of ferrite (α-iron) and cementite (Fe_3C) see (Table 7.1). However, phases should generally be bigger than 100-nm diameter for reliable measurements of plasticity due to the bluntness of available indenters.

Elastic Properties

Whereas determining the plastic properties of distributed phases is tractable in many cases, this is not true of elastic properties. The elastic deformation associated with an indentation extends to much greater distances than the plastic deformation, and it is therefore very difficult to generate data that is not influenced by neighboring phases. In many cases the best way to determine an estimate for the elastic modulus of a precipitate is to make tests at a range of indentation sizes and extrapolate the results to zero contact depth. Failing this the use of modeling to extract the data from the indentation response of the composite is required [7].

7.2.2 Effect of Sample Size and Geometry

A similar argument can be used regarding the assessment of the elastic and plastic properties of small components using nanoindentation. Equation (7.2) may be used to determine the thickness of a component that may be assessed without a contribution from plastic deformation of the substrate and also provide some guidance for the minimum spacing of an indentation from the edge of the sample—this should typically be at least 10 times the indentation depth. Elastic properties of very small components are generally influenced by the fixturing, and the use of extrapolation or modeling methods is necessary in most cases. This is amplified in the next section.

7.2.3 Effect of Surface Roughness

Of serious concern when measuring the nanoindentation response of real engineering surfaces is the effect of surface roughness. In its simplest form the effect of roughness is to increase the experimental scatter in the data as tests undertaken at the top of asperity peaks usually give different results to those from the bottom of valleys. This is due to the different amounts of constraint from the deformed region around the indent. The region at the top of an asperity provides less constraint compared to the bottom of a valley and thus a lower value of hardness or contact modulus is measured. If the surface roughness is randomly distributed about the centerline and the positions of the indents are randomly distributed, the average values determined will be the same as those obtained from tests on smooth surfaces where the size of the impression is small compared to the wavelength of the roughness. However, surfaces with appreciable skewness can lead to considerable variations from the smooth surface result. Similarly if there is some bias in the positioning of the indents, for example, by always testing the tops of asperities, there can be a considerable change from the values measured on a flat surface. For accurate assessment of the hardness and contact modulus of engineering surfaces, it is therefore necessary to obtain some idea of the surface profile prior to testing.

7.2.4 Measurement of Interfacial Phases

Given that high-resolution positioning and imaging of the indentation is possible using modern nanoindentation equipment, there has been a drive to measure the mechanical properties in the vicinity of interfaces or interfacial phases. For instance, a mapping approach identifies changes in hardness in the vicinity of copper grain boundaries, and hardness enhancements due to twin boundaries have been reported [11]. Whereas useful results can be generated for plasticity, showing the increase in hardness associated with dislocation density increasing near boundaries, the assessment of elastic properties is more problematic. This is particularly true if it is desired to use nanoindentation to measure the elastic properties of a thin interfacial layer, such as the hydroxy-catalysis glue bond used to join the fused silica mirror suspension components in a gravity wave detector [12]. In this case the bond line is 80 nm thick [visualized in the AFM topography image in Fig. 7.2(a)], and the modulus of the glue line is less than that of the fused silica it is joining. Indentations can be made in the glue line [Fig. 7.2(b)], but it is not possible to position them exactly at the bond centerline due to the limitations of the scanner system.

The AFM image in Figure 7.2(a) revealed about 4 nm of polishing relief, which complicates the nanoindentation analysis since the test surface is below the surrounding fused silica. Analysis of the geometry indicated that the total indenter displacement must be less than 20 nm to avoid its sloping sides resting on this material. The sharper cube corner indenter was used to maximize the chance of avoiding this. A number of indents were made and their positions accurately determined by AFM on this sample. The Young's modulus was determined by the method of Oliver and Pharr [3] and showed some variation with position; thus the situation was analyzed by finite-element modeling to assess the extent of the elastic deformation. A general-purpose finite-element program

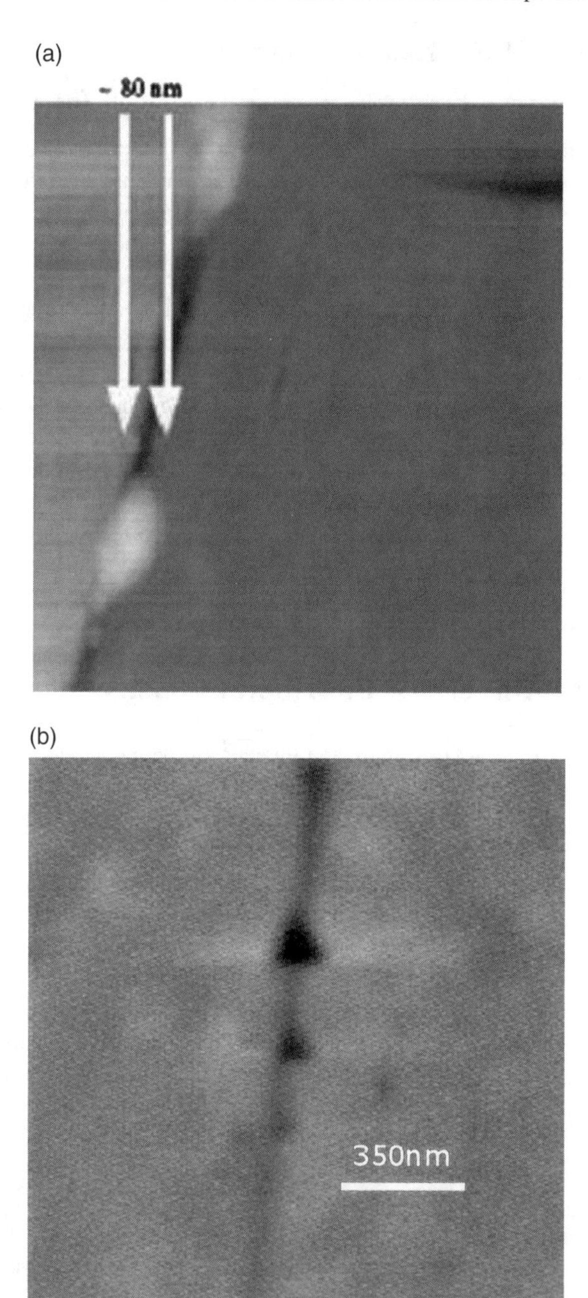

Figure 7.2 (a) AFM image of hydroxy-catalysis glue bond showing a width of 81±4 nm, and (b) AFM image of the bond made with the cube corner tip used to make the indents showing the good alignment of the indents with the bond.

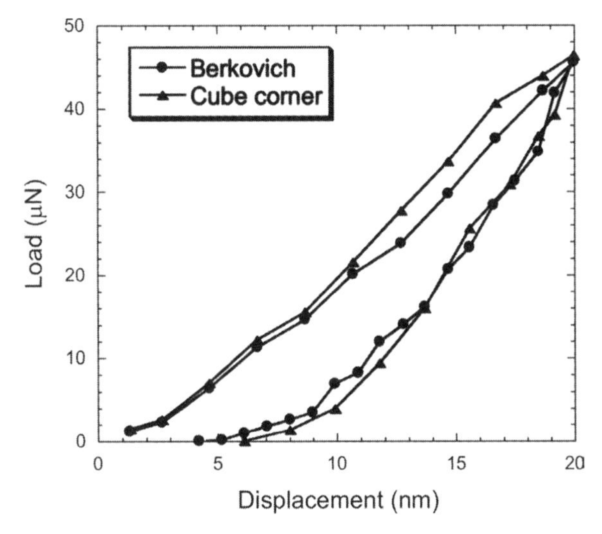

Figure 7.3 Load–displacement curve obtained from finite-element modeling.

ANSYS, version 6.1, was used for modeling load–displacement curves [13] with results shown in Figure 7.3. Since a three-dimensional (3D) model produces similar results to a two-dimensional (2D) model [14], a 2D plane strain model can be used to reduce computation time. The elastic properties of silica (E=73 GPa, υ=0.18) and diamond (E=1141 GPa, υ=0.18) were fixed for all models, and the properties of the glue layer were varied. The diamond was assumed to be elastic in all tests. Figure 7.4 shows typical simulations for the Berkovich [Fig. 7.4(a)] and cube corner indenters [Fig. 7.4(b)] used in this study. In both cases the indenter is aligned centrally with the glue line, and there is considerable elastic deformation in the surrounding fused silica. However, a residual indent is left showing that plastic deformation occurs in the glue bond, and the use of the Oliver and Pharr method for data analysis might thus be thought to be valid because an elastic-plastic indentation has been produced. Interpretation for gray scale in Figure 7.4 is as follows: Stress is zero in the dark gray regions some way away from the contact and rises as the indenter is approached, and the shades of gray for each contour represent regular increases in Von Mises stress of 2 MPa as indicated by the scale at the bottom.

Finite-element predictions of the load–displacement curves for each of the indenters (Fig. 7.3) show that for indentations less than 20 nm deep the rounding of the tip dominates behavior, and there is no difference in using either a Berkovich or cube corner tip. A 50-nm tip radius represents the best commercially available tips produced today. Since a well-developed indentation is produced, the Oliver and Pharr method [3] can be applied to the data produced by finite-element modeling to compare with experimental data. However, despite the fact that the plasticity is confined within the glue bond, the elastically stressed region is somewhat larger and a considerable part is accommodated by the surrounding fused silica. Thus the values of contact modulus determined by use of the Oliver and Pharr method will overestimate the properties of the glue bond. This effect is more significant the closer the indent is positioned to the edge of the bond.

(a)

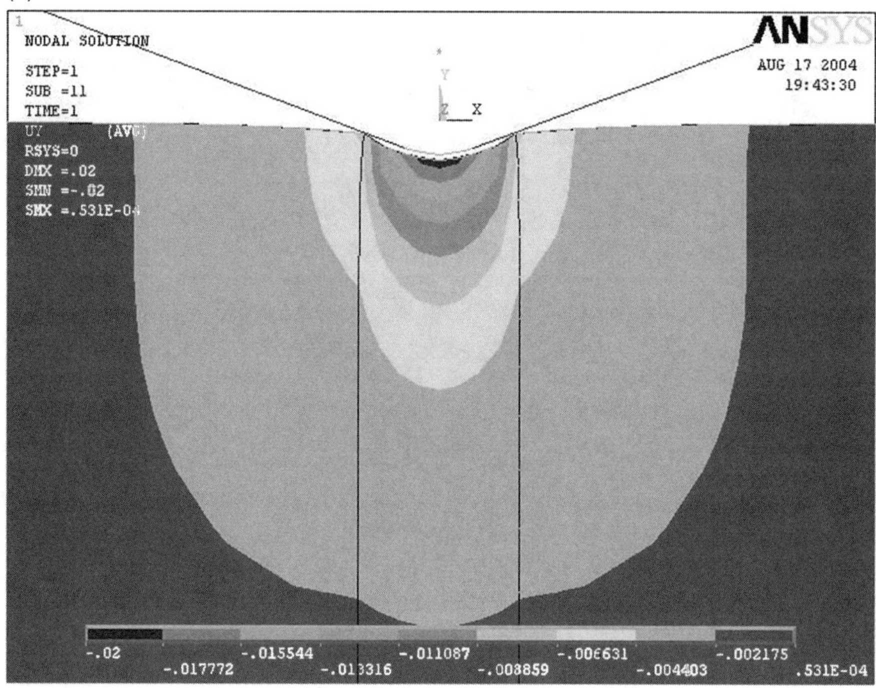

(b)

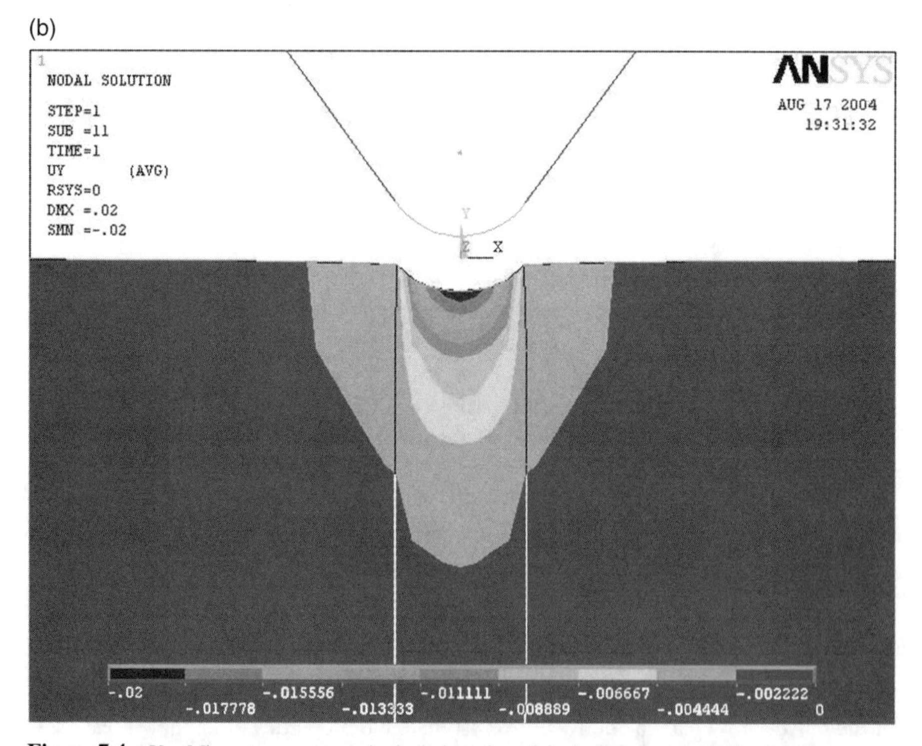

Figure 7.4 Von Mises stress contours in the indentation of the hydroxy-catalysis glue bond: (a) Berkovich indenter and (b) cube corner indenter. The stress is zero in the dark gray regions some way away from the contact and rises as the indenter is approached—the shades of gray for each contour represent regular increases in Von Mises stress of 2 MPa as indicated by the scale at the bottom.

The effect of indenter offset from the bond centerline is pronounced as shown in finite-element (FE) predictions [Fig. 7.5(a)]; experimental measurements are compared to finite-element predictions in Figure 7.5(b). The key observation is that the lowest measured modulus is still higher than the modulus of the glue bond due to the influence of the surrounding fused silica. Thus, when testing the properties of an embedded particle or boundary layer using nanoindentation, it should be recognized that although the plastic properties (hardness) may be measured with good accuracy, if the impression is small enough, the elastic properties will be influenced by the surrounding media and modeling is necessary to extract accurate data.

7.2.5 Effect of Pileup on Measured Properties

Pileup has a significant effect on the measured hardness and contact modulus of many technologically important materials. In particular, the properties of many steels are influenced by pileup, and the properties obtained by the Oliver and Pharr analysis tend to be overestimates. An example is the case of carburized EN40B steel used in gears [15].

Figures 7.6(a) and 7.6(b) show the variation of hardness and elastic modulus, respectively, as a function of indentation load using two different analysis methods to determine the contact area: the contact area as determined by the standard Oliver and Pharr method and the direct measurement of contact area using AFM. In the absence of pileup these results would be expected to be identical, but when pileup occurs the direct measured contact area is larger and the mechanical properties given by the Oliver and Pharr analysis and those based on area calculated by AFM differ by a factor of proportionality. The average factors, which are referred to in this chapter as $k_H = H_{\text{pileup}}/H_{\text{Oliver–Pharr}}$ (the ratio between hardness with the pileup effect included and the hardness calculated by the Oliver–Pharr method) and $k_E = E_{\text{pileup}}/E_{\text{Oliver–Pharr}}$ (the ratio between the elastic modulus with the pileup effect included and the elastic modulus calculated by the Oliver–Pharr method) have been calculated experimentally as $k_H = 0.82 \pm 0.044$ and $k_E = 0.75 \pm 0.021$.

The proportionality factors can also be determined theoretically from the dimensions of the impression and pileup. If r_c is the radius of a conical indenter and r_p is the pileup radius, it can be shown that $r_c/r_p = 0.66$ [15]. Thus

$$k'_H = \frac{H_{\text{pileup}}}{H_{\text{Oliver-Pharr}}} = \left(\frac{r_c}{r_p}\right)^2 = 0.56 \tag{7.4}$$

$$k'_E = \frac{E_{\text{pileup}}}{E_{\text{Oliver-Pharr}}} = \frac{r_c}{r_p} = 0.75 \tag{7.5}$$

The experimental and the theoretical approach give similar results for the elastic modulus correction factor, $k_E = k'_E = 0.75$. However, different results are obtained for the hardness correction factor: $k_H = 0.82$ and $k'_H = 0.56$. Hardness as described by a power law of the radii ratio r_c/r_p, therefore, is more sensitive to the precise radii values. Also,

(a)

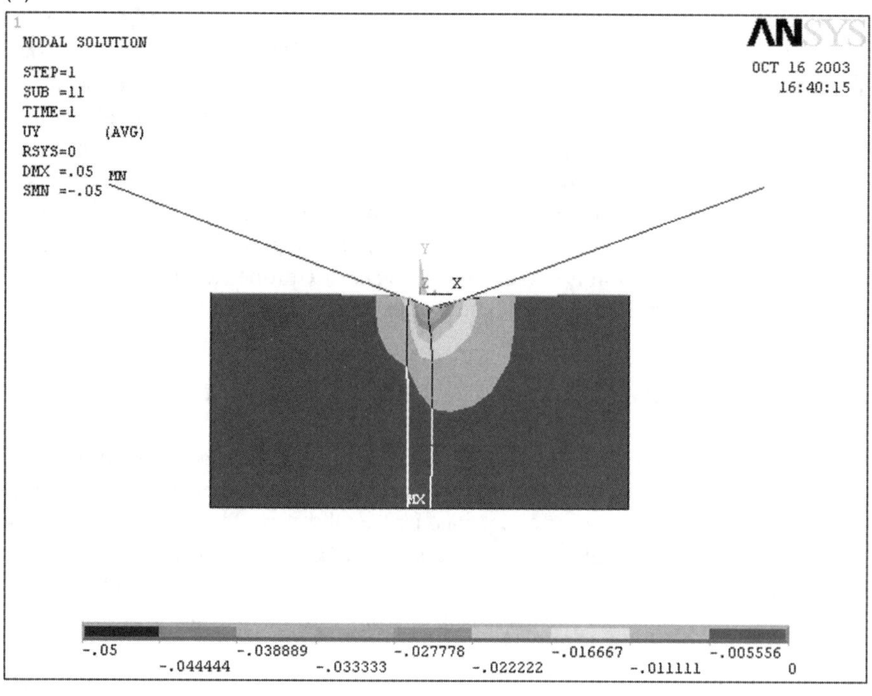

(b)

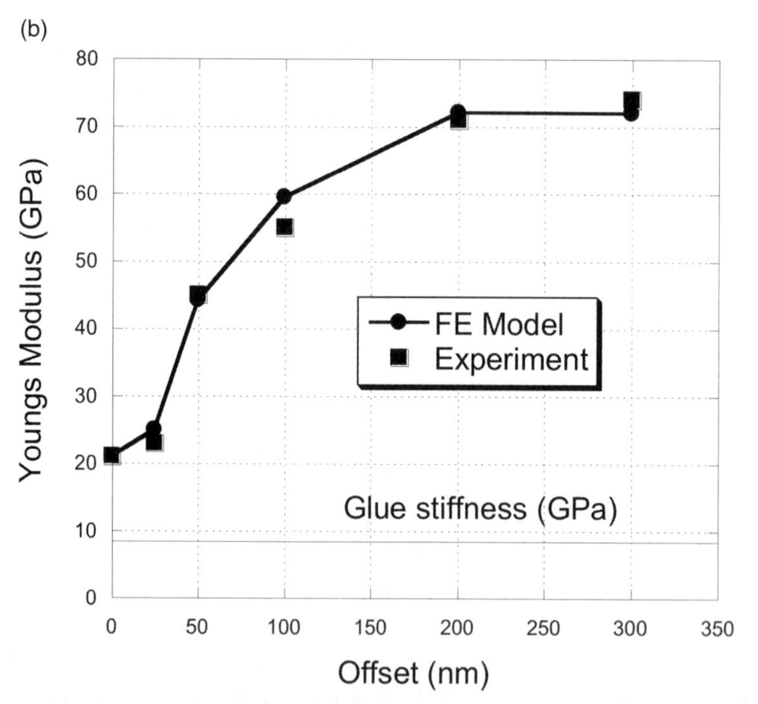

Figure 7.5 (a) Finite-element prediction of the Von Mises stresses for an indentation with a 50-nm offset from the centerline of an 80-nm glue bond. The gray contour regions represent a change in stress of 5 MPa as indicated by the bottom scale. (b) Comparison of Young's modulus determined by the Oliver and Pharr analysis for FE and experimental data.

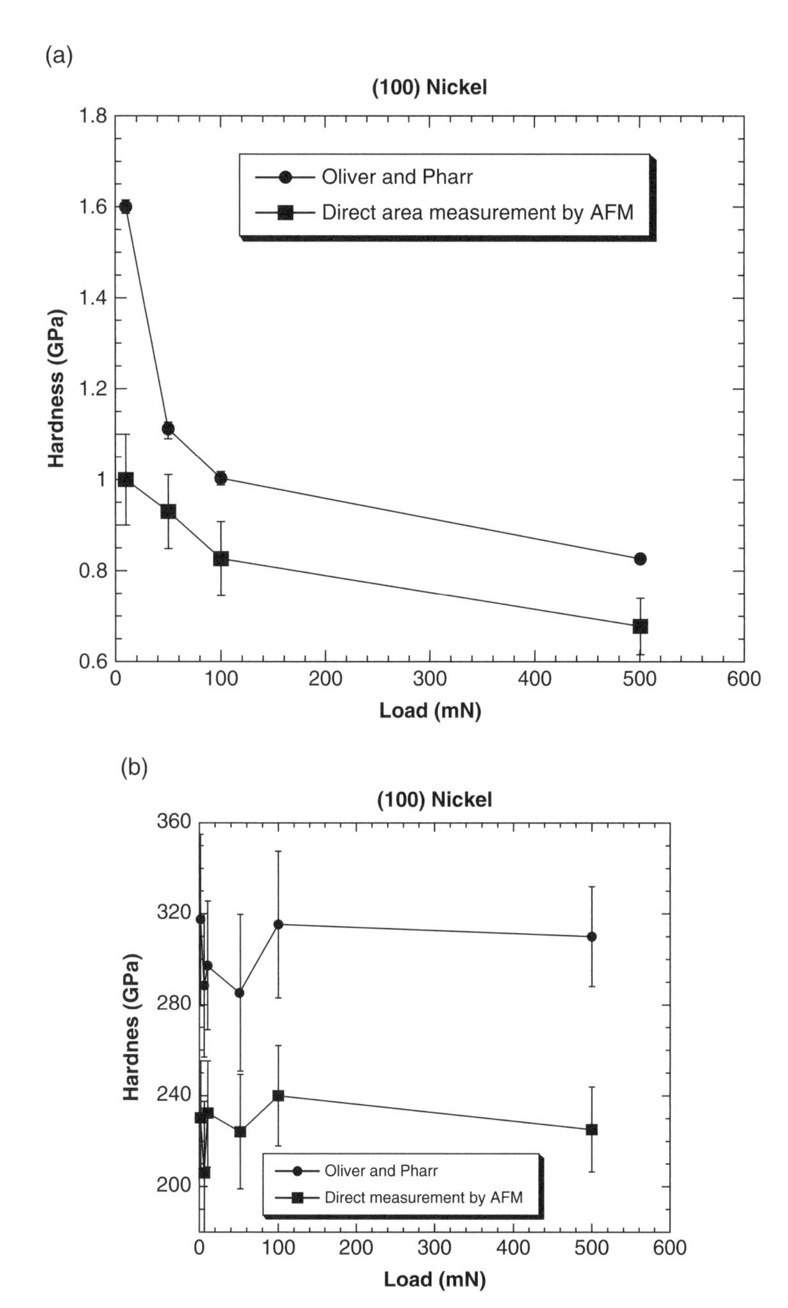

Figure 7.6 Variation of (a) hardness and (b) contact modulus as a function of contact depth for single crystal nickel comparing the Oliver and Pharr analysis method to data determined from direct measurement of the indentation area by post facto AFM. The Oliver and Pharr analysis overestimates both properties due to the effect of pileup.

the geometric model does not take into account any elastic recovery of the material. Finally, the constraint from the surrounding material will be less in the piled up zone than in the bulk material below it, and this will tend to reduce the hardness.

Thus, for practical purposes it is better to use an experimentally measured pileup correction factor if accurate values of hardness are required. For gold it has been found that $k_E=0.92$ and $k_H=0.85$, whereas for gold nitride the values are 0.6 and 0.92, respectively, despite the similarities in material properties [16]. The difference is in the mechanics and geometry of pileup formation.

7.2.6 Time-Dependent Deformation

When testing materials that show significant time-dependent deformation such as viscoelastic polymers, the conventional nanoindentation analysis methods cannot be applied since the deformation created during loading continues after the unloading process starts. This results in a nose on the load–displacement curve [Fig. 7.7(a)] as the depth of an indentation continues to increase in the early stages of unloading. The slope of the unloading curve at peak load is negative in such cases and the Oliver and Pharr analysis method is invalid.

One solution to this problem is to use load or displacement control during the indentation cycle where you are actively controlling or feedbacking of load (load control) or displacement (displacement control) to counteract the effects of time-dependent deformation. Displacement control is preferred for polymer testing, whereas load control is more reliable for assessing creep in metals. Figure 7.7(b) shows a displacement control load–displacement curve obtained from a polycarbonate instrument case. Values of hardness and contact modulus can be obtained using the Oliver and Pharr analysis and agree reasonably well with values obtained from conventional hardness and tensile tests. The indentation values for contact modulus will depend on test speed in both cases, and testing must be performed at comparable strain rates if such a comparison is to be attempted. Another advantage of the displacement control process is that pull-off forces can be measured that give an indication of the adhesive properties of the material surface.

An alternative approach is to using an oscillating tip to measure the properties of the surface. Oliver and Pharr [3] introduced a dynamic technique called continuous stiffness measurements (CSM) where a small-amplitude oscillation of relatively high frequency is superimposed to the direct current (DC) signal that controls the indenter load. This way, we can obtain the contact stiffness all along the load–displacement curve and not only in the upper part of the curve as in quasi-static methods. The technique was originally developed for metals and ceramics and was aimed at providing indentation data as a function of depth in a single indentation cycle. Its applications are the study of hardness and contact modulus, of various materials as a function of depth, although the results are affected by time-dependent deformation.

To address this, Syed et al. [17] developed a very similar technique, joining in the same machine the continuous stiffness technique and the dynamic measurement analysis (DMA) widely used in the study of polymers to measure viscoelastic

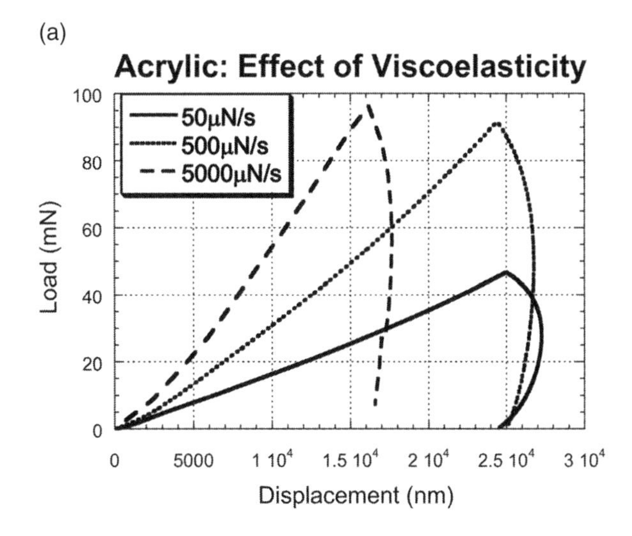

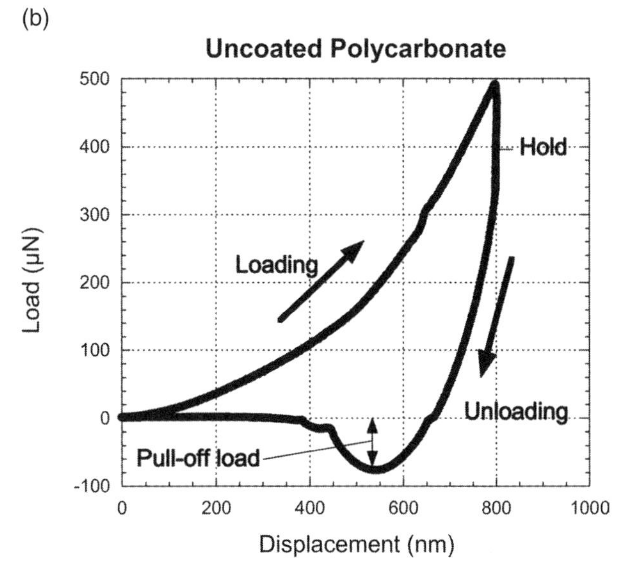

Figure 7.7 Load–displacement curves for (a) open-loop indents in acrylic at different loading rates and (b) a displacement control indent in polycarbonate. Without displacement control viscoelastic deformation leads to a nose in the unloading curve that generates a negative slope of the unloading curve at peak load. The extent of the nose depends on loading rate.

properties through the storage and the loss moduli of the sample. This is the so-called dynamic stiffness measurement (DSM) method.

The storage (E') and the loss moduli (E'') are

$$E' = \frac{k_s \sqrt{\pi}}{2\sqrt{A_C}} \tag{7.6}$$

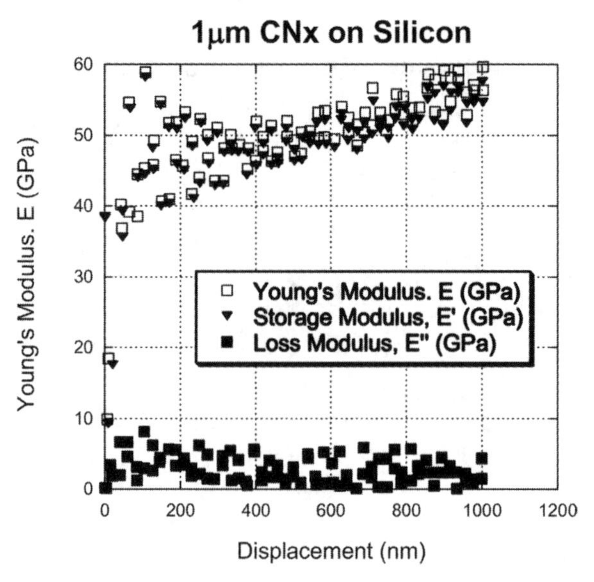

Figure 7.8 Variation of Young's modulus, storage modulus, and loss modulus with contact depth for CNx coatings on silicon.

$$E'' = \frac{\omega C_S \sqrt{\pi}}{2\sqrt{A_C}} \tag{7.7}$$

where k_s is the contact stiffness, C_S is the damping coefficient, and A_C is the contact area ($24.5\,h^2$ for an ideal Berkovich indenter but corrected for tip blunting by the method of Oliver and Pharr [3]). The stiffness and damping coefficient of the indenter head need to be determined by free oscillation of the head assembly at different frequencies in air. These are then coupled with the stiffness and damping coefficient of the contact to assess the material under test. Details of the approach can be found in Ref. [17].

Figure 7.8 shows the storage and loss moduli for a CN*x* coating on silicon compared to the conventional contact modulus (or reduced modulus E_r) determined from quasi-static measurements obtained on the same sample—it is expected that the conventional contact modulus should be given by the square root of the sum of the squares of storage and loss modulus. The loss modulus is low but increases toward the surface where the properties of the coating dominate the measured response. For more viscoelastic materials such as polypropylene, the loss modulus can be a more substantial proportion of the storage modulus.

7.3 COATINGS

The use of nanoindentation for the assessment of coatings is widespread and has been reviewed recently [18].

7.3.1 Contribution of the Substrate

As with the testing of bulk materials and distributed phases, considerable care has to be taken to ensure that the substrate does not contribute to the measurement if coating-only data is desired. A commonly reported rule, the substrate effect or so-called 10% rule, is where coating-only hardness can be measured if the depth of the indent is less than 10% of the coating thickness. Equation (7.3) can be used to assess the validity of this rule—for a typical hard TiN coating where $E = 600$ GPa and $H = 22$ GPa so that $E/H \sim 27$, the radius of the plastic zone is about four times the maximum displacement (i.e., the sum of elastic and plastic deflections of the test surface) and six times the contact depth of the indent (i.e., its plastic depth), which is consistent with the 10% rule.

Measurement of the elastic properties of a coating independent of the substrate is a much more difficult task since elasticity is a long-range property as discussed previously. Strictly speaking, the coating and substrate must be treated as springs in series, and there will always be a substrate contribution in the contact modulus results. However, as the coating thickness and the sharpness of the tip both increase, the contribution from elastic deformation in the substrate is reduced and eventually will be negligible. The precise conditions when this occurs are still a subject of ongoing research (eg. [19]). Factors affecting the critical indenter penetration for measuring coating properties have been recently reviewed [20].

7.3.2 Coating Plasticity

An extra concern in testing thin hard coatings on a softer substrate is whether the coating will plastically deform before the substrate. If the substrate deforms first, a visible impression will be formed, and the coating will be flexed into the impression and may fracture but is not likely to plastically deform significantly. In such circumstances a hardness modeling approach is necessary to determine anything meaningful about coating properties [7]. An estimate of whether the substrate will deform first can be obtained by considering the shear stress distribution below a Hertzian contact [21].

7.3.3 Modeling of the Hardness of Coating–Substrate Systems

It is straightforward to model the indentation of a coating–substrate system using modern finite-element codes if the mechanical properties of the coating are well known, but it is rather more difficult to extract such properties from indentation data if they are not known. The process is often iterative with many simulations run in an endeavor to match a given load–displacement curve. The solutions are not usually unique, and matching other data such as the shape of the pileup around an indentation may also be necessary. This complex and time-consuming process is not consistent with routine coating development program requirements so there is a need to extract useful information from

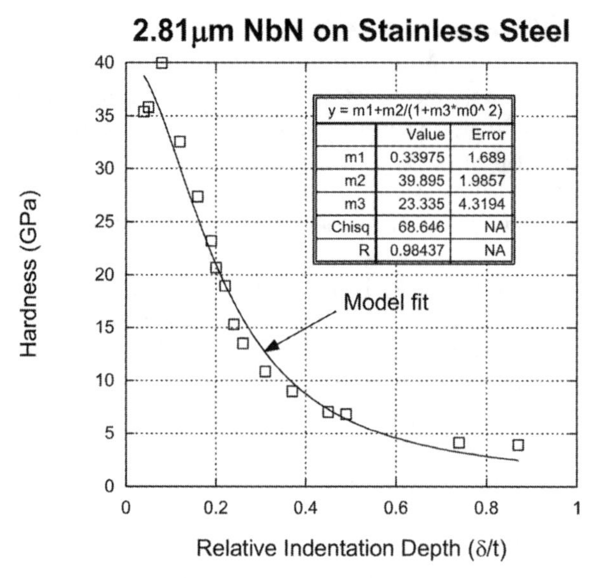

Figure 7.9 Variation of hardness with relative indentation depth for 2.81-μm NbN on stainless steel showing a fit of the model of Korsunsky et al. to the experimental data [29].

measured indentation data using relatively simple models. This usually involves fitting the variation of hardness or contact modulus with contact depth to determine a property for the coating that is independent of substrate effects.

Early models of composite hardness based on apportioning hardness contributions to different layers [22, 23] using deforming area [24] or deforming volume arguments [25] do not describe very well the available experimental data in most circumstances. However, in the volume law-of-mixtures hardness model introduced by Burnett and Rickerby [26, 27] and extended by Bull and Rickerby [28], it was realized that deformation of the harder material (coating or substrate) could constrain deformation in the softer one, reducing the deforming volumes in the softer layer and generally increasing the composite hardness. The use of this volume law-of-mixtures hardness model incorporating indentation size effects for both coating and substrate [28] allows a good description of experimental data but requires a complicated fitting approach. For this reason a simplified model based on work of indentation was developed by Korsunsky and co-workers [29, 30]. In its most simplistic version

$$H_c = H_s + \frac{H_f - H_s}{1 + k\beta^2} \tag{7.8}$$

where H_s is the substrate hardness, H_f the coating hardness, H_c the composite hardness, k is a constant, β is the relative indentation depth (δ/t) where δ is the indenter displacement and t is the coating thickness. Figure 7.9 shows the good quality of the fit of Eq. (7.8) to experimental data obtained by nanoindentation testing of a 2.8-μm physical vapour deposition (PVD) NbN coating on steel. The constant k in Eq. (7.8) controls how the hardness changes in the region where $\beta < 1$.

In order that a hard material constrain deformation in the softer one in contact with it, a strong interface is necessary to transmit stress form one layer to another. Therefore, the properties of the interfacial layer will control the hardness measured both in terms of the energy needed to deform this interface (effectively a measure of the amount of constraint) and the reduced energy associated with deforming a smaller volume of softer material (or increased energy deforming a greater volume of hard material if this were to occur). Any predictive model for hardness of the coating–substrate system must account for this.

When modeling coating–substrate system hardness, it has been customary to start with the basic definition of hardness, H, as a pressure, but an alternative but equivalent definition of hardness is [31]

$$H = \frac{W}{V} \tag{7.9}$$

where W is the plastic work of indentation and V is the deforming volume. Any mechanism that contributes to energy dissipation in the indentation cycle is automatically included in the work of indentation, which is just the sum of these contributions [7]. In instrumented indentation testers where load and indenter displacement are continuously monitored, W can be measured directly. However, a direct measurement of V is not possible, and, if the deforming volume is to be used as a basis for modeling, it needs to be related to something that is measurable, such as the indenter displacement. Since the deforming volume is approximately hemispherical, Eq. (7.2) can be used to determine V for bulk materials.

The modeling of a coating substrate system is relatively straightforward. Two interfaces must be considered. Thus

$$H = \frac{V_s H_0(s) + V_f H_0(f) + A_s \gamma_s + A_i \gamma_i}{V} \tag{7.10}$$

where V_f and V_s are the deforming volumes of coating and substrate, $H_0(f)$ and $H_0(s)$ are the bulk hardnesses of coating and substrate, A_s is the surface area, A_i is the interfacial area within the plastic deforming region, and γ_s and γ_i are the energies of deformation of the surface and interface, respectively. Given the plastic zone radius, R, Eq. (7.10) may be rewritten as

$$H = \left(\frac{3t}{2R} - \frac{1}{2}\frac{t^3}{R^3} \right) H_0^F + \left(1 - \frac{3t}{2R} + \frac{1}{2}\frac{t^3}{R^3} \right) H_0^S + \frac{3\gamma_s}{2R} + \left(\frac{3}{2R} - \frac{3t^3}{2R^3} \right) \gamma_i \tag{7.11}$$

Here the radius of the plastic zone is assumed to be the same in the coating and substrate. For a hard coating on a soft substrate, such as TiN on steel, the radius of the plastic zone in the substrate can be more that three times that in the substrate at the same test load. As plastic deformation expands from the coating into the substrate, there must be a

Table 7.2 Modeling Parameters for Coating Layers in Solar Control Coatings on Architectural Glass

Material	Contact Modulus, E_r (GPa)	Hardness (GPa)
SnO_2	131	14.0
ITO	133	12.0
Ag	83	2.4
ZnO	117	15.0
TiOxNy	122	9.0
Float glass substrate	79	6.1

rapid expansion of the plastic zone since at large loads the substrate will dominate indentation behavior. As the indenter penetration increases, there is a smooth growth of the plastic zone radius from the coating-dictated size to that of the substrate. Deviations from the hemispherical shape can be accounted for by the interfacial energy, γ_i.

This approach has been used to assess the properties of individual layers in a multilayer solar control coating stack on glass deposited by physical vapor deposition [31]. These coating stacks are typically based on a 10-nm silver layer surrounded by antireflection and barrier layers that are oxides and the total multilayer thickness is about 100 nm. To assess an individual layer by nanoindentation, it must be deposited onto the correct underlayers to a thickness of at least 100 nm to ensure that plastic deformation starts in the coating rather than the substrate when using commercially available indenters. The measured data must be corrected for these underlying layers using the above approach. Once values for the individual coating layers are known, the same modeling approach may be used to predict the variation of hardness, the penetration depth for the multilayer stack architecture, and this can be compared to experimental data. Good agreement was obtained using the parameters in Table 7.2. The hardness was found to be relatively constant with coating thickness [32], which implies that indentation size effects are not that important in these coatings. This is because most of the layers are completely amorphous. The exception for the oxide layers is ZnO, which shows some crystallinity, and the observed hardness of this layer increases as the coating thickness is reduced or the contact depth decreases. This fits with an indentation size effect model based on geometrically necessary dislocations or discrete deformation events [33].

The complexity of the processes occurring during the deformation of a coating–substrate system makes it difficult to produce a generic model for the behavior of the coating substrate system that can be used to extract coating properties from system property data. For this reason a simplified approach is adopted in ISO14577 Part 4. To determine coating properties from nanoindentation, the hardness and contact modulus is measured at a range of indentation loads and plotted as a function of contact depth. A linear fit to the data is extrapolated to zero contact depth to give the properties of the coating. This approach works well if there are no large changes in coating properties with contact depth—when the coating and substrate have very different properties, any results generated by this method should be treated with caution.

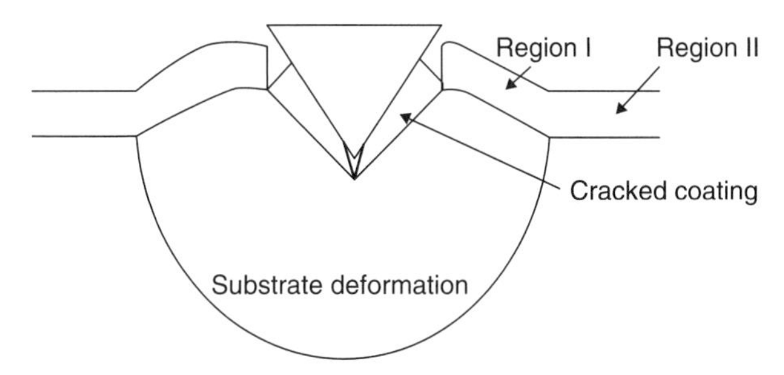

Figure 7.10 Schematic of the cross section of an indention with through-thickness cracking by a sharp tip. High tensile stress at the wedge tip causes fracture and the bending of the coating around the indenter causes failure at the indent edge. The upthrust of the coating and the plastic deformation of substrate is exaggerated in this figure.

This smooth transition can be modified by fracture of the coating caused by bending of the coating into the indentation caused by plastic deformation of the substrate. There are two types of fracture associated with indentations in brittle coated systems:

1. Through-thickness fracture, which generally runs normal to the coating–substrate interface for thin coatings but may run at a lower angle to it as the coating thickness increases. Through-thickness fracture is exacerbated by the bending stresses that arise once plastic deformation of the substrate occurs and the coating is bent into the impression or over the material piled-up around it.
2. Interfacial fracture, which occurs at or near the coating–substrate interface or the weakest interface in a multilayer stack. Interfacial detachment may occur around the impression during loading and in the contact zone during unloading.

Once fracture occurs, the properties of the coating–substrate system will be affected. For instance, during a through-thickness fracture event in a brittle coated system, the following changes of the mechanical response of the coated system may occur:

1. The stiffness of the coating decreases.
2. Plastic deformation of the substrate is more likely or even dominates.
3. Redistribution of the elastic and plastic strain may occur. The stored elastic energy in a cracked coating is released. Thus, during further deformation, this part of coating may be elastically deformed rather than continuing to plastically deform.
4. Any membrane stress is released.

Figure 7.10 displays the schematic of cross section of an indentation with through-thickness cracking by a sharp tip illustrating the processes that might occur. This has a critical influence on any technique to extract toughness from indentation data as is discussed in the next section.

7.3.4 Coating Fracture Due to Indentation with a Sharp Tip

When a brittle coating on hard substrate is indented by a very sharp tip, say a cube corner tip, radial cracks easily initiate at the indenter edges. When through-thickness cracking occurs, the compressive stress and membrane stress (in region I in Fig. 7.10) of the uncracked coating is released. The stiffness of the cracked coating and its adjacent uncracked coating decreases and more load is supported by the substrate. Thus plastic deformation of the substrate is likely to play an increasing role during further indentation. The cracked coating that was plastically deformed will still support some of the load imposed by the indenter; after the stored elastic energy is released by through-thickness fracture, it may be elastically deformed again prior to carrying on plastic deformation during further indentation. After through-thickness cracking the bending stress caused by conforming to the substrate will also be released. The wedging effect for a sharp tip will enhance the opening displacement of radial cracks.

Figure 7.11 shows the comparison of load–displacement (P–δ) curves between a 400-nm TiO_xN_y single layer on glass tested under displacement control with peak displacements of 100 and 400 nm. There is no evidence for fracture at the lower penetration either in the load–displacement curve [Fig. 7.11(a)] or in an AFM scan of the indent [Fig. 7.11(b)], but clear load drops are visible in the load–displacement curve for the deeper penetration [Fig. 7.11(c)]. There is also some evidence of uplift next to the bigger indent [circled in Fig. 7.11(d)], which is evidence for coating detachment since this material does not show appreciable pileup and the substrate shows neither significant plastic deformation nor the significant constraint of plastic deformation in the coating, which would enhance what pileup there is.

For very low load tests, especially for tests that intend to eliminate the substrate influence, it can be expected that the bending effect imposed by the plastic deformation of substrate is not significant. Further, given the sharp geometry of the cube corner tip (with a small tip radius as well), the high stress intensity should result in the appearance of radial cracking. Also, it has been shown that the wedging effect is most significant for an acute probe [34], which makes the radial crack rather than picture frame crack more likely to occur.

Various methods to extract fracture toughness and adhesion information from nanoindentation have been developed, and these have been recently reviewed [32, 35, 36]. The energy dissipated in fracture of the coatings can be assessed from the load drops in Figure 7.11, and the crack driving force and fracture toughness can be determined from this data if the size of the crack and the Young's modulus of the coating is known [37]. For radial fracture, the crack horizontal dimension is approximately equal to the indentation radius (no evidence of extended radial cracking was observed by in situ AFM) and the vertical dimension is equal to the coating thickness. Since there are three indenter edges, this area must be multiplied by 3 to get the total crack area, which assumes uniform fracture around the indentation.

The energy dissipated in fracture is determined from a plot of work of indentation as a function of displacement during loading. There are discontinuities in this plot when fracture events occur, and a simple estimate of the fracture-dissipated

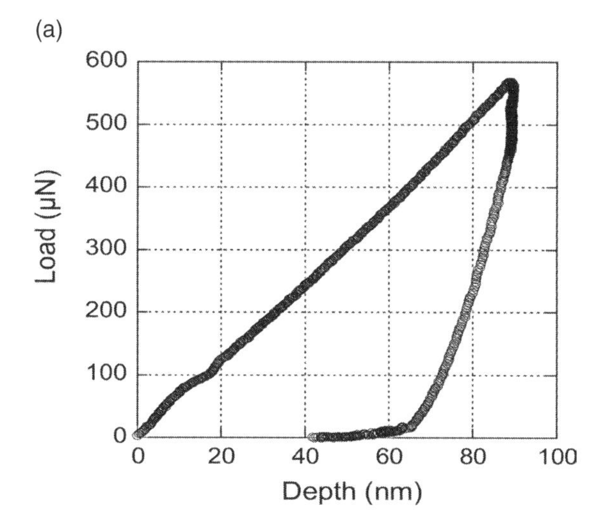

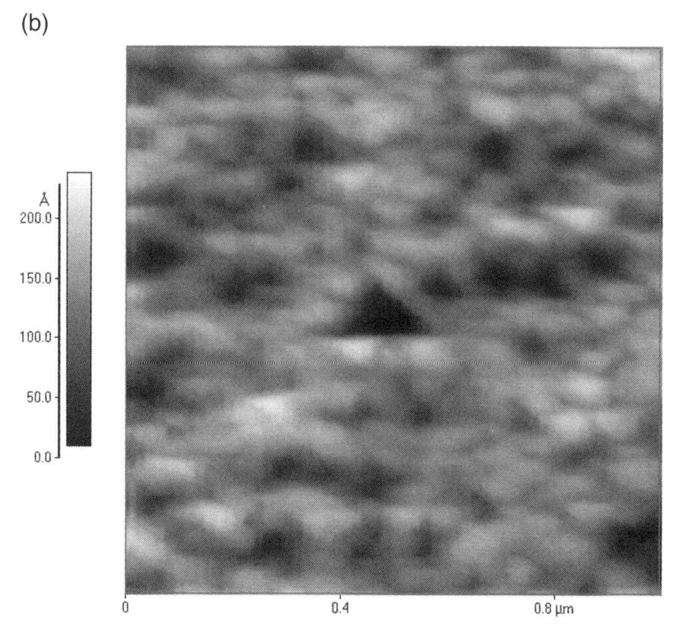

Figure 7.11 (a, c) Load–displacement curves and (b, d) AFM images of a 400-nm TiO_xN_y coating on a glass substrate (a, b) without cracks and (c, d) with cracks, respectively. The coating was indented by a cube corner tip under displacement control at 100-nm maximum displacement (a, b) and 400-nm maximum displacement (c, d). The circle in (d) marks an area of uplift associated with through-thickness and interfacial fracture. In (c), position A is regarded to be the position where plastic deformation extends to the softer substrate, which is reasonable since the estimated plastic deformation zone radius exceeds 400 nm; points B and C are the start point and endpoint of through-thickness cracking, respectively; D and E are the start point and endpoint of interfacial fracture.

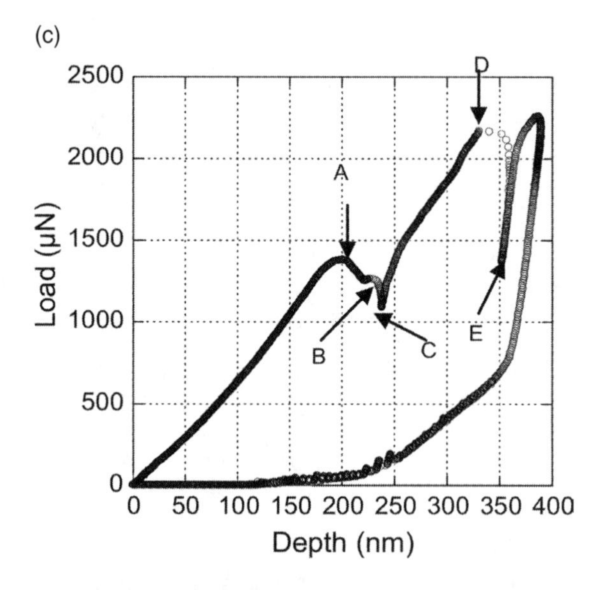

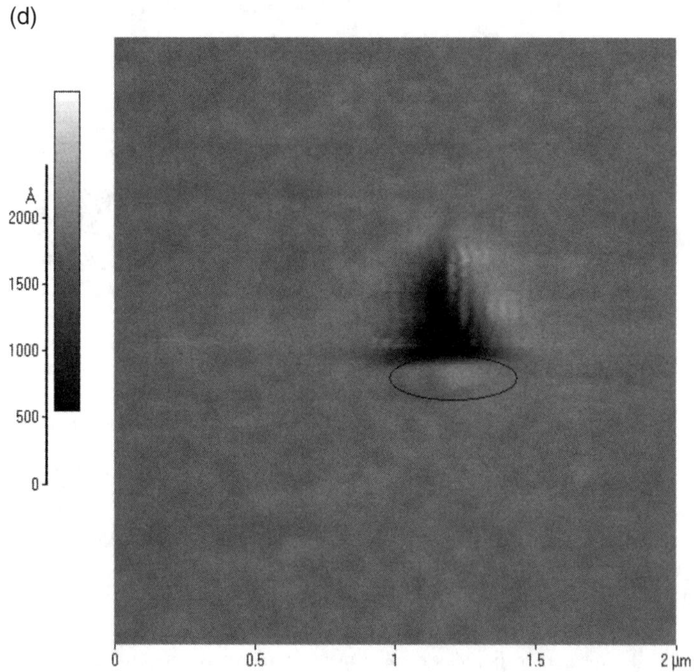

Figure 7.11 (*cont'd*)

energy is given by the size of this discontinuity. However, the fracture of the coating changes the manner in which energy is dissipated in the plastic deformation of the coating substrate system, and this must be accounted for in any calculation [37]. The method to determine fracture-dissipated energy is explained in Figure 7.12. First, the total work versus displacement curve W_t–dp is extrapolated from the

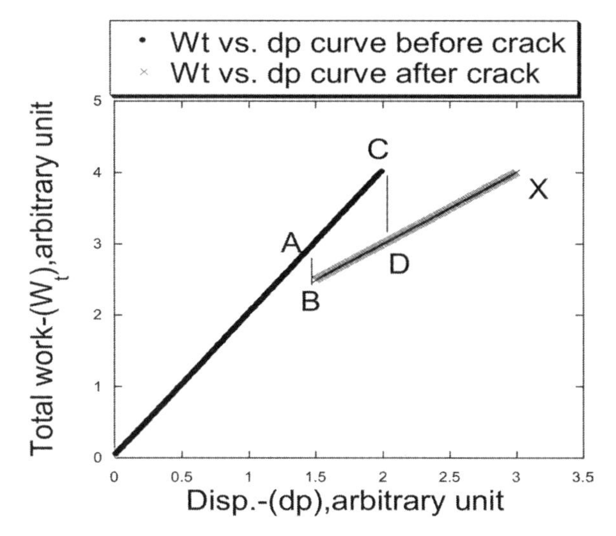

Figure 7.12 Schematic of the extrapolated total work vs. displacement curve before and after cracking to determine the fracture dissipated energy CD-AB.

cracking start point *A* to the cracking end point *C*, to get the work difference *CD* after fracture; then the W_t–dp curve after cracking is extrapolated backward to the cracking start point, thus obtaining the work difference *AB* at the onset of fracture. *AB* represents the work caused by different elastic-plastic deformation behavior of the material before and after fracture, whereas *CD* represents the total work difference caused by the presence of cracking, which consists of the change of elastic-plastic deformation behavior between the uncracked system and cracked system plus the fracture-dissipated energy (not including contribution from relaxation of residual stress in the coating). The difference between the two (i.e., *CD* minus *AB* in Figure 7.12) will be the fracture-dissipated energy. This method has been successfully applied in many different coated systems [37–39].

From Table 7.3, it can be seen that the results from this approach are reasonable for several brittle coatings on glass, whereas the conventional indentation fracture method (CIM) developed for bulk materials just returns the toughness of the substrate glass in most cases.

Applying these methods to coatings that make up the individual layers in solar control coatings demonstrates that for coatings thicker than 100 nm there is no size effect in fracture toughness and [32]. This is due to the fact that the defect distribution in the coatings probably controls fracture behavior.

7.3.5 Nanoindentation in Industrial Coating Applications

Nanoindentation is widely used in industrial research and development in the coatings industry and for statistical process control testing in some specific applications (e.g., in the hard disk industry). In some cases it is being used to provide

Table 7.3 Energy Release Rates and Toughness Values Calculated for the Solar Control Coating Components Investigated in This Study Based on the Radial Through-Thickness Fracture

	Energy Release Rate of Coating G_{IC} (J/m^2) Calculated by the W_t–dp Model	Toughness of Coating K_{IC} (MPa$\sqrt{m}$)	
		W_t–dp Model	Estimated by CIM
400 nm TiO_xN_y top layer, single-layer stack	24.4 ± 1.4	1.8 ± 0.2	0.9 ± 0.1
240 nm ITO top layer, mutilayer stack	36.3 ± 8.2	2.2 ± 0.3	0.9 ± 0.1
400 nm ITO top layer, mutilayer stack	32.7 ± 4.4	2.1 ± 0.2	0.7 ± 0.1
400 nm TiO_xN_y top layer, mutilayer stack	24.1 ± 7.8	1.8 ± 0.2	1.0 ± 0.1
400 nm SnO_2 top layer, mutilayer stack	29.3 ± 9.8	1.9 ± 0.3	1.3 ± 0.1

design data when there is no other way to obtain the measurement (e.g., determining the elastic properties of small-molecule layers used in organic light-emitting diodes that do not exist in the bulk state). The use of nanoindentation is usually required when the coating thickness is of the order of a few microns or less as it enables the measurement of coating-only properties according to the procedures laid down in ISO14577 Part 4. To achieve good coating properties the thickness of the coating must be known and measurements should be made at a range of contact scales (peak loads) such that the contact depths of the impressions are substantially less than the coating thickness. The ISO method requires that a linear extrapolation of the measured data be made to zero contact depth to get the properties of the coatings. For plastic properties it is usually assumed that the indenter penetration must be less than 10% of the coating thickness, whereas for elasticity the penetration must be much smaller, and it may not be possible to get any measured values that are not influenced by the substrate in some way.

Practically this approach works very well for hard ceramic coatings on softer metallic substrates (e.g., TiN on steel) and can also give reasonable results for soft metallic coatings of harder substrates such as copper on silicon. One key consideration is the surface roughness of the coating, which can be too large in the as-processed state leading to a large scatter in the measured data. In such cases polishing the coating to achieve a roughness of less than 50 nm can dramatically improve the quality of the data—this approach is necessary in the case of arc-evaporated TiN coatings where macrodroplet formation can dramatically increase experimental scatter. If such precautions are undertaken, the nanoindentation technique can be used to assess the effects of coating architecture (i.e., composition and layer arrangements) on properties and highlight issues of poor process control—for example, see [40] for an assessment of commercial TiCN coatings.

When testing metallic or ceramic coatings, it is usual to use time alone to control the indentation cycle and no feedback control systems are used. For some polymer coatings such as the lacquers on food cans or the protective layers on the cases of consumer electronics, this is not sufficient since time-dependent deformation is an issue (viscoelasticity). In such cases the use of displacement control can greatly increase the reliability of the data produced and makes it sufficient for process development or statistical process control (SPC) purposes. The added advantage of the displacement control process is that the unloading process is also very well controlled, and accurate measurements of pull-off forces are possible at the end on the unloading cycle [Fig. 7.7(b)], which enables an assessment of the stickiness of surfaces processed in different ways to be made at the same time as assessment of the mechanical properties of the material.

The precision stages associated with most nanoindentation systems allow the controlled positioning of indentations at particular positions within a component. This is particularly useful for failure analysis studies in complex systems such as microelectronic devices. For instance, it is possible to accurately position indentations on bonding pads for the connecting wires to semiconductor chips—these are often multilayer coated structures, and the nanoindentation technique can be used to assess if the mechanical properties of individual layers meet the desired specification. In some cases the indentation may be performed at high enough loads to generate through-thickness fracture or interfacial failure, which are key factors in the mechanical failure of the wire binding joint. Mechanical stages can generally position indentations with an accuracy of 1 μm, but higher precision can be achieved by mounting the nanoindentation head on an AFM scanner; with such systems it is possible to place indentations in metallization lines in the semiconductor industry that are only 200 nm in width. Positioning at the micron level of accuracy is also useful in other industries such as when testing printed metal heating elements or sensor structures.

When feature sizes are very small, it is not only the accuracy of positioning that is important but also the scale of deformation associated with the indentation test. For instance, when testing thin coatings, the contact depth should be less than 10% of the coating thickness if possible, but the indent should also be positioned at a suitable distance from the edge of the coated feature. For very thin films (<200 nm) the tip end radius of the diamond indenter becomes critical. Commercially available indenters typically have a tip end radius of greater than 100 nm and even with very careful calibration may not give reliable results at contact depths less than 20 nm, making determining of coating only properties almost impossible without using the ISO14577 extrapolation method. It is essential to ensure that some plastic deformation occurs at this small contact scale if the Oliver and Pharr method is to be used for assessment. Furthermore, if coating properties are to be assessed, it must be clear that some plastic deformation has occurred in the coating—when depositing hard ceramic coatings on polymeric substrates, this may not be the case and some modeling of the contact is necessary if good data is to be produced. An example of this is in the thin alumina barrier layers, which may be used to protect organic light-emitting diodes on polymer substrates [41].

Increasingly, scratches and wear tests are being performed using nanoindentation systems because they allow precise control of scratch and wear scar dimensions and the possibility for assessing the topography and extent of damage when compared to more traditional tribological tests. For instance, it is possible to simulate the types of scratches the are produced on coated architectural glass during production and delivery by performing scratches in the nanoindenter. In this case it is necessary to change the indenter from diamond, which is generally used, and both glass shards and polymer balls have been used to simulate observed field behavior. In fact, it is the scratches from the poly methyl methacrylate (PMMA) spheres that are sprayed onto the surface, leading to coating detachment that is the most important failure mode in solar control coatings on glass deposited by physical vapour deposition [42].

7.4 CONCLUSIONS

Nanoindentation studies can be used to determine the mechanical properties of bulk materials, coatings, and interfacial phases. Direct measurement of plastic properties at high spatial resolution is possible, but the assessment of elastic properties is more difficult due to the longer range of elastic stress distributions. For this reason a modeling approach is necessary to fully understand elastic behavior. Fracture toughness can be determined from the energy dissipated during indentation if well-defined fracture events can be associated with features in the nanoindentation load–displacement curve.

ACKNOWLEDGMENTS

Many of my former Ph.D. students have contributed to this work including Jinju Chen, Eva Gutierrez-Berasetegui, Isabel Arce-Garcia, Adrian Oila, Sorin Soare, Jon Tuck, Martin McGurk, Reza Rastegar Tohid, and Krishna Belde.

REFERENCES

1. Wolf, B. *Cryst Res Technol* **35** (2000): 377–399.
2. Fischer-Cripps, A. C. *Nanoindentation*, Springer: New York, 2002.
3. Oliver, W. C., and Pharr, G. M. *J Mater Res* **7** (1992): 1564–1583.
4. Pharr, G. M. *Mater Sci Eng* **A253** (1998): 151–159.
5. Cheng, Y. T., and Cheng, C. M. *Mater Sci Eng* **R44** (2004): 91–149.
6. Lawn, B. R., Evans, A. G. and Marshall, D. B. *J Am Ceram Soc* **63** (1980): 574–581.
7. Bull, S. J. *J Vac Sci Techno* **A19** (2001): 1404–1414.
8. Johnson, K. L. *J Mech Phys Sol* **18** (1970): 115–126.
9. Chen, J., and Bull, S. J. *J Mater Res* **21** (2006): 2617–2627.
10. Oila, A., Bull, S. J., Shaw, B. A., and Aylott, C. J. *J Eng Tribology (Proc IMechE Part J)* **219** (2005): 77–83.
11. Lu, L., Schwaiger, R., Shan, Z. W., Dao, M., Lu, K., and Suresh, S. *Acta Materialia* **53** (2005): 2169–2179.
12. Sneddon, P. H., Bull, S., Cagnoli, G., Crooks, D. R. M., Elliffe, E. J., Faller, J. E., Fejer, M. M., Hough, J., and Rowan, S. *Classical and Quantum Gravity* **20** (2003): 5025–5037.

13. ANSYS Inc., Canonsburg, PA, 1999.
14. Liu, J., Mwanza, M. C., Reed, P. A. S., and Syngellakys, S. *ANSYS Application Note* (2002).
15. Oila, A., and Bull, S. J. *Zeitschrift fur Metallkunde*, **94** (2003): 793–797.
16. Šiller, L., Peltekis, N., Krishnamurthy, S., Chao, Y., Bull, S. J., and Hunt, M. R. C. *Appl Phys Lett* **86** (2005): 221912.
17. Syed, Asif, S. A., Wahl, K. J., and Colton, R. J. *Review of Scientific Instruments* **70** (1999): 2408–2413.
18. Bull, S. J. *J Phys D: Appl Phys* **38** (2005): R393–R413.
19. Bull, S. J. *J Vac Sci Technol* **A30** (2012): 01A1690.
20. Chen, J., and Bull, S. J. *Vacuum* **83** (2009): 911–920.
21. Johnson, K. L. *Contact Mechanics*, Cambridge University Press: Cambridge, 1985.
22. Buckle, H. *Metal Rev* **4** (1959): 49–100.
23. Buckle, H. in *The Science of Hardness Testing and Its Research Applications*, Westbrook, J. H., and Conrad, H., Eds., ASME: Metals Park, OH, 1973 pp. 453–494.
24. Johnson, B., and Hogmark, S. *Thin Solid Films* **114** (1984): 257–269.
25. Sargent, P. M. Ph.D. thesis, University of Cambridge, 1979.
26. Burnett, P. J., and Rickerby, D. S. *Thin Solid Films* **148** (1987): 41–50.
27. Burnett, P. J., and Rickerby, D. S. *Thin Solid Films* **148** (1987): 51–65.
28. Bull, S. J., and Rickerby, D. S. *Surf Coat Technol* **42** (1990): 149–164.
29. Korsunsky, A. M., McGurk, M. R., Bull, S. J., and Page, T. F. *Surf Coat Technol* **99** (1998): 171–183.
30. Tuck, J. R., Korsunsky, A. M., Bull, S. J., and Elliott, D. M. *Surf Coat Technol* **127** (2000): 1–8.
31. Stillwell, N. A., and Tabor, D. *Proc Phys Soc* **78** (1961): 169–178.
32. Bull, S. J. *Comptes Rendus Mecanique* **339** (2011): 518–531.
33. Nix, W. D., and Gao, H. *J Mech Phys Sol* **46** (1998): 411–425.
34. Morris, D. J., and Cook, R. F. *International Journal of Fracture* **136** (2005): 237–264; 265–284.
35. Chen, J., and Bull, S. J. *J Phys D: Applied Phys* **40** (2007): 5401–5417.
36. Chen, J., and Bull, S. J. *J Phys D: Applied Phys* **44** (2011): 034001.
37. Chen, J., and Bull, S. J. *Thin Solid Films* **494** (2006): 1–7.
38. Chen, J., and Bull, S. J. *Mater Res Soc Symp* **890** (2006): Y0202.
39. Chen, J., and Bull, S. J. *Thin Solid Films* **517** (2009): 2945–2952.
40. Bull, S. J., Bhat, D. G., and Staia, M. H. *Surf Coat Technol* **163–164** (2003): 499–506.
41. Bull, S. J. *J Vac Sci Technol* **A30** (2012) 01A1690.
42. Belde, K. J., and Bull, S. J. *J Adhesion Sci Technol* **22** (2008): 121–132.

Chapter 8

Scanning Probe Microscopy for Critical Measurements in the Semiconductor Industry

Johann Foucher

LETI–CEA, Grenoble, France

8.1 INTRODUCTION

The main objective of this chapter is to focus on the measurement of critical dimensions (CD) for the semiconductor industry. During electronic device fabrication, a CD measurement step occurs after each technological step that may induce a morphological modification of the fabricated patterns. With shrinkage of dimensions following Moore's law, metrological steps are becoming increasingly critical, and CD measurement accuracy is currently a major concern in semiconductor manufacturing and quality control. Consequently, intrinsic limits of each metrological technique, likely to provide dimensional information on manufactured patterns, must be known.

In this chapter we will first define the critical parameters that must be controlled by the semiconductor industry in order to guarantee sufficient yield and process quality in device manufacturing. In the second part, we will briefly introduce and define the limitations of the two main production techniques [CD scanning electron microscope (CD-SEM) and scatterometry: optical CD measurement] that are currently used in production for CD process control. Then, we will introduce and discuss the interest of using scanning probe microscopy (SPM) for CD purpose in the semiconductor industry, originally developed in the 1990s [1] to complement conventional techniques. Last but not least, we will present a new concept of CD metrology, so-called hybrid metrology, which is supposed to integrate SPM technology into the global production process control schematic in order to push away the current limitations of conventional CD techniques.

Scanning Probe Microscopy in Industrial Applications: Nanomechanical Characterization, First Edition. Edited by Dalia G. Yablon.

8.2 CRITICAL DIMENSION IN THE SEMICONDUCTOR INDUSTRY

What is CD? CD is the *critical dimension* of an object made on a given substrate. For the most advanced devices, the substrate is a 300-mm wafer made of crystalline silicon. Currently, the next generation of wafers that is scheduled for the industry to reduce costs is a 450-mm-diameter wafer. The fabricated patterns on substrates can have any kind of shape as a function of the device design. For each pattern we can associate a critical dimension as shown in Figure 8.1 which shows a typical cross section view of three lines. Therefore, we can characterize different nanoscale objects associated with their own CD. Usually, the CD is associated with the desired feature dimensions along all three dimensions required to obtain the right device performances. Typical devices already studied or produced for advanced electronic devices include a high-k metal gate, nanowires, magnetoresistive random-access memory (MRAM), and FinFet devices; we are entering into a new era of complexity for material stack, device functionalities, multiple technological steps, and subsequently more complicated CD metrology steps.

We have entered a new metrology area that is so particular that a complicated technological schematic must be used to address the semiconductor industry roadmap requirements. As shown in Figure 8.2, we must address in parallel the requirements of device dimension shrinkage, pattern roughness reduction (because the absolute value is no more negligible compared to the average pattern CD), and fabrication of two- (2D) and three-dimensional (3D) devices. From such a technological schematic, we can infer another schematic associated to CD metrology that consists of being able to develop a

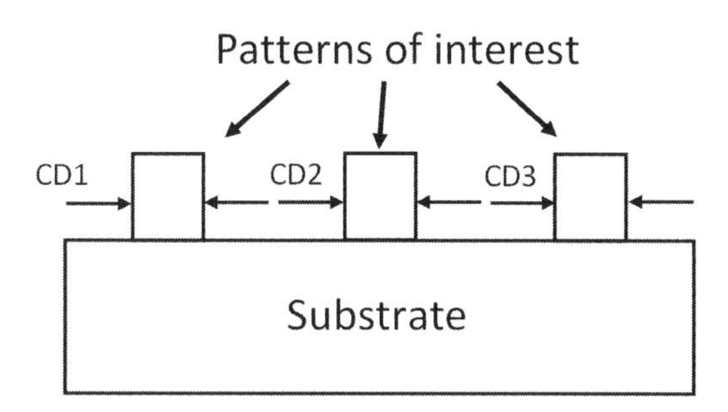

Figure 8.1 Definition of the critical dimension of line structure.

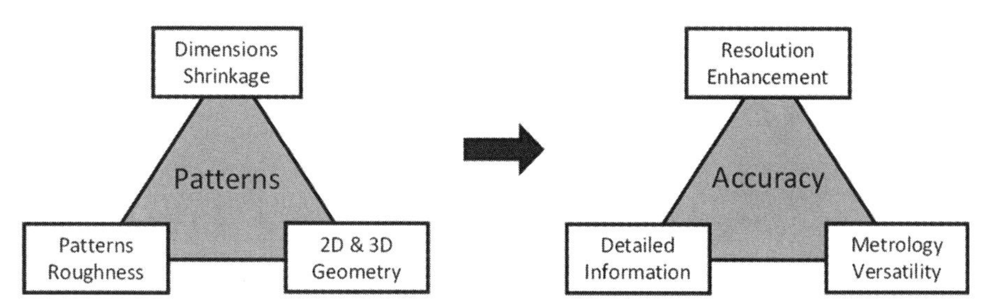

Figure 8.2 Technological challenges impact on CD metrology step.

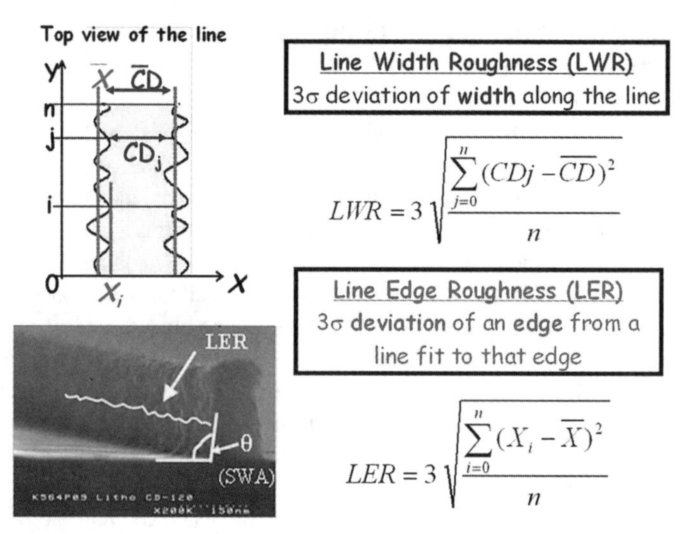

Figure 8.3 Line edge roughness (LER) and LWR (line width roughness), sidewall angle (SWA) definitions.

metrology protocol to address in parallel the accuracy requirements of resolution enhancement, detailed information measurements, and therefore the capability to measure one-dimensional (1D), 2D, and 3D patterns, so-called metrology versatility.

Such parallel challenges lead to obsolescence of traditional CD measurement parameters (as shown in Fig. 8.1, i.e., a single dimension value) and subsequently to move on to multiple parameters measurements as shown in Figure 8.3. These include, for example, a pattern's height (in the range of 100 nm), a pattern's sidewall angle (SWA) that could vary as a function of applications (tapered, vertical, or reentrant profiles), sidewall roughness along pattern's height, and line edge and line width roughness (LER, LWR) in the range of one or several nanometers. Moreover, CD at different height locations must be perfectly controlled within 1 nm of accuracy for the most aggressive devices. If such requirements are not met, production yields will degrade and research and development (R&D) cycle time will continue to increase over future technological nodes, which corresponds to each device generation over an 18-month period. As a result, since nanometer accuracy in three directions is required for both R&D and production, we can legitimately say that there is a key role for SPM to play.

8.3 CD METROLOGY TECHNIQUES FOR PRODUCTION

8.3.1 CD-SEM

Scanning electron microscopy is a powerful technique to observe surface topography. SEM is mainly based on detection of secondary electrons emerging from the studied surface when a very narrow beam of primary electrons scans it. Images with resolution usually smaller than 3 nm and a large depth of field may be obtained.

An in-line version of SEM for production, or so-called CD-SEM, consists of top view analysis of individual or multiple patterns across a wafer at low voltage (typically in the range of 300–800 V as a function of materials type). Due to this nondestructive behaviour, CD-SEM is compatible with production environment and with an in-line fabrication process control. The major advantages of CD-SEM technique are: (1) easy to implement, (2) repeatable (currently in the range of 5 Å to 1 nm at 3σ), and (3) well-known technique for decades. CD-SEM requires specific recognition patterns to locate the patterns to be measured in order to accurately and repeatably position the beam. From the corresponding intensity profiles for each measured patterns, typical CD-SEM algorithms apply filters, derivatives, and thresholds in order to extract related CD for each pattern of interest.

Currently, CD-SEM equipment is limited in accuracy because it is calibrated using only pitch standards (which are enough for accurate pitch measurements) where pitch represents a period of a repeated nanostructure made of any material type. However, SEM is not calibrated with CD standards, which is nearly impossible since one specific standard is needed for each material–height–CD configuration (leading to an almost infinite number of standards!). Therefore, the main drawback of CD-SEM technology is the limitation of top-view analysis, which hides major information coming from the pattern's sidewalls. Though the information is present, unfortunately, it is integrated in the global intensity profile so that conventional algorithms are unable to separate accurately secondary electron contributions coming from different areas of interest such as top CD, sidewall angle, bottom CD, and height. Consequently, it is impossible to get accurate information on a pattern's height, sidewall angle values, or pattern's footing information. We will see later in this chapter that an alternative solution can potentially overcome the current CD-SEM technology limitations.

8.3.2 Scatterometry

Scatterometry is an indirect optical method to measure geometrical parameters of diffraction gratings. The patterns of interest are repeated periodically on a plane substrate. The technique is based on theoretical analysis of light diffracted by the measured objects. It analyzes the optical reflectivity and compares it to the theoretical reflectivity computed out from a grating model. This technique has been used since the 1980s [2]. As explained previously in the chapter, more stringent requirements on CD metrology have pushed the development of this technique in order for it to be applied to the semiconductor industry [3] to overcome CD-SEM limitations. Scatterometry enables 2D characterization of gratings (height, CD, SWA), and it is faster and more reproducible than CD-SEM. Consequently, the technology has been well received by the industry.

Although scatterometry has been quickly introduced at the manufacturing level, the technique is suffering from major drawbacks including strong limitations for advanced devices fabrication process control. Here, we have listed some major current issues or challenges:

- The scatterometry technique is intrinsically very sensitive to the material stack configuration under the patterns of interest, which means that with the increasing complexity of material stack, modeling development time and validation is drastically increasing and uncertain without any calibration data coming from other technologies [e.g., transmission electron microscopy (TEM), SPM, etc.].
- The introduction of metal (Ti, W, Hf, etc.) in CMOS transistor gate stack degrades measurement quality because of high reflection index.
- Although, the theory says that scatterometry technique can measure independently various parameters (angle, height, CD), the reality does not fully confirm it. Indeed, as a function of tool optics configuration (ellipsometry, reflectometry, goniometry), the parameters of interest are always more or less correlated. Moreover, the correlation will be a function of material stack, dimensions, optical setup, and so forth. The immediate impact is the uncertainty of the global process control, which degrades and therefore does not guarantee sufficient reliability as a function of dimension shrinkage node after node.

8.3.3 CD Metrology Current Status

Subsequently, the introduction of new material stacks, more sophisticated design rules, and complex 3D architectures in semiconductor technology has led to major metrology challenges by posing stringent measurement precision and accuracy requirements for CD, feature shape, and profile. Current CD metrology techniques used in production such as CD-SEM and scatterometry have their inherent limitations that must be overcome to fulfill advanced roadmap requirements. The approach of hybrid automated CD metrology seems necessary. Using multiple tools in unison is an adequate solution when adding their respective strengths to overcome individual limitations. In a more mathematical vision, a data fusion algorithm to gather information—first to create new information that will be more efficient and finally achieve much more accurate inferences—has great appeal. Such a solution should give the industry a better nanocharacterization solution than the conventional approach. However, to be successful, such an approach must include consideration of accuracy (through the notion of true value) that can be possible only with the introduction of reference CD metrology. We will therefore introduce in the next section the CD atomic force metroscopy (AFM) technology (SPM technique) dedicated to 3D measurements of advanced patterns for the semiconductor industry.

8.4 OBTAINING ACCURATE CD IN THE SEMICONDUCTOR INDUSTRY

The concept of measurement uncertainty (MU) and guidelines of its implementation have been introduced and developed in the early 1990s [4–6]. The *ITRS Metrology 2007* edition has adopted the concept and changed metrology performance

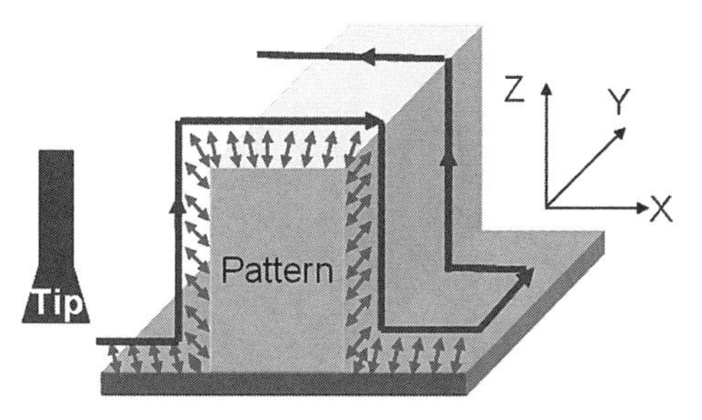

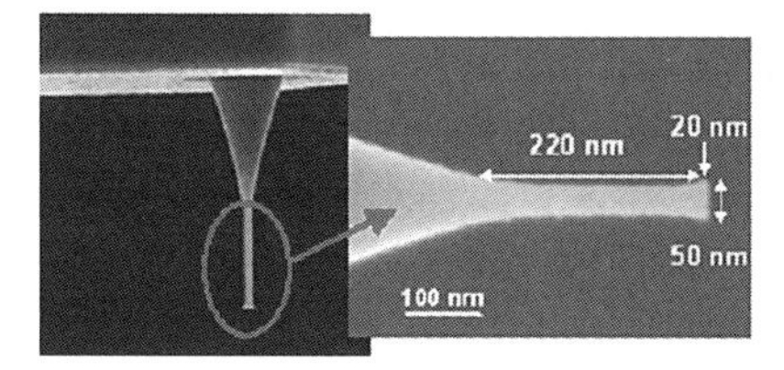

Figure 8.4 CD-AFM basic principle.

metric from precision to uncertainty [7] The new metric, MU, has precision as one of the many uncertainty components. This is a deep change in the way of thinking, proposing, and developing nanometrology solutions for the future of the semiconductor industry.

8.4.1 Three-Dimensional SPM

Atomic force microscope scanning in *CD mode* using a flared probe (cf. Fig. 8.4) is capable of generating a 3D profile of the scanned 3D patterned sample in a nondestructive manner and through piezotube usage. This is also referred to as CD-AFM (or 3D-AFM), and the tool is capable of making measurements at precise locations (typically within 200 nm of precision) on a range of samples ranging from a small sample up to 450 mm industrial wafers for the next generation of electronic devices. This technique provides valuable information about the sample such as the depth of trenches, height of a line, physical CD of a trench, or a line and sidewall angle. The 3σ precision of the tool and technique is subnanometer for CD and height measurements and subdegree for the sidewall angle measurement. These attributes of the technique make it a preferred reference metrology technique in the semiconductor industry [8, 9]. CD-AFM can be used to calibrate workhorse metrology tools such as scatterometry and CD-SEM. The CD-AFM measurement process is much slower compared to scatterometry and CD-SEM, and it is therefore more suitable for calibrating these workhorse metrology techniques or other

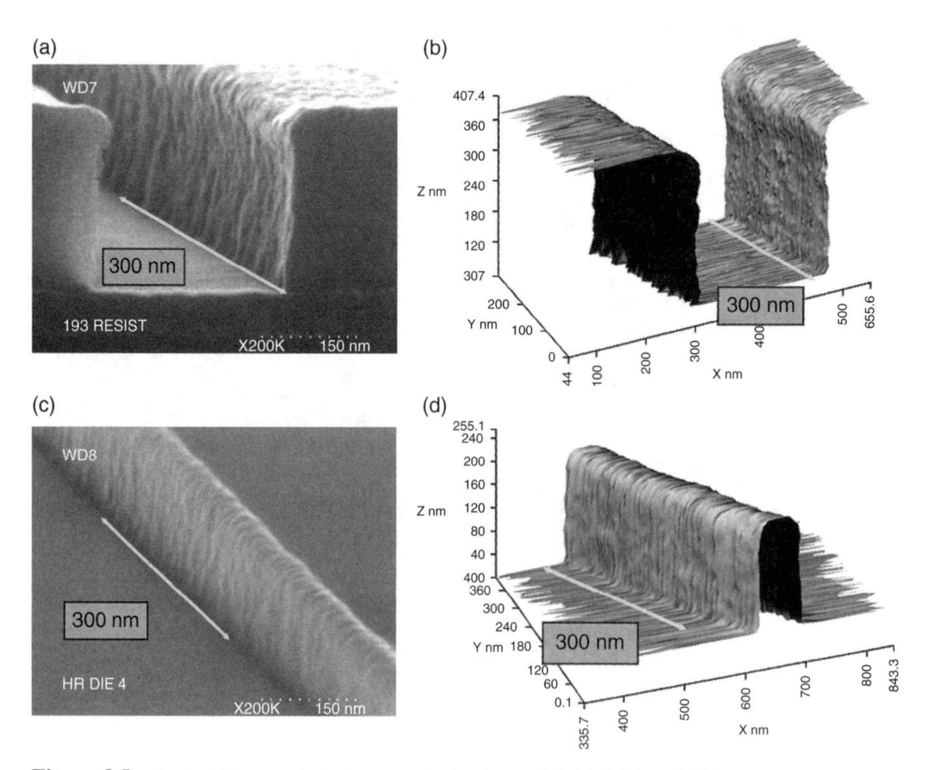

Figure 8.5 Typical high-resolution images obtained with CD-AFM (b) and (d) in comparison to SEM images (a) and (c).

future techniques. However, in a certain way, where CD-SEM and scatterometry are limited, the CD-AFM technique can replace them in a high-volume manufacturing environment if the periodicity of the measurements are smartly tuned (specific protocols have to be defined in such cases). CD-SEM measurement accuracy of resist patterns is compromised due to the well-known phenomenon of electron-beam-induced shrinkage and electron proximity effect that start to appear mostly below 100 nm of the trenches dimensions. CD-AFM measurement conducted prior to CD-SEM measurement of resist pattern can correct for the resist shrinkage occurring during the CD-SEM measurement. CD-AFM can also take into account and correct for electron proximity effects [10–12]. As shown in Figure 8.5, the main advantage of CD-AFM is the true nondestructive three-dimensional capabilities, which enable the method to be introduced in the semiconductor industry at the production level and as a reference technique for all materials, which is not the case today. However, measurement success of the CD-AFM significantly depends upon the probe geometry, that is, its size and shape. Figure 8.6 shows key parameters of the CD-AFM probe or tip. These include tip width (C), tip edge radius (D), lateral reach (B), and effective length (A).

Pattern dimensions are scaling with the successive technology nodes, and scaling of space dimension is of particular interest for the success of CD-AFM measurements.

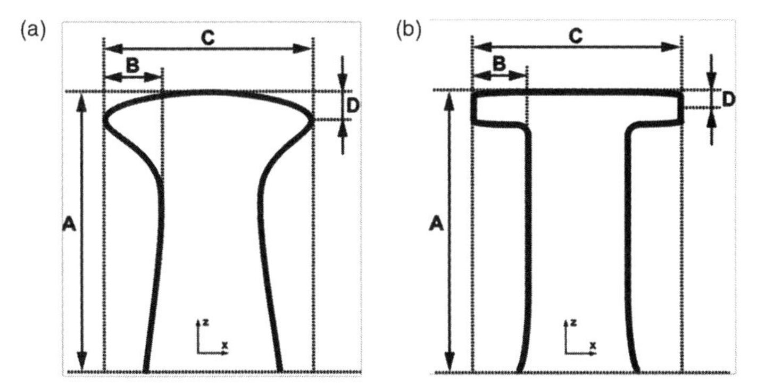

Figure 8.6 Schematic illustration of tip design and parameters of a typical flared silicon-based (a) and an EBIP-fabricated T-shape-like CD-AFM tip made of HDC/DLC (b), where *A* is the effective tip length, *B* is the lateral reach overhang, *C* is the tip width, and *D* represents the vertical edge height.

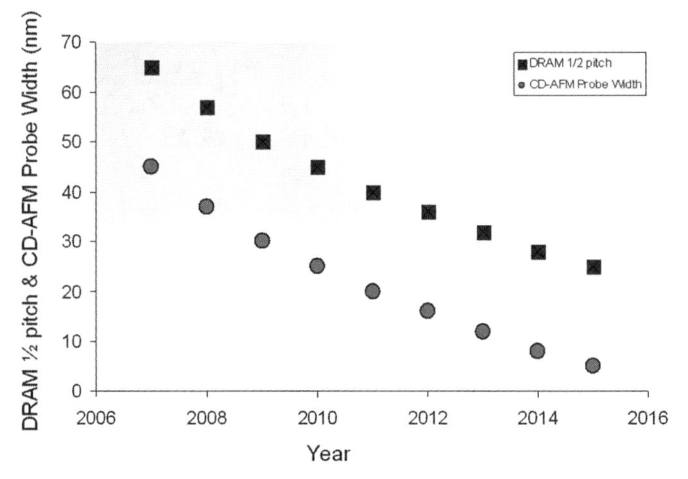

Figure 8.7 Roadmap outlook for CD-AFM probe width.

In a state-of-art-the CD-AFM tool, the CD-AFM probe requires a feature spacing of at least 15–20 nm apart from its diameter to be able to scan productively in CD mode. This dependence requires a scaling of CD-AFM probe with the technology as shown in Figure 8.7. A desired probe should have small tip width, small edge height, customized lateral reach, and effective length as a function of patterns of interest. These are opposing requirements and pose a fabrication challenge in the scaling of CD-AFM probes. As the CD-AFM probe width scales down, the mechanical strength of the probe also degrades. Commonly, the probes are made of silicon. Scanning patterns on an etched wafer wears the probe fast and therefore requires a carbon coating [13]. The carbon coating process adds to the probe width by 5–10 nm. Figure 8.7 shows a plot of dynamic random-access memory (DRAM) half-pitch taken from the ITRS 2007 roadmap and the corresponding CD-AFM probe width that would be required to make its corresponding successful 3D measurement.

8.4.2 Challenge for the AFM Tip

Need for Technology Breakthrough

Generally, the performance of an AFM tip depends on its sharpness, shape, mechanical stability, and durability, as well as its surface properties (e.g., hydrophilic or hydrophobic), primarily determined by its chemical composition. Aiming for highly accurate measurements on advanced structures and lithography patterns, therefore, requires extremely sharp AFM tips with a sufficiently small tip radius combined with enhanced mechanical stability and high wear resistance. These properties play a crucial role especially in the field of CD-AFM metrology, with a tip design like the one schematically illustrated in Figure 8.6a. Critical tip parameters such as tip width, vertical edge height, and lateral reach overhang are subject to changes due to wear effects in the course of repeated CD-AFM measurements. This decreases resolution over tip lifetime and, as a consequence, reduces throughput during quality control measurements.

Here, we present and discuss high-density diamond-like carbon tips (HDC) induced through an electron beam process that could generate a T-shape-like design (Fig. 8.6b), which is unachievable with current silicon-based technology, and compare their performance to standard silicon tips used for CD-AFM purposes. In contrast to state-of-the-art silicon CD-AFM tips, repeated CD-AFM measurements performed with HDC tips reveal 3D measurements at constant tip edge radius and therefore constant high resolution.

Example of Manufacturing and Processing

Critical dimension AFM tips made from amorphous carbon are fabricated by means of gas-assisted electron beam deposition according to the well-established nanofabrication of supersharp scanning probe microscopy tips. Initially, a substrate (e.g., a standard silicon cantilever) is placed into a low-pressure chamber providing a defined atmosphere, which contains carbonaceous precursors. Carbon deposition is then initiated by exciting the process gas with an electron beam. By selectively directing the focused electron beam to the initial deposition spot, a defined T-shape-like HDC tip is fabricated in a two-step process as shown in Figure 8.6b.

Experimental Setup

In order to investigate the performance of HDC-manufactured tips and compare their wear effects on CD measurements to commercially available silicon tips, tip characterization and tip shape reconstruction were conducted. Therefore, tip width and tip shape of both HDC and Si tips were determined using, respectively, a silicon vertical parallel structure (VPS) and a silicon flare characterizer (IFSR or isolated flare silicon ridge) schematically illustrated in Figure 8.8. Initially, the tip width was investigated with the VPS characterizer corresponding to the standard production application by scanning three consecutive silicon lines and performing eight scans at one characterizer site.

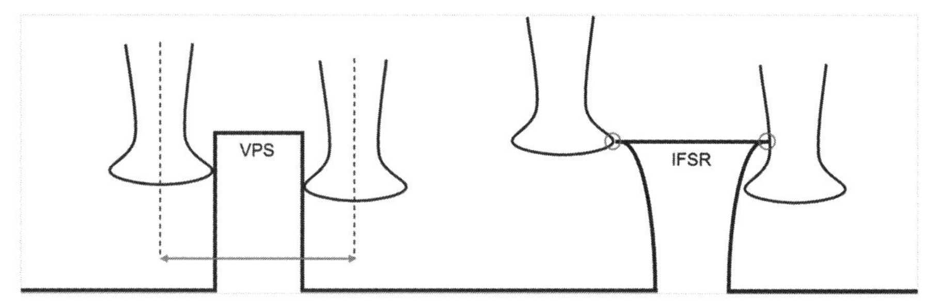

Figure 8.8 (Left) Schematic illustration of a VPS characterizer to determine tip width, where the measured value corresponds to the VPS width plus 2 times the half tip width (gray arrow). (Right) IFSR characterizer for the determination of the tip shape.

Tip shape reconstruction was conducted in the same way with the IFSR characterizer. This procedure was repeated continuously and tip width and tip edge height were reconstructed individually after each scanning measurement.

Results: Customizability of tip Dimensions The customizable HDC fabrication process of CD-AFM tips described in Section 8.4.2 allows for independently controlling tip shape dimensions, that is, tip length, tip width, reentrant profile distance, and tip edge radius. Such a process has been validated for 25-nm tip width with various tip lengths (90, 130, and 180 nm).

Results: Process Control: SEM vs. AFM3D

In order to control the fabrication process stability, HDC-manufactured AFM tips were inspected using SEM and AFM3D characterization. Table 8.1 summarizes lengths and widths of the various 25-nm tips, respectively, measured with SEM and AFM3D. Comparing the values extracted from the quality control process based on both approaches (indicated as offset in Table 8.1) reveals a tip width variation of +2.3 nm to −2.2 nm while tip lengths vary by less than 20 nm.

Tip Performance

The tip wear of a T-shape-like 3D-AFM tip made of HDC is shown together with the tip wear of a state-of-the-art 3D-AFM tip made of silicon in Figure 8.9, where the relative tip width is plotted as a function of the number of measurements. The measured values of an HDC-manufactured tip (CDR50-EBD Model) are represented in blue and the corresponding values of a standard silicon-based tip are shown in orange (both tips feature a nominal tip width of 50 nm and a nominal tip edge height of 10 nm). The black dotted horizontal line indicates the relative tip width of 0.90, that is, the measurement cycle at which the initial tip width is degraded by 10%. While the CDR50-Si tip reaches this point after ~15 measurements (dashed orange vertical line), the CDR50-EBD tip exhibits a 10% tip width wear after ~75 measurements (dashed blue vertical line)

Table 8.1 Resulting Values and the Corresponding Offset Extracted from SEM and AFM3D Measurements*

		SEM		AFM3D		OFFSET (SEM–AFM3D)	
		Length (nm)	Width (nm)	Length (nm)	Width (nm)	Length (nm)	Width (nm)
CDR25-90	1	88	25	90.3	24.5	−2.3	0.5
	2	90	27	98.1	28.0	−8.1	−1.0
	3	98	27	108.5	23.6	−10.5	3.4
	AVG	**92.0**	**26.3**	**99.0**	**25.4**	**−7.0**	**1.0**
	3σ	**15.9**	**3.5**	**27.4**	**7.1**	**12.7**	**6.9**
CDR25-130	6	130	25	121.0	23.0	9.0	2.0
	7	130	24	131.0	21.7	−1.0	2.3
	8	124	26	126.0	23.3	−2.0	2.7
	AVG	**128.0**	**25.0**	**126.0**	**22.7**	**2.0**	**2.3**
	3σ	**10.4**	**3.0**	**15.0**	**2.6**	**18.2**	**1.1**
CDR25-180	9	181	25	N.A.	N.A.	N.A.	N.A.
	10	166	25	N.A.	N.A.	N.A.	N.A.
	11	162	27	163.5	29.2	−1.5	−2.2
	AVG	**169.7**	**25.7**	**163.5**	**29.2**	**−1.5**	**−2.2**
	3σ	**30.0**	**3.5**	–	–	–	–

*Where tip lengths and widths of three individual HDC/DLC AFM tips with a nominal tip width of 25 nm at a tip length of 90, 130, and 180 nm, respectively, were measured and characterized. Average numbers and standard deviations of 3σ are noted in bold.

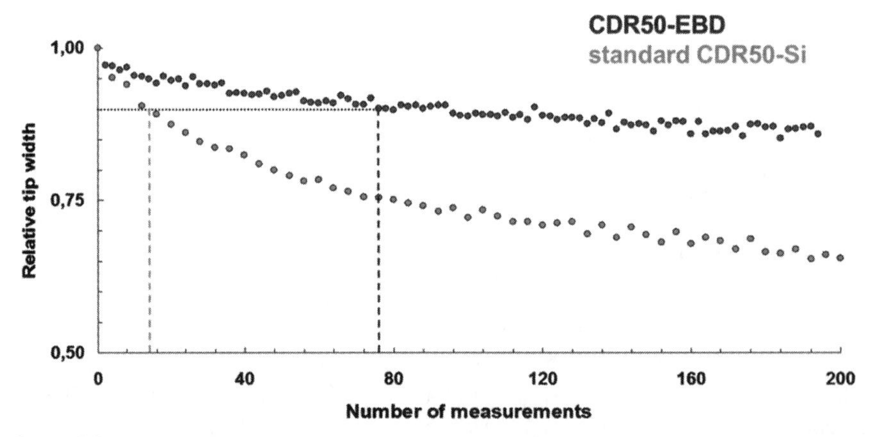

Figure 8.9 Relative tip width of an EBIP-manufactured CD-AFM tip (blue, CDR50-EBD) and a commercial state-of-the-art Si-based CD-AFM tip (orange, CDR50-Si) as a function of the number of measurements. The gray horizontal line represents the 10-percentage reduction of the tip width related to the initial tip width. The dashed vertical line points to a 10-percentage reduction after ~15 measurements for the standard CDR50-Si, while the CDR50-EBD exhibits the corresponding tip width wear after ~75 measurements (dashed blue vertical line). For color details, please see color plate section.

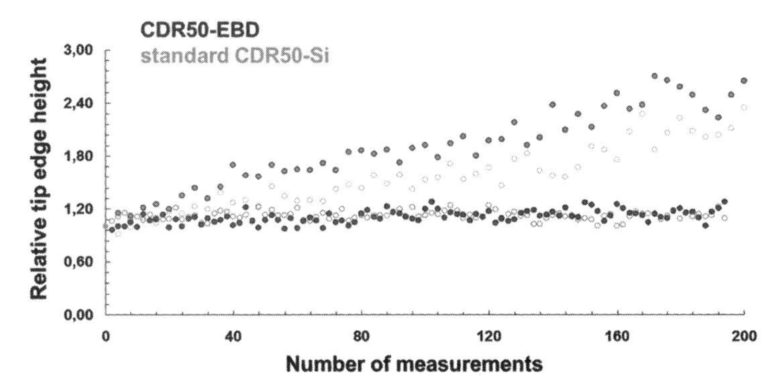

Figure 8.10 Relative tip edge height of a HDC/DLC CD-AFM tip (blue, CDR50-EBD) and a standard Si-based CD-AFM tip (orange, CDR50-Si) as a function of the number of measurements. Filled circles and open circles correspond to the right tip edge height and the left tip edge height, respectively. For color details, please see color plate section.

clearly pointing to a tip lifetime improvement by a factor of 4 of the CDR50-EBD compared to the standard CDR50-Si tip [14].

The observed tip wear not only reduces the tip width but also has an effect on the tip shape, in particular on the tip edge height as shown in Figure 8.10 for the corresponding right tip edge height (open circles) and left tip edge height (filled circles) of a CDR50-EBD tip (blue symbols) and a CDR50-Si tip (orange symbols): while the CDR50-EBD tip with its T-form-like shape exhibits a nearly constant tip edge height over the whole range of measurements, the flare-shaped CDR50-Si tip is subjected to a considerable increase of the tip edge height already after a few measurement cycles.

8.4.3 AFM Application Enhancement

Alternative Mode for the Semiconductor Industry

A new CD-AFM mode has appeared recently [15] and has started to be implemented in the semiconductor industry. The particularity of this mode is not the usage of flared tip but the AFM head rotation associated with the removal of piezotube, which is replaced by the decoupling of Z with X-Y motion to eliminate the crosstalk issue. As shown in Figure 8.11, the pattern of interest is scanned through three different angles in order to reconstruct a full 3D profile. Therefore, a conventional sharp conical tip can be used that does not require a special tip design like, for example, a flared tip technology. Subsequently, the user can benefit from a very aggressive conical tip radius (typically in the range of 2–5 nm), which allows access to high-resolution pattern sidewall information. Such detailed information is nearly impossible with the flared tip technology. The main remaining drawback for this technology is the measurement of tiny spaces that is limited by the AFM head angle. Nevertheless, new hardware is under development to allow smaller angles tilt (typically less than 9°) that could be enough to get access to accurate and complementary sidewall information for high-density pattern. This technique is

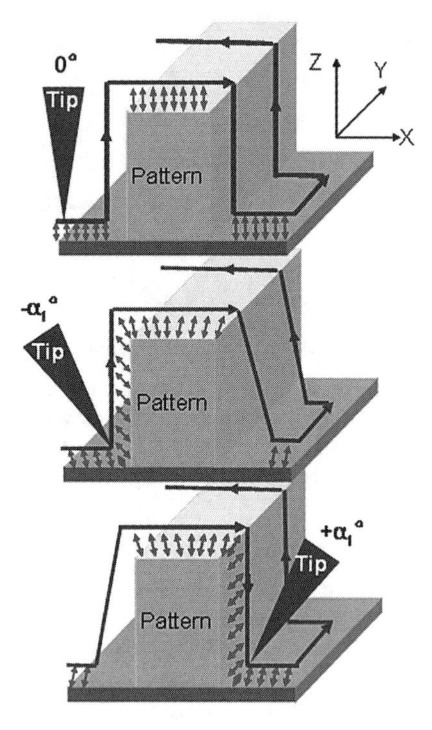

Figure 8.11 New CD-AFM mode using the AFM head rotation instead of flared tip.

very complementary to the flared tip technology in order to fulfill advanced roadmap requirements since it allows more accurate detail information that is mandatory to know and control in a near future.

Major Role for the CD-AFM Technology

In the semiconductor industry, CD-AFM is primarily used as a *reference* metrology technique. Key applications of CD-AFM include calibration of the CD-SEM and scatterometry methods. Calibrated CD-SEM data is used during process development and optical proximity correction (OPC) model calibration [12]. Thousands of CD-SEM measurements are acquired for input to the OPC model and its verifications. These measurements span variations in design and process space resulting in complex profile variations. It is important to cover the full range of profile variations in the calibration, otherwise measurement inaccuracies will result in OPC model inaccuracies and eventually printing inaccuracies on the wafer. As we move toward increasingly smaller printing, the accuracy requirements only become more stringent. Calibration based on limited sampling may not be practical anymore, and so increased scope of AFM application is inevitable. In this context, CD-AFM technique must cover two important areas:

- Fast calibration (critical radius or tip width (CR) and roughness) of noncalibrated techniques (such as CD-SEM for OPC model input or scatterometry for measurement uncertainty calibration as a function of material stack variations)
- Fast R&D process development for better process windows definition that will allow cycle time reduction, a major challenge for future nodes production

8.5 HYBRID METROLOGY AS A FINAL SOLUTION TO OVERCOME CD-AFM, CD-SEM, AND SCATTEROMETRY INTRINSIC LIMITATIONS

According to Jean-Marie Lehn (Nobel Prize for Chemistry in 1987) who prefaces a 2003 book entitled *Les Nanotechnologies* [16]: "Beyond dimensions shrinkage, the new era of complexity is initiated." Indeed, nanotechnology is the union of physics, chemistry, biology, and industrial sciences that meet up to the atomic or molecular scale. All metrology techniques must be compatible with such a union of scientific disciplines. To that end, and on the basis of all previous conclusions observed in this chapter, a versatile platform has to be developed to simplify multiple metrology platforms software in order to enhance R&D work and to optimize production process control.

As a function of the technology step the engineer is working on, metrology requirements can vary, although globally the constraints remain the same, for example: microscopic uniformity within a die, within wafer uniformity, wafer to wafer uniformity, lot to lot uniformity and material modification under electron exposure and plasma exposure. As shown in Figure 8.12, the previously listed requirements drastically impact process control quality that subsequently induces pessimistic ITRS (International Technological Roadmap for Semiconductor) perspectives. We can infer that we are clearly entering into the "coming of age era" of accurate and versatile in-line metrology that creates new needs and issues. Some of these needs and issues are listed here:

- Many driving forces are increasing the need for more control (more tuning knobs)
- Time consuming (e.g., TEM)
- Accurate CD metrology protocol (e.g., CD-AFM parameters)
- Parameter correlations (e.g., scatterometry)
- No more unique CD metrology solution (e.g., CD-SEM or scatterometry)
- Accurate 2D or 3D measurements (e.g., for FinFet measurements, nanowires)
- Understanding pattern's profile transfer from layer to layer (e.g., double patterning process)

ITRS roadmap front-end (2008)

Year of production	2010	2013	2016	2019	2022
MPU M1 half pitch (nm)	45	32	22	16	11
MPU Physical gate length (nm)	18	13	9	6.3	4.5
L gate 3σ variation (nm)	2.16	1.56	1.08	0.76	0.54
Max allowable etch 3σ (trim and gate) (nm)	1.08	0.78	0.54	0.38	0.27
Eq. oxide thickness - FDSOI MPU-metal gates (nm)	0.7	0.5	0.5	0.5	0.5

Semiconductor equipment must uniformly process nearly a trillion transistors uniformly (entire 300 mm wafers)

Figure 8.12 International Technological Roadmap for Semiconductor 2008 (3σ variation for gate level).

From all these parameters and new challenges, we can infer that hybrid metrology will be requested in the near future by the industry in order to use multiple techniques together to overcome individual technique limitations and, subsequently, push ahead the limits of CD metrology without major investments on tools themselves. In the end, multiple benefits for the end-users will issue from such an approach:

1. Reduction of uncertainty
2. Reduction of cycle time for process development
3. Yield enhancement
4. Standardize internal software interface for data processing
5. Improve tool efficiency without new tool investment
6. Customer can keep its own strategy

8.5.1 Contour Metric as a Pillar for 3D Metrology

In order to measure multiple kinds of devices that require 1D, 2D, or 3D measurements with a high level of accuracy in all dimensions, we must define a standard metric common to all devices and metrology techniques. In this section, we want to focus on the contour metric. In Figure 8.13a, we show a single nanowire

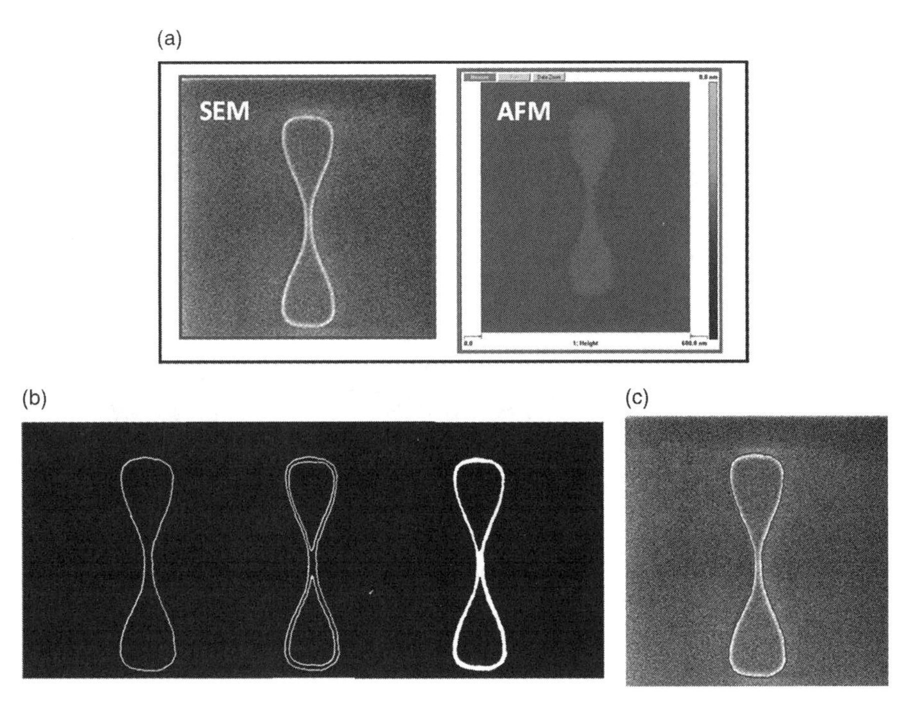

Figure 8.13 (a) Example of a single nanowire device measured by two techniques: CD-SEM and CD-AFM; (b) extraction of *X* and Y contour information from CD-SEM and CD-AFM images; and (c) contour superimposition on an original CD-SEM image. For color details, please see color plate section.

device that is measured by two different techniques, the first is the CD-SEM technique and the second is the CD-AFM technique. Subsequently, from these two images we can extract 2D contours in *X* and *Y* directions shown in Figure 8.13b. We can, therefore, create a new single metrological object or multiple metrological objects as a function of the mathematical filters used in the image analysis (e.g., Canny, Laplace, or Sobel matrix). Each of these new contours can be superimposed on the raw image, as shown in Figure 8.13c so that we can infer the real dimensions of the fabricated structure since AFM data coupling with SEM will combine various contour information with *Z* altitude, which is unavailable with single CD-SEM image analysis. In our approach, we propose to define multiple contours for one specific pattern measured by multiple techniques. Then, from these multiple contours we can extract various useful information based on data fusion that will be defined in the next section.

8.5.2 Data Fusion for Hybrid Metrology

Data fusion is generally defined as the use of techniques that combines data from multiple sources and gathers that information into discrete, actionable items in order to achieve inferences, which will be more efficient and narrowly tailored than if they

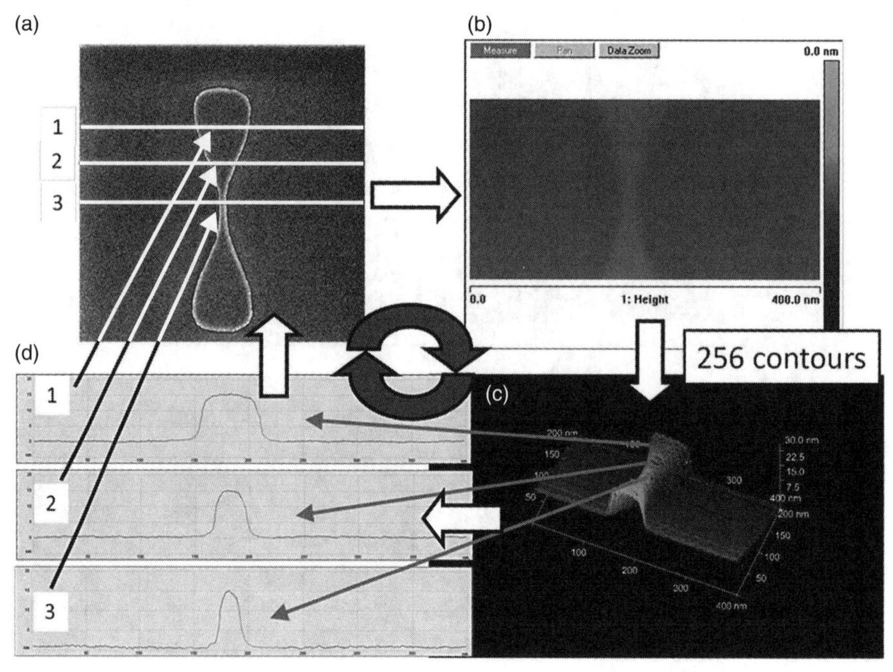

Figure 8.14 Example of a potential data fusion method applies to CD metrology between CD-SEM (a) and CD-AFM (b) and (c) through multiple contour extractions (d). For color details, please see color plate section.

were achieved by means of disparate sources. In our daily life, data fusion is commonly used is several areas such as geospatial information systems, business intelligence, oceanography, discovery science, business performance management, intelligent transport systems, loyalty card, bioinformatics, intelligence services, and wireless sensor networks.

From all these examples of application areas, we can clearly see that today, data fusion methodologies are almost everywhere at different degrees of complexity. In our field of interest, that is, CD metrology for the semiconductor industry, we think that data fusion can have great potential for global accuracy enhancement to reduce R&D cycle time through optimized process window definition and to improve production yield through easier hotspot detection that are typically hidden by conventional CD methodology with large uncertainties. A typical application for data fusion is hybrid metrology, which has started to be explored by different authors [17–19]. In our approach we would like to go further and propose a generic methodology that could be applied to both R&D and production. We show in Figure 8.14 a potential data fusion example application to a silicon nanowire device and illustrated by the implementation of a CD-AFM measurement into a top-view SEM image through multiple contours extraction and fusion.

Data fusion can be divided in two distinguishable parts as shown in Figure 8.15. The first part, the so-called calibration loop, consists of the definition of a high-level

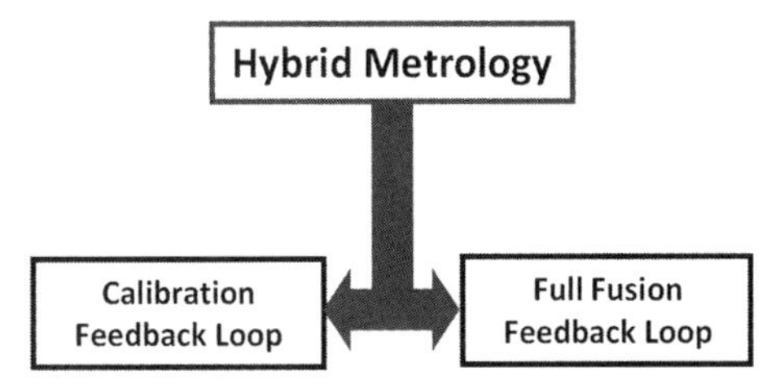

Figure 8.15 Hybrid metrology methodology made of two loops: a calibration loop and full data fusion loop.

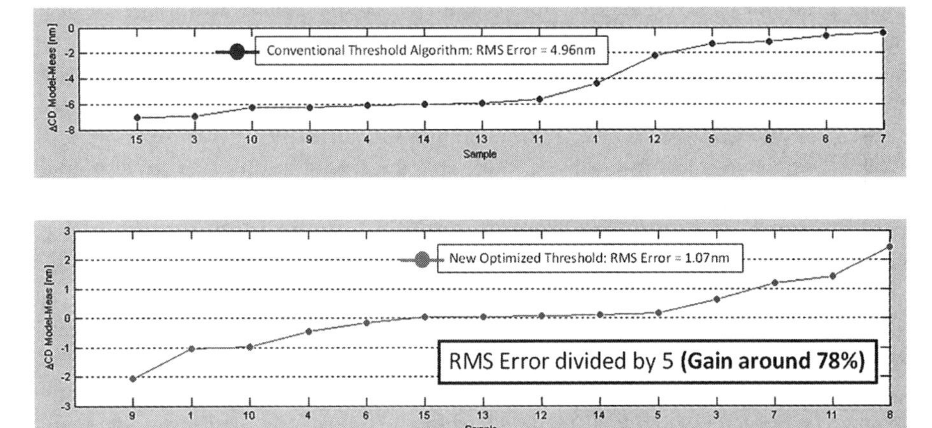

Figure 8.16 CD-SEM new optimized threshold impact on RMS error in comparison to conventional constant threshold approach. It leads to 78% improvement on RMS error by using a reference technique calibration.

calibration protocol. The second part, the so-called full fusion loop, consists of the creation of one image from two single measurements coming from two different techniques. The new image will offer the capability to sum the strengths of each technique and therefore will propose a nonnegligible added value in the final CD metrology protocol. An example for a set of CD-SEM images is shown in Figure 8.1, which is considered as the production tool we would like to calibrate. In this hybrid metrology module, we have integrated reference features (height, CD, sidewall angle, LER…) coming from CD-AFM or TEM images in order to recalculate the new contour or new threshold in order to extract the correct CD. Such a module is capable of extracting a CD bias function that must be applied to each image. As shown in Figure 8.16, it can be possible to improve the initial RMS error of a CD-SEM tool by nearly 80% (from 5 nm of error down to 1 nm) with, for example, a new threshold calculation algorithm. All patterns were made of 193 resist material with CD from 60 nm up to 200 nm. The conventional CD-SEM algorithm was a constant threshold algorithm taken at 80%. Such a data fusion methodology allows the user to take into account at the same time the electron proximity effect and resist material patterns shrinkage during the measurements which introduce CD bias up to several nanometers.

8.6 CONCLUSION

In this chapter, we have tried to show that the introduction of SPM (through CD-AFM technology) for CD metrology in the semiconductor industry is critical to introduce accuracy at the manufacturing level. Many driving forces are increasing the need for more accurate process control, including the complexity through new devices, the introduction of new materials, and tight specifications for metrology. Subsequently, SPM cannot answer all the requirements and must be used together with other techniques. Our view is that the contour metric will play a major role for most devices and CD metrology techniques. It will become the future metric of interest for sub-14-nm node production. However, its usage must be part of a global new CD metrology methodology based on data fusion. The application of such methodology in the semiconductor industry can be called hybrid metrology.

To initiate such an approach, various application modules have to be developed such as robust contour extraction for various CD metrology techniques, specific calibration loops, and multiple data fusion algorithms dedicated to heterogeneous and homogeneous fusion. One example of a CD-SEM calibration exercise has been shown in this chapter through the implementation of reference features coming from CD-AFM or TEM technology in order to reanalyze CD-SEM images. We have shown an improvement of CD-SEM accuracy by 80% compare to the conventional constant threshold methodology. Such a module is an example of one typical module that can be implemented on a versatile metrology platform compatible with any kind of CD metrology techniques.

One specific advantage of the data fusion approach is that all techniques (CD-SEM, scatterometry, TEM, CD-AFM, and newly emerging techniques) are exactly at the same level to be implemented on such a future versatile platform since the final goal is to extract useful information from each individual technique.

Finally, we would like to conclude this chapter with the statement that SPM technology must remain a workhorse technique for the semiconductor industry mostly because of its great versatility.

REFERENCES

1. Martin, Y., and Wickramasinghe, H. K. *Appl Phys Lett* **64**(19), (1994): 2498.
2. Kleinknecht, H. P., and Meier, H. Linewidth Measurement on IC Masks and Wafers by Grating Test Patterns, *Appl Opt* **19**(4), (1980): 525.
3. Azordegan, A., Bunday, B., Banke, B., Archie, C., Solecky, E., Allgair, J., and Silver, R. Unified Advanced Optical Critical Dimension (OCD) Scatterometry Specification for sub-65 nm Technology (2007 version), Technical Report, Sematech, Inc., January 2007.

4. ISO, Guide to the Expression of Uncertainty in Measurements, 1993 (corrected and republished in 1995), International Organization for Standards: Geneva.
5. Taylor, B., and Kuyatt, C. Guidelines for Evaluating and Expressing the Uncertainty of NIST Measurement Results, *NIST Technical Note 1297*, 1994 Edition.
6. ANSI/NCSL Z540-2-1997, American National Standard for Expressing Uncertainty—U.S. Guide to the Expression of Uncertainty in Measurement, National Conference of Standards Laboratories: Boulder, CO, 1997.
7. International Roadmap for Semiconductors (ITRS), http://www.itrs.net/Links/2007ITRS/2007_Chapters/2007_Metrology.pdf.
8. Dixson, R., and Guerry, A. Reference Metrology Using a Next-Generation CD-AFM, *Proc. SPIE* 5375, 633 (2004).
9. Dixson, R., Orji, N. G., Fu, J., Cresswell, M., Allen, R., and Guthrie, W. Traceable Atomic Force Microscope Dimensional Metrology at NIST, *Proc. SPIE* 6152, 61520P (2006).
10. Foucher, J., Pargon, E., Martin, M., Farys, V., Bécu, S., and Babaud, L. The CD Metrology Perspectives and Future Trends, *Proc. SPIE* 7140, 71400F (2008).
11. Foucher, J., Pargon, E., Martin, M., Reyne, S., and Dupré, C. Paving the Way for Multiple Applications for the 3D-AFM Technique in the Semiconductor Industry, *Proc. SPIE* 6922, 69220F (2008).
12. Foucher, J., Faurie, P., Foucher, A-L., Cordeau, M., and Farys, V. The Measurement Uncertainty Challenge for the Future Technological Nodes Production and Development, *Proc. SPIE* 7272, 72721K (2009).
13. Liu, H-C., Osborne, J. R., Dahlen, G. A., Greschner, J., Bayer, T., Kalt, S., and Fritz, G. Recent CD AFM Probe Developments for sub-45 nm Technology Nodes, *Proc. SPIE* 6922, 69222J (2008).
14. Foucher, J., Schmidt, S. W., Penzkofer, C., and Irmer, B. Overcoming Silicon Limitations: New 3D-AFM Carbon Tips with Constantly High-Resolution for sub-28 nm Node Semiconductor Requirements, *Proc. SPIE* 8423, 842318 (2012).
15. Hua, Y., Buenviaje-Coggins, C., Lee, Y. H., Lee, J. M., Ryang, K.-D., and Park, S.-I. New Three-Dimensional AFM for CD Measurement and Sidewall Characterization, *Proc. SPIE* 7971, 797118 (2011).
16. Wautelet, M. *Les Nanotechnologies*, Book Dunod, 2003.
17. Rana, N., and Archie, C. Hybrid Reference Metrology Exploiting Patterning Simulation, *Proc. SPIE* 7638, 76380W (2010).
18. Foucher, J., Faurie, P., Dourthe, L., Irmer, B., and Penzkofer, C. Hybrid CD Metrology Concept Compatible with High Volume Manufacturing, *Proc. SPIE* 7971, 79710S (2011).
19. Vaid, A., Yan, B. B., Jiang, Y. T., Kelling, M., Hartig, C., Allgair, J., Ebersbach, P., Sendelbach, M., Rana, N., Katnani, A., Mclellan, E., Archie, C., Bozdog, C., Kim, H., Sendler, M., Ng, S., Sherman, B., Brill, B., Turovets, I., and Urensky, R. A Holistic Metrology Approach: Hybrid Metrology Utilizing Scatterometry, CD-AFM and CD-SEM, *Proc. SPIE* 7971, 797103 (2011).

Chapter 9

Atomic Force Microscopy of Polymers

Andy H. Tsou[1] and Dalia G. Yablon[2]

[1]*ExxonMobil Research and Engineering, Annandale, NJ*
[2]*SurfaceChar LLC, Sharon, MA*

9.1 INTRODUCTION

Polymers are macromolecules resulting from polymerizing or linking of repeating building units of small molecules known as monomers. Although polymeric materials comprise a major portion of the materials around us, the study of polymers in a field called polymer science has only existed for about 80 years. Recognition of the existence of macromolecules of very high molecular weight did not occur until the 1920s [1]. During the 1920s and 1930s, polymer-based industries predominantly produced natural or modified natural materials. Between World Wars I and II, the development of synthetic rubbers was pursued primarily in the United States and in Germany, resulting from the concerns of potential disruption in the natural rubber supply [2]. Synthetic polymers started being produced in the 1940s. Since then, many synthetic polymers have been invented, and we cannot live a modern life today without them. Polymers now account for roughly 80% of the global chemical industry outputs. The chemical industry was a $3.7 trillion global enterprise in 2008 with more than 70,000 industrial chemical products. Commercial synthetic polymers are many, with polyethylene being the largest volume product. Industrial customers of polymers include rubbers, plastics, textiles, apparels, petroleum refining, pulp and paper production, and many others. Synthetic polymers are now in our everyday life as they are used to make milk containers [high-density polyethylene (HDPE)], diapers [polypropylene (PP)], soda bottles [polyethylene terephthalate (PET), trash bags [linear low-density polyethylene (LLDPE)], coffee cup [polystyrene (PS)], and tire tread [styrene–butadiene rubber (SBR)], among others.

Scanning Probe Microscopy in Industrial Applications: Nanomechanical Characterization, First Edition. Edited by Dalia G. Yablon.

Since the emergence of polymer science in the 1930s, its fundamental challenge is the development of the structure–property relationships for polymers. The ultimate goal has always been to identify the polymer structures needed to attain the desired properties and performance requirements of a given material. The various aspects of polymer structures that need to be defined include the identity of the monomers, the length of the polymer or the number of the monomer, and the backbone linking structure, which could be linear, block, star, comb, or dendritic. However, the backbone structure of a polymer affects the polymer chain assembly and the resulting morphology. Additionally, the polymer processing conditions employed to fabricate polymers into the final useful articles affects the assembly process and the resulting morphology. Thus, it is the morphology of a polymer product that ultimately controls its performance, and it would be impractical to establish the structure–property relationship of a polymer without developing the understanding of structure-processing–morphology-property relationships.

Even a simple linear polymer such as polyethylene (PE) without any comonomer possesses a complicated morphology. Linear polyethylene forms a semicrystalline morphology with a structural hierarchy going from an orthorhombic unit cell, to folded crystalline stacked lamellae with amorphous phase in between, to twisting lamellae assembled into spherulites if it is allowed to crystallize statically [3]. However, under blown-film conditions, linear PE can form a biaxially oriented fibril morphology with significant performance improvements over PE formed by compression molding and static annealing, resulting in higher in-plane mechanical stiffness, higher elongation to break, and across the plane impact toughness. In order to understand the backbone structural effects of a polymer on its morphology development under static and dynamic processing conditions, it is critical to have the capability to examine and to quantify the morphology evolvement in polymers.

Many additives, including stabilizers, plasticizers, and curatives, have been used ever since the first use of commercial natural polymers in product applications. The synergy that is derived from a polymer blend incorporating additives has long been recognized leading to the employment of polymer blends in polymer products. For example, in a commercial tire, almost every layer consists of blends of many polymers. Synthetic rubbers, first commercialized in the 1940s, have been used even from the beginning in blends with natural rubber. The inherently low modulus of most polymers and their viscoelastic nature can lead to die swell and complicate their processing and shaping into the final fabricated products. Hence, many polymer products also contain stiff fillers, such as carbon black, silica, calcium carbonate, or talc, to name a few, in order to raise the product modulus and to suppress its viscoelasticity. Performance of these polymer products depends on their submicron polymer blend domain sizes, filler dispersion, and filler and additive phase partition [4]. The ability to characterize the complex morphologies of these polymer blends and composites is essential in the establishment of structure–property relationships for blends and composites and in optimizing the material design, selection, and processing for delivering final polymer products with desirable performance.

Before the invention of atomic force microscopy (AFM) in 1986 [5], transmission electron microscopy (TEM), and scanning electron microscopy (SEM), were the commonly used microscopic methods to examine polymer morphology [6]. Due to the lack of materials contrast in polymers, staining and etching are sometimes required to deliver the contrast in these methods. Additionally, polishing, and a conductive coating are also necessary to enhance the image quality in addition to cryo-microtomy to produce a thin section and/or smooth surface. For those polymers that require osmium or ruthenium staining for contrast enhancement, subsequent cryo-sectioning of the stained samples to arrive at a section that is <100 nm thick and with a "proper" staining concentration for TEM contrast is necessary. Due to the complexity in sample preparation and the limitation in the magnification field, TEM has been utilized only in very limited situations to examine polymer morphology and to assist in the development of polymer structure–property relationships.

Early AFM used contact mode where a sharp tip attached to a cantilever was brought into contact with the sample surface under a constant interaction force [7]. The force is measured through the deflection of the cantilever and controlled by adjusting the cantilever's vertical deflection. Contact mode allows surface morphology imaging with the highest lateral resolution of 0.2–0.3 nm and lattice resolution under certain conditions [8]. However, by maintaining a constant tip–sample contact, a very high lateral force results with shear stresses reaching 100 GPa that can damage most polymers during AFM contact mode scanning. Typical plastics have break strengths that are below 10 GPa, and typical strengths of neat rubbers or elastomers are below 100 MPa. Only limited applications of contact mode AFM were applied early in polymers [9–11].

Tapping mode AFM, first introduced in 1993 [12], with a cantilever oscillating at or close to its resonance frequency, allows the tip to come into intermittent contact with the sample. There are two possible feedback mechanisms in tapping mode. The most widely used is amplitude modulation mode, where a constant vibrational amplitude is maintained. However, frequency modulation, where the resonance frequency is maintained constant, is also possible. In both these forms the surface topography can be imaged while the lateral forces are significantly reduced. Soon after the introduction of tapping mode AFM, the difference between the phase of the drive signal and that of the cantilever response (the phase lag or the phase lead) was found to depend on the material properties and can be used to provide material contrast [12–14] (see Chapter 4 for more detail on phase imaging). This tapping phase contrast depends on the cantilever vibrational parameters and is related to the cantilever damping due to the viscoelastic and adhesive properties of the sample surface [15–17].

Although the change in phase signal during tapping mode is difficult to be quantitatively related to material properties, it offers strong material contrast that does not require sample staining, etching, or ablation. Once tapping phase imaging was recognized to deliver excellent polymer contrast, the AFM, especially tapping phase AFM, went into an explosive growth in polymer applications. The extensive usage of AFM in all polymer research and product development areas came from not just the relatively low instrument cost of an AFM relative to a TEM or SEM, but also the ease of the sample preparation. To examine the bulk morphology in a polymer, polymer blend, or a composite, the sample only needs to be cryofaced to expose the bulk as the surface. Comprehensive reviews of the usage of AFM in examining polymer morphologies were provided back in 1997 and 2001 [18, 19].

Since 2001, due to the extensive enhancements in AFM instrumentation-cantilever modeling, data acquisition and analysis, and the introduction of many newer modes of operation-AFM may have finally arrived at its namesake with "atomic" resolution that can precisely measure the "force" for nanomechanical characterization of viscoeleastic polymers. AFM has now become one of the standard microscopic techniques utilized in examining polymer morphologies as indicated by more than 2700 publications since 2001 on AFM of polymers. Correspondingly, there are 13,674 patents and patent applications that mentioned the use of AFM on polymers with only 815 patents discussing applying AFM on polymers from 1991 to 2001. In this review, we cover the developments and applications of AFM methods on polymers since 2001. We first discuss our present understanding regarding tapping phase contrast mechanism in polymers and the enhancements of tapping phase (background in Chapter 4), force modulation (background in Chapter 5), and pulsed force mode AFM (background in Chapter 6). We will then discuss the recent innovations, developments, and applications of several AFM methods on polymers, including force–volume AFM, peak force AFM, multifrequency dynamic AFM, and quantitative nanomechanical measurements based on contact resonance. The emphasis will be on qualitative and quantitative mapping of polymer morphologies based on nanomechanical contrast; chemical force microscopy and single-chain dynamics probe microscopy on polymers will not be discussed here.

9.2 TAPPING PHASE AFM

Tapping phase AFM was found to be very useful for multicomponent polymer systems that do not exhibit strong density contrast in electron microscopy, such as polyolefin blends [20], elastomer blends and compounds [21, 22], and semicrystalline polymers [23–25]. Although all these polymer systems do possess contrast due to chemical composition, the spectroscopic imaging methods had and still have spatial resolution limitations to examine submicron domains and additive phase partition [26]. Tapping phase contrast in these systems comes from the local differences in energy dissipation as a consequence of the differences in material properties of adhesion, viscous damping, and modulus [27–30].

Although phase contrast is difficult to be related to the material properties quantitatively due to its dependence on the cantilever vibrational parameters, tip, contact area, and material properties, one can still establish polymer contrast rank ordering empirically. Based on a relatively soft cantilever (with a spring constant of 1–5 N/m and a corresponding resonance frequency between 50 and 100 kHz) operating at 50% setpoint amplitude (i.e., where the amplitude setpoint is 50% of the free vibration amplitude of the cantilever in air), contrast ordering in elastomers has been demonstrated [21, 22]. Using this empirically derived phase contrast order, tapping phase imaging successfully resolves the five phases present in a typical tire white sidewall

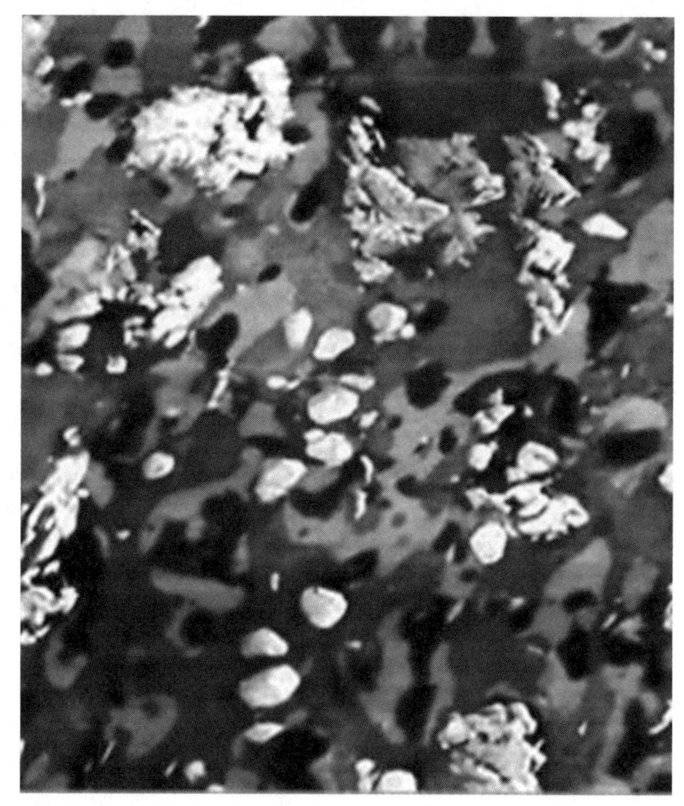

Figure 9.1 A 5-by-5-μm tapping phase image of a tire sidewall compound where the black phase is NR, the brown phase is EPDM, the yellow phase is CIIR, the individual round white particle aggregate is silica, and the large white flaky particle is TiO_2 (titanium dioxide). For color details, please see color plate section.

rubber compound, as shown in Figure 9.1: natural rubber (NR), ethylene propylene diene rubber (EPDM), chlorobutyl rubber (CIIR), silica particles, and TiO_2 flakes. It should be noted that with a relative soft cantilever (spring constant of 1–3 N/m), no phase contrast is established between hard silica and titanium dioxide particles. However, they can still be distinguished based on the particle shape. Additionally, it can be clearly shown that NR is encapsulated by CIIR, whereas almost all particles are inside the EPDM phase.

Tapping phase AFM is, in fact, the most effective microscopic technique in evaluating and quantifying the filler phase partition in rubber compounds. As shown in Figure 9.2, bright white carbon black (CB) particles have the tendency to cluster inside the EPDM domains in blends of EPDM and butyl rubber (IIR). The ability to quantify the filler partition in rubber compounds is critical to understanding the resulting compound performance. Based on finite-element simulation shown in Figure 9.3, it was found that the computed (red curve) and measured (blue curve) stress–strain curves of a CB-filled blend of brominated isobutylene-*p*-methylstyrene rubber (BIMS) and butadiene rubber (BR) can only be matched by considering the

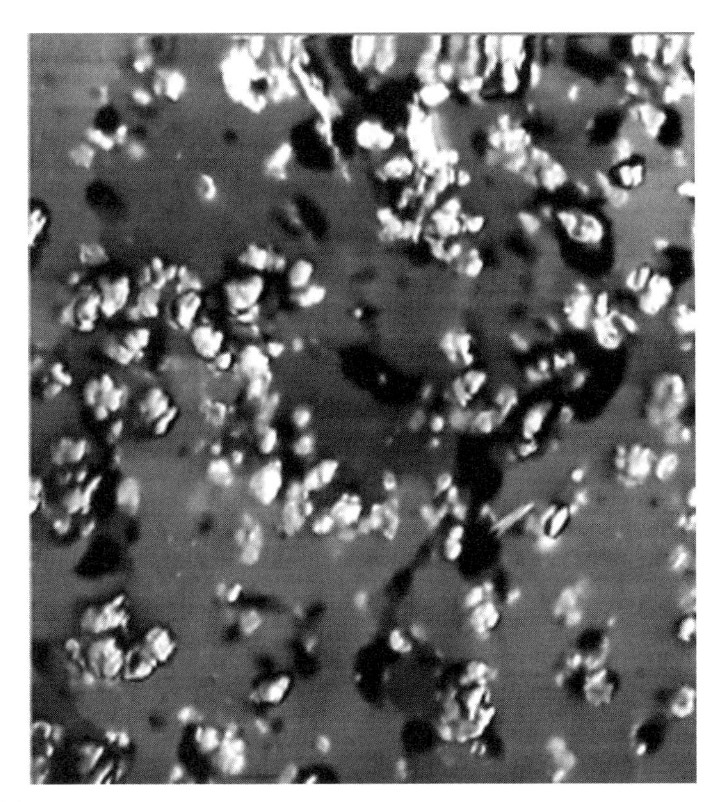

Figure 9.2 A 2.5-by-2.5-μm tapping phase image of an IIR (butyl rubber) and EPDM blend illustrating the preferential partition of the bright white CB particles into the dark EPDM phase. (White particles: CB; yellow matrix phase: IIR; dark areas: EPDM.) For color details, please see color plate section.

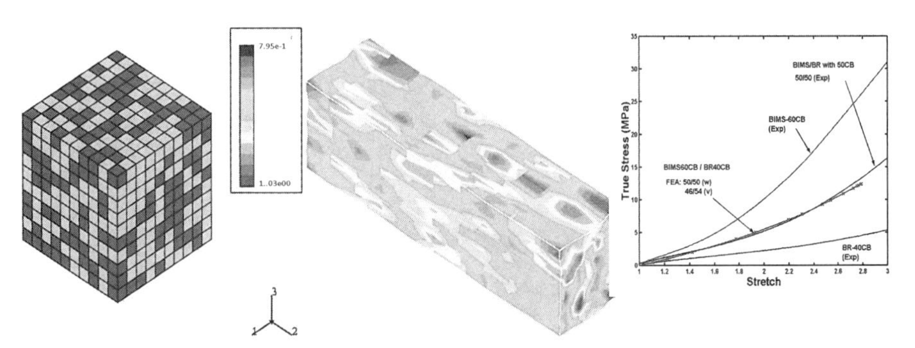

Figure 9.3 Finite-element simulation of a CB-filled blend of BIMS and BR (50/50 by weight). Image processing of tapping phase AFM micrographs quantified the phase partition of CB with 60% CB in BIMS and 40% CB in BR. Monte Carlo simulation was used to create the random blend morphology shown. For color details, please see color plate section.

AFM measured CB phase partition [31]. With the strong phase contrast that can be delivered from tapping phase AFM, morphology quantification through image processing of tapping phase micrographs was possible, thus allowing the development of quantitative morphology–property relationships [20, 22, 32, 33].

Impressive resolution has been achieved on polymer features with tapping mode. Using a whisker tip [34] reveals crystallographic arrangement in polydiacetylene. A specialized version of tapping mode called torsional resonance mode, where the torsional resonance of the cantilever is measured instead of the flexural resonance has led to recent exciting high-resolution results on the morphology of PE films. The combination of enhanced sensitivity and reduced contact force of torsional resonance mode using a specialized whisker tip delivers spatial resolution down to 3.7Å, [35] leading to the direct observation of buried chain folds and chain end defects in the crystalline phase of a semicrystalline PE.

Ever since the first use of tapping phase AFM to examine the polymer morphology, attempts have been made to extract viscoelastic properties from the phase signal, including using the energy conservation [15] and the virial theorem [36] to calculate the conservative and dissipative responses. Advantages of using tapping phase AFM for quantitative mechanical characterization are high measurement speed and spatial resolution while maintaining low contact forces. However, the practical measurement of the highly nonlinear interaction forces present in tapping mode is challenging, not to mention the presence of multiple energy dissipation mechanisms and potential phase contrast inversion. In order to understand the origin of phase contrast, a simulation of amplitude-modulated (tapping) cantilever in contact with a viscoelastic polymer has been conducted [37] with the assumptions of Hertzian contact, elastic contact area, and Kelvin–Voigt viscoelastic material model. Although this simulation managed to explain some experimental observations, it does have limitations due to the elastic contact assumption and the simple Kevin–Voigt model used.

Kiracofe and Raman proposed a new viscoelastic contact model through the simulation software VEDA (Virtual Environment for Dynamic AFM) [38, 39] that accounts for the tip contact area dependence on strain rate and loading history [40] with a three-element viscoelastic model or a complete generalized Maxwell–Prony series viscoelastic model for the material. Recently, a loss tangent measurement method has been developed for viscoelastic polymers with loss tangent values derived from the cantilever phase and normalized amplitude in tapping phase AFM [41] for operating under certain conditions. This simplified expression for the loss tangent was verified with numerical solutions of a cantilever interacting with a viscoelastic polymer represented by a Voigt model. As shown in Figure 9.4, a loss tangent map of a blend of PP and BIMS can be obtained using this method, however the ability of this method to accurately measure absolute loss tangent values still needs significant work. Furthermore, if the loss tangent of a material varies significantly within the range of frequencies from the tapping frequency to frequencies above by a decade, an average loss tangent value would be obtained instead. Finally, this average measured loss tangent value is sensitive to the tapping mode operation conditions and requires a number of careful calibrations.

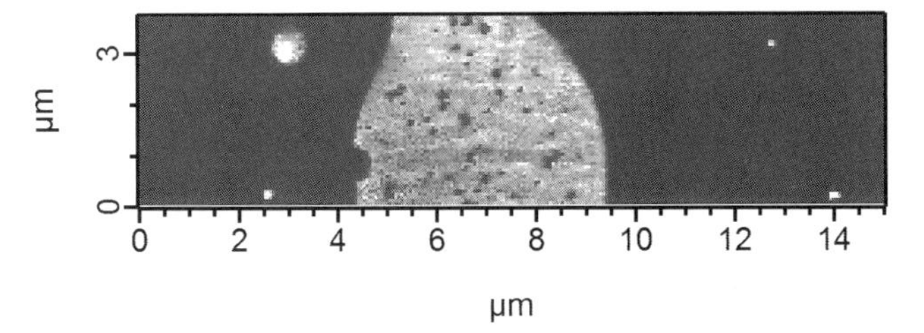

Figure 9.4 Loss tangent map obtained from tapping phase AFM for a PP–BIMS blend.

9.3 NANOINDENTATION

Instrumented nanoindentation methods with abilities to precisely measure and control millinewton loads and nanometer displacement were introduced right around the invention of AFM for measuring mechanical properties, moduli, and hardness of metals, ceramics, and polymers in films and in coatings on the microscopic scale [42–45]. These so-called nanoindenters do not have nanometer spatial resolution but, rather, have the ability to control and measure vertical displacement in nanometers. (See overview in Chapters 1 and 7 for more detailed information on nanoidentation.) A typical nanoindenter employs a Berkovich (pyramidal), cube corner, or conical tip of about 100 nm in radius. Based on the method developed by Oliver and Pharr [46, 47], the elastic modulus is calculated from the unloading portion of the indent to avoid the viscoelastic and plastic interferences. Later, a sinusoidal oscillation superimposed on the indent in the direction of indentation but at a frequency much greater than the indentation rate was introduced, allowing the continuous measurements of viscoelastic properties of a solid polymer during indentation in a measurement of dynamic nanoindentation.

A typical instrumented nanoindenter does not have imaging capabilities unless it is coupled or attached to an AFM; however, many recent commercial nanoindenters contain some form of low-resolution "AFM-type" imaging with the admittedly large nanoindenter tip. Without AFM, a dynamic nanoindenter with 0.1-µN load resolution can measure the cure state distributions of rubbers and rubber compounds down to 0.1 MPa modulus value but with micron-level spacing [48]. However, it is difficult for a commercial nanoindenter to deliver nanonewton or subnanonewton load resolution that an AFM can provide.

As an alternative, AFM-based nanoindentation, or AFM equipped with an indenter tip, has been used to measure the mechanical properties of polymers [49–54]. One advantage of AFM-based indentation over an instrumented nanoindenter is that the contact area and depth can be obtained directly by imaging. The mechanical properties can be computed from the force–distance curve following the same procedure developed by Oliver and Pharr. Typically, a sapphire cantilever with a Berkovich-shaped diamond tip is used that has a spring constant >100 N/m and a tip radius <30 nm. This type of force–distance measurement is a single-point measurement that requires precise calibrations of cantilever stiffness, tip shape, and piezo displacement in addition to the carefully planned experiments of controlled penetration depth, holding time, and unloading rate for the correct calculation of the elastic moduli [55]. The major disadvantage of AFM based nanoindentation, as discussed in prior chapters, is the flexibility of the cantilever resulting from its multiple degrees of freedom, making quantitative measurements challenging. However, in order to map the elastic, viscoelastic, and adhesive properties of polymers with internal heterogeneity, other AFM techniques such as force modulation, pulsed force, peak force, and force–volume are necessary. These techniques will be discussed next.

9.4 FORCE MODULATION

Force modulation is discussed in detail in Chapter 5. Briefly, the tip is in contact with the sample while scanning under a constant cantilever deflection at low frequencies (far below the cantilever resonance). The tip vibrates at a frequency that is below the cantilever resonance but above the instrument resonance [56–58], typically around 1–20 kHz. The vertical directional compliance or the response amplitude of the cantilever is related to the elastic properties of the material. However, in order for the cantilever to respond to the material properties, its force constant has to be comparable to that of the material. Therefore, a 3000-N/m force modulation cantilever was used for measuring the moduli of carbon fibers [56], while a 0.17-N/m cantilever was selected for soft organic thin films [58]. The modulation amplitude needs to stay small so as to maintain the constant contact with the material examined and is typically in the range of 1–20 nm.

All methods discussed in the following sections, including force modulation, pulsed force, peak force, and force–volume AFM, involve force measurements. The contact force depends on the contact area; contact area depends on the penetration depth of the AFM tip. Since the contact area cannot be measured directly, a contact mechanics model is required to relate contact area to the tip penetration and then to the load. Three contact mechanics models are commonly used, and they are Hertz [59], DMT (Derjaguin–Muller–Toporov) [60], and JKR (Johnson–Kendall–Robert) [61] theories, and their differences in contact area dependences on load have been analyzed by Carpick and Salmeron [62]. These contact mechanics theories are discussed in detail in Chapter 2. Briefly, a Hertzian contact is a pure elastic contact without any adhesion, whereas the DMT theory considers the long-range adhesion on hard materials, and the JKR theory incorporates the short-range adhesion on soft materials. Direct explicit expressions relating penetration/contact area to load can be derived based on Hertzian and DMT contacts, and, hence, they are the most frequently used methods, although a measured adhesive force is a required input for the use of DMT theory.

Without the ability to measure the contact adhesion force during force modulation AFM operation, a Hertzian contact is assumed for computing material moduli from force modulation response amplitudes [56–58]. Although strong adhesion effects of rubbers were recognized, force modulation AFM has been applied earlier to semiquantify the cure states of submicron rubber domains [21, 63] due to the needs in understanding the cure state of rubber domains thermoplastic vulcanizates (TPVs) and rubber compounds. Some internal standards were used to overcome the rubber adhesion effects [21, 63]. Recently, force modulation AFM was applied to examine the crystalline and amorphous

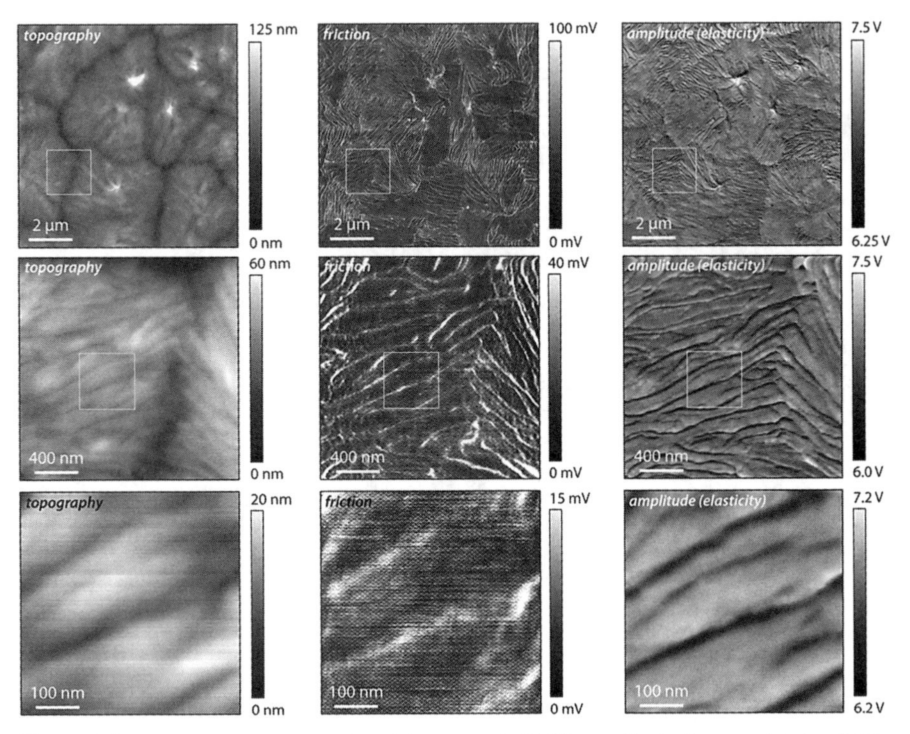

Figure 9.5 Topography, friction, and modulus maps of a HDPE cast film by force modulation AFM. Data courtesy of Jason Bares and Robert Carpick (University of Pennsylvania).

morphologies of an HDPE cast film using a silicon cantilever having 0.28-N/m force constant modulated at 1-kHz frequency, 17-nm amplitude, and average applied load at 5 nN. The topography, friction, and elasticity maps of this HDPE cast film are shown in Figure 9.5 where the correlation between the friction and elasticity measurements is clearly demonstrated where areas between the PE lamellae, thought to be the soft amorphous phase, are exhibiting higher friction and lower amplitude (elasticity).

9.5 PULSED FORCE IMAGING

Force modulation AFM is a dynamic contact method, whereas pulsed force, peak force, and force–volume AFM are based on single-point force—distance measurements (force spectroscopy). Pulsed force mode, introduced in 1997, was the first development in digitizing the force–time curves for nanomechanical contrast mapping of polymers using dynamic AFM [61, 65]. By modulating the tip frequency and amplitude, typically at around 1–10 kHz (substantially below the cantilever resonance) and hundreds of nanometers, forces exerted on the cantilever during the consecutive indentations were determined. Several points (trigger forces within the search windows) from the force–time curves can then be used to generate the topography, adhesion, and stiffness maps with the maximum repulsive force typically used for the stiffness mapping and the maximum pull-off force applied for the adhesion mapping [66].

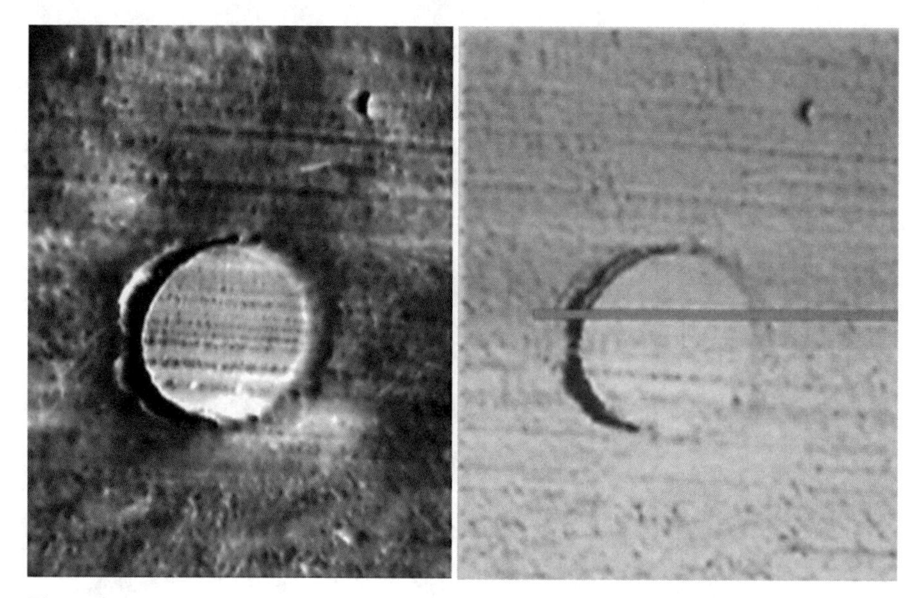

Figure 9.6 Topography (left) and stiffness (right) maps of a PP–PS blend.

Pulsed force AFM has found wide applications in adhesive application due to its high adhesion sensitivity through the pull-out force measurements. Before the introduction of pulsed force AFM, soft tapping or tapping where attractive tip–sample interactions dominated was used for examining adhesives. However, this method of operation has the tendency for phase inversion (resulting from the cantilever resonance shifts due to the tip contamination, shape change, or bistability in tip-sample interaction). Pulsed force AFM has been used to measure the adhesive properties of surface modified silicon [67], polyurethane coatings [68], and tackifier containing ethylene vinyl acetate (EVA) [69], to name a few. As shown in Figure 9.6, little stiffness contrast is observed between a PS dispersion in a PP matrix (right image), although the PS modulus is about 0.2 GPa higher than that of PP at 500 Hz, the frequency of the pulsed force measurement. Using pulsed force AFM on a PP–BIMS blend, both strong stiffness (left) and adhesion (right) contrast can be obtained between the PP domain (left part of image) and BIMS domain (right part of image), as shown in Figure 9.7. Pulsed force imaging is qualitative but can be quantitative if the force versus time curves can be converted to force versus distance curves offline after calibrations. It is possible to continuously map the force–distance curves at every contact point while scanning, though this is slow; this is one possible manifestation of force–volume AFM.

9.6 FORCE–VOLUME AFM

Force–volume AFM involves the collection of force–distance curves at each pixel in an image followed by offline data analysis to measure the elastic and adhesive properties over a larger area [70, 71]. Force–volume imaging is typically slow due to the large amounts of data to be collected. As shown in Figure 9.8, a 4×4 array of force–distance

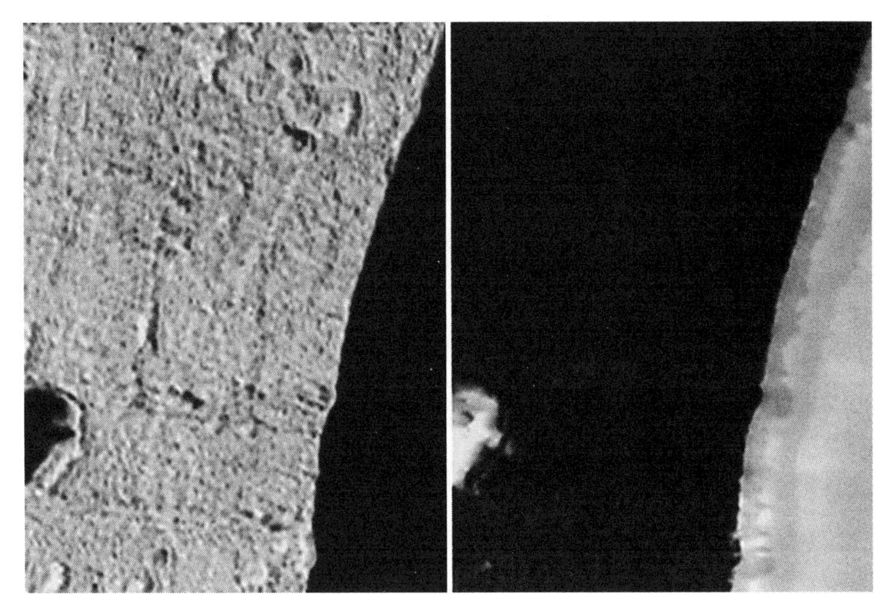

Figure 9.7 Stiffness (left) and adhesion (right) maps of a PP–BIMS blend, where PP is the material on the left.

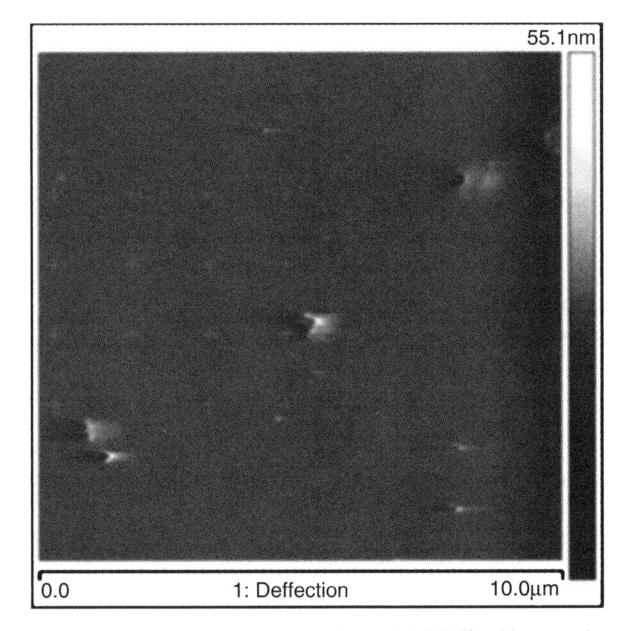

Figure 9.8 A 4×4 force curve measurement array shown in a BIMS rubber sample.

curves were measured on a BIMS rubber sample. The particles shown in the image are zinc oxide, which is one of the curative components. This BIMS rubber was cured using zinc oxide and stearic acid curatives. All force curves were obtained specifically on the rubber domains and not on the ZnO particles. The force curves thus obtained on these

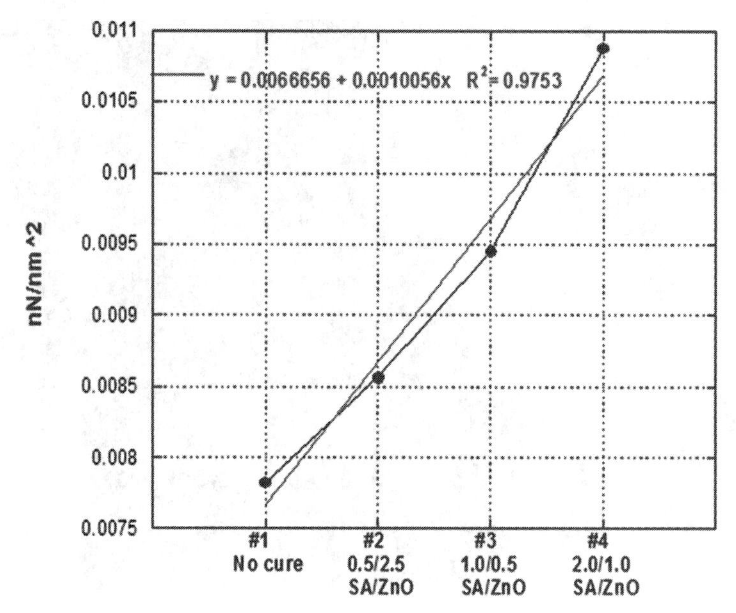

Figure 9.9 Effective modulus measured from force–distance curves for BIMS rubbers with varying curative amounts.

rubber samples were fit with a Hertz model to determine the rubber modulus. As shown in Figure 9.9, the rubber modulus increases with increasing curative content and its value is around 1 MPa, similar to the tensile modulus measured from the bulk sample.

If significant contact adhesion is present during the force–distance curve measurements, a procedure based on the combination of JKR contact mechanics and a two-point method [72] has been developed for force–volume AFM [73]. This procedure has been successfully applied to generate modulus and adhesive energy maps of carbon nanotube-(CNT)-filled NR.

9.7 HARMONIX AND PEAK FORCE QNM IMAGING

HarmoniX AFM [74, 75] was introduced in 2008 as a quantitative mechanical property mapping method for AFM. HarmoniX still employs tapping mode, where the cantilever oscillates at a single resonance. It uses specialized hammer-shaped tips with a tip offset on the side. Due to the offset tip, each tap excites torsional motion with the same harmonic components, but without the attenuation experienced by the cantilever's flexural mode. These harmonic components are then used to build a time-resolved force curve for each moment of impact; the force curves can then be fit to various models to extract mechanical properties [76].

This method has been applied to measure the bound rubber storage modulus (at about 60 kHz) of a CB-filled hydrogenated nitrile butadiene rubber (HNBR) that has a bound rubber thickness being less than 20 nm [77] and the elastic moduli of soft and hard segments of polyurethanes with lateral resolution less than 30 nm [78]. The interpretation of the force curves extracted from the HarmoniX signal can, however, be quite difficult if the torsional sensor does not have sufficient bandwidth, if the vertical (or flexural) motion is significant, or if the cantilever overtones coincide with the multiples of the drive frequency [79]. (The use of Harmonix to explore mechanical properties of low dielectric constant coatings is explored in Chapter 14.)

Peak force quantitative namomechanical (QNM) imaging is based on peak force tapping, which despite its name, is not a tapping-mode-based method. Instead, peak force tapping is a force–distance curve method where the tip oscillates at a low frequency, typically 1–2 kHz, and the peak force or maximum tip–sample interaction force is used for feedback control. With peak force tapping, force curves are thus collected at each pixel point in an image, with real imaging speed acquisition time. Peak force QNM is then an additional capability on top of peak force tapping that fits the force curves to a variety of models in real time and extracts various mechanical property channels such as DMT modulus, adhesion, deformation, and dissipation, together with topography in real time [80]. In order to quantitatively measure the materials moduli using the DMT fitting to the retract force curve, it is necessary to know the tip shape, cantilever stiffness, and the relationship between cantilever amplitude voltage and actual vertical displacement. Great care with a variety of calibrations such as cantilever force curve and tip area must be taken in peak force QNM in order to get quantitative mechanical data; even a slight mismatch in cantilever spring constant to the material can cause significant deviations in the measured mechanical properties. Finally, with the advancements in noise reduction, data acquisition, pattern recognition, and data processing speed, peak force imaging has higher force resolution with the ability for real-time force curve acquisition and mechanical property calculations, which have been successful on a variety of polymer materials [81].

As shown in Figure 9.10, modulus, dissipation, and deformation maps were obtained with peak force QNM for an impact copolymer (ICP) with ethylene–propylene (EP) rubbers dispersed in an isotactic polypropylene (iPP) synthesized in a serial reactor. It is clear from Figure 9.10 that the EP rubber is qualitatively softer, has higher deformation, and dissipates more energy in contact. The peak force QNM method was also applied to a binary blend of a plastic PP and an elastomeric BIMS with BIMS as the dispersed phase. As indicated in Figure 9.11, the BIMS elastomer behaves as an elastomer with lower modulus and higher adhesion/deformation/dissipation. With lower contact force–area and peak force feedback control, peak force imaging is also fairly easy to operate and protects the tip from damage.

Peak force quantitative nanomechanical mapping has been applied successfully to 12 different polymeric materials [81] and to polyurethane [78], with moduli measured in reasonable agreement with expected values for the majority of the polymers. The ones that were not in agreement have been attributed to the DMT contact model not being the appropriate model for the particular sample during peak force imaging [81].

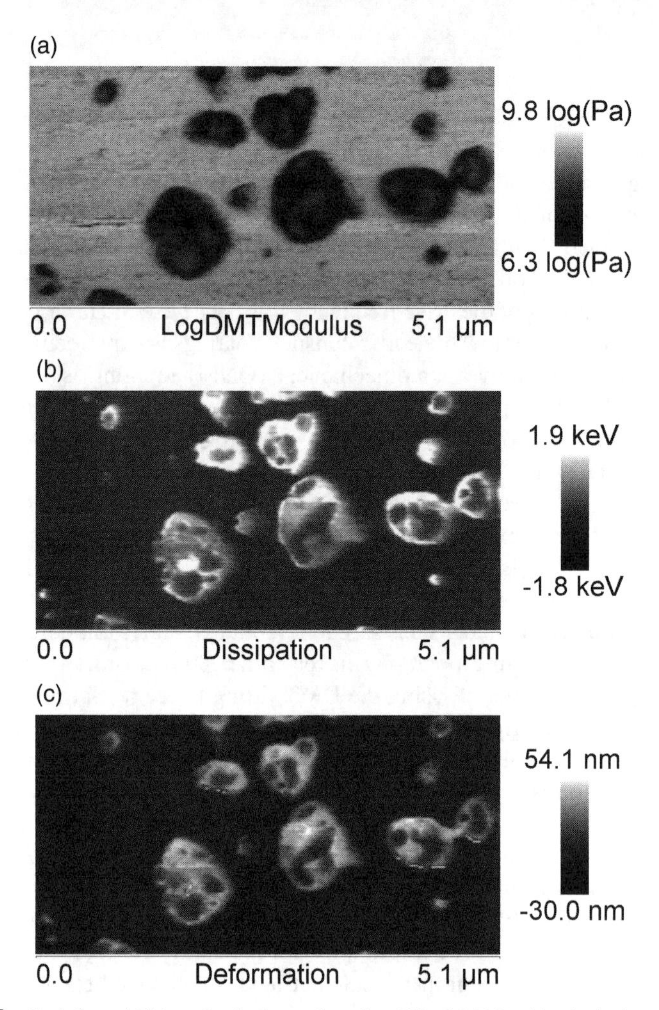

Figure 9.10 Peak force AFM mechanical mapping of an ICP with E-P rubber inclusions in a PP matrix shown in (a) log DMT modulus channel, (b) dissipation channel, and (c) deformation channel.

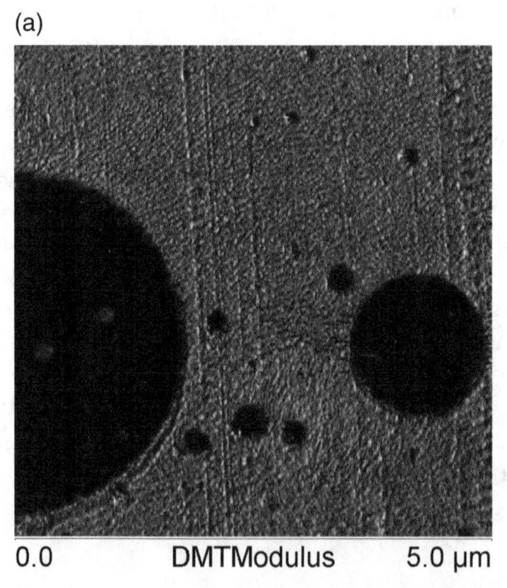

Figure 9.11 Peak force AFM mechanical mapping of a PP–BIMS blend with BIMS being the inclusions in the PP matrix showing (a) DMT modulus channel, (b) adhesion channel, (c) dissipation channel, and (d) deformation channel.

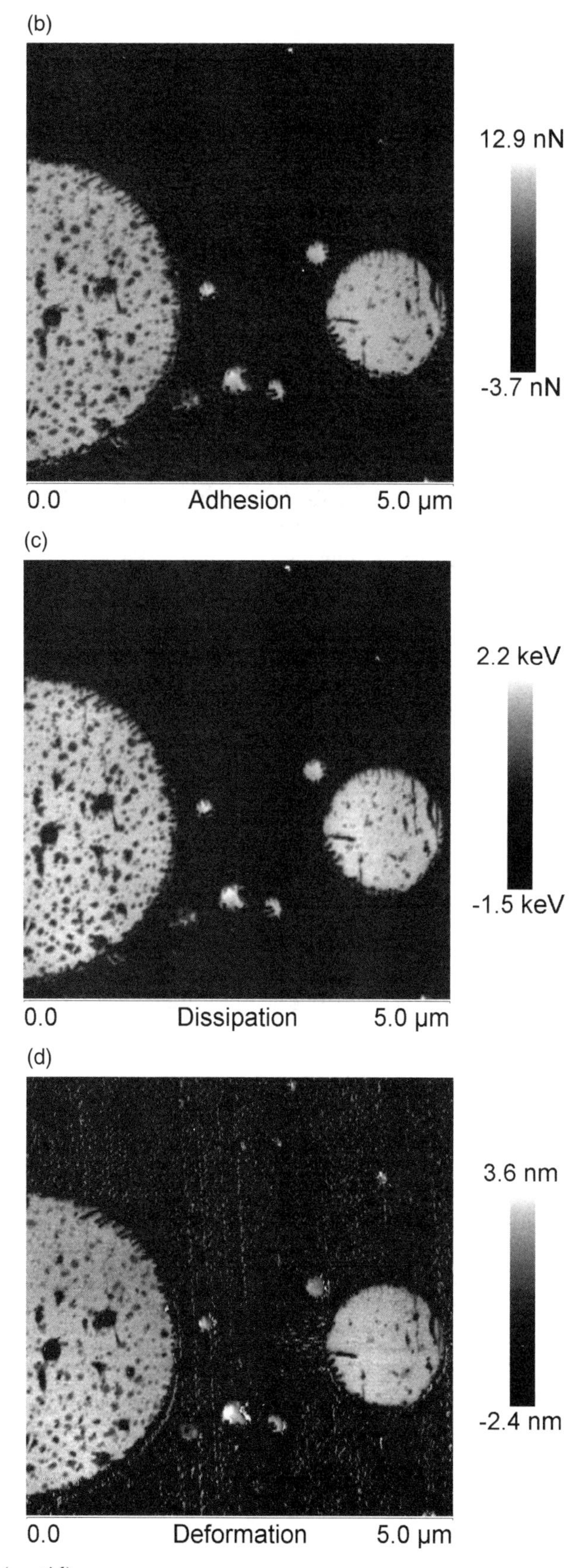

Figure 9.11 (*cont'd*)

9.7.1 Multifrequency AFM

A cantilever used in AFM has many flexural and even torsional resonant frequencies. Typically, tapping mode AFM excites only the fundamental or first resonance frequency with its oscillation amplitude modulated during scanning. However, other higher frequencies of the cantilever can be accessed including higher harmonics (integer multiples of the fundamental) and higher eigenmodes. Soon after the development of tapping phase AFM, enhanced phase contrast and cantilever sensitivity were observed on materials when the cantilever was excited at a higher eigenmode instead of just the first eigenmode [82–84]. This enhanced contrast has been attributed to the sensitivity of higher harmonics to the nonlinear tip–sample interactions [85, 86]. Rodriguez and Garcia [87] simulated the AFM cantilever driven at its two lowest vibrational modes and found that the second resonant mode oscillation can be sensitive to the Hamaker constant of the material by the nonlinear coupling of the two oscillating modes. It was then suggested that the composition sensitivity of an AFM can be significantly enhanced by the simultaneous excitation of the cantilever's first two resonant frequencies [88].

Thus, simultaneous excitation at multiple cantilever frequencies has become an area of active research over the past few years. Both observing a separate harmonic (integer multiple) of the drive frequency simultaneously with the fundamental frequency or simultaneous excitation of a cantilever at two frequencies coinciding with the cantilever eigenmodes [87, 89] have been reported. In this case, the AFM controller is equipped with multiple lock-in amplifiers to excite the lever at multiple frequencies.

The ability of bimodal AFM, where the cantilever is excited at two distinct eigenmodes, to discriminate material properties in the higher order mode is still an active area of research. However, enhanced contrast in the higher order mode has been observed [90–94] on a ternary blend of polypropylene, polyethylene, and polystyrene as shown in Figure 9.12. Note that in Figure 9.12(a), the first mode phase (conventional phase from tapping mode) image does not discriminate the three materials. However, the second-order mode phase data in Figure 9.12(b) successfully identifies the three materials in the blend.

Contact Resonance

An emerging method for probing mechanical properties of viscoelastic materials is contact resonance. Contact resonance is described in detail in Chapter 5. Briefly, similar to force modulation, it is a dynamic contact method where the cantilever is in contact with the sample, and the tip–sample contact is oscillated at resonance. The advantage of a linear system such as this one, opposed to a nonlinear system like tapping mode where the tip intermittently strikes the sample, is that it can be modeled as a damped harmonic oscillator in order to extract quantitative material properties.

Contact resonance has characterized stiff materials for the past decade, but in the past couple of years has been applied to polymer systems. The storage and loss moduli

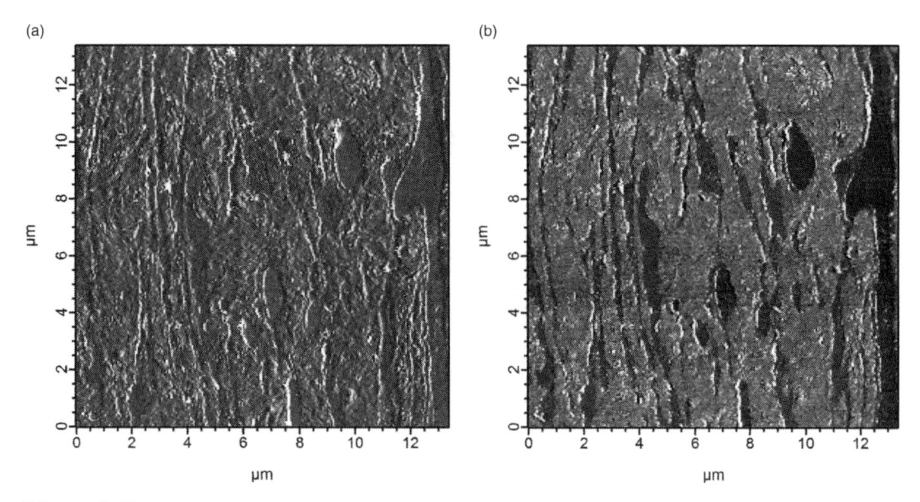

Figure 9.12 (a) First mode phase and (b) second mode phase image shown of a blend of 60% polyethylene (PE), 20% polypropylene (PP), and 20% polystyrene (PS). For color details, please see color plate section.

of materials can be successfully imaged with contact resonance [95, 96]. A contact resonance image is shown of a ternary blend of PP, PS, and PE in Figure 9.13. The topography image is shown in Figure 9.13(a). The contact resonance image successfully differentiates the three materials in the E' and E'' plots shown in Figure 9.13(b) (PE is lower E' than PP or PS) and Figure 9.13(c) (PS has the lowest E'', followed by PE and then PP), respectively.

9.8 SUMMARY

Atomic force microscopy has clearly been demonstrated to be a powerful tool to characterize a variety of polymer materials in the past 25 years. While phase imaging is the main characterization tool, several new advanced methods have emerged over the past decade with significant promise to improve on the ability to discriminate polymer materials and measure their properties quantitatively. However, there is still significant research to be done in polymer characterization applications. Polymers exhibit certain specific challenges to AFM: Materials can exhibit a wide range of properties (e.g., tires, impact copolymers); many polymer materials such as elastomers have significant adhesion properties that can contaminate the tip and confound the measurement; and contact mechanics models do not adequately model many of the polymer materials due to adhesion and viscoelasticity. We believe with the continued progress in the hardware, software, and modeling capabilities, these problems can be resolved for true nanomechanical characterization and discrimination of polymer-based materials, making scanning probe microscopy (SPM) arguably the most powerful tool for this application in the arsenal of microscopy tools.

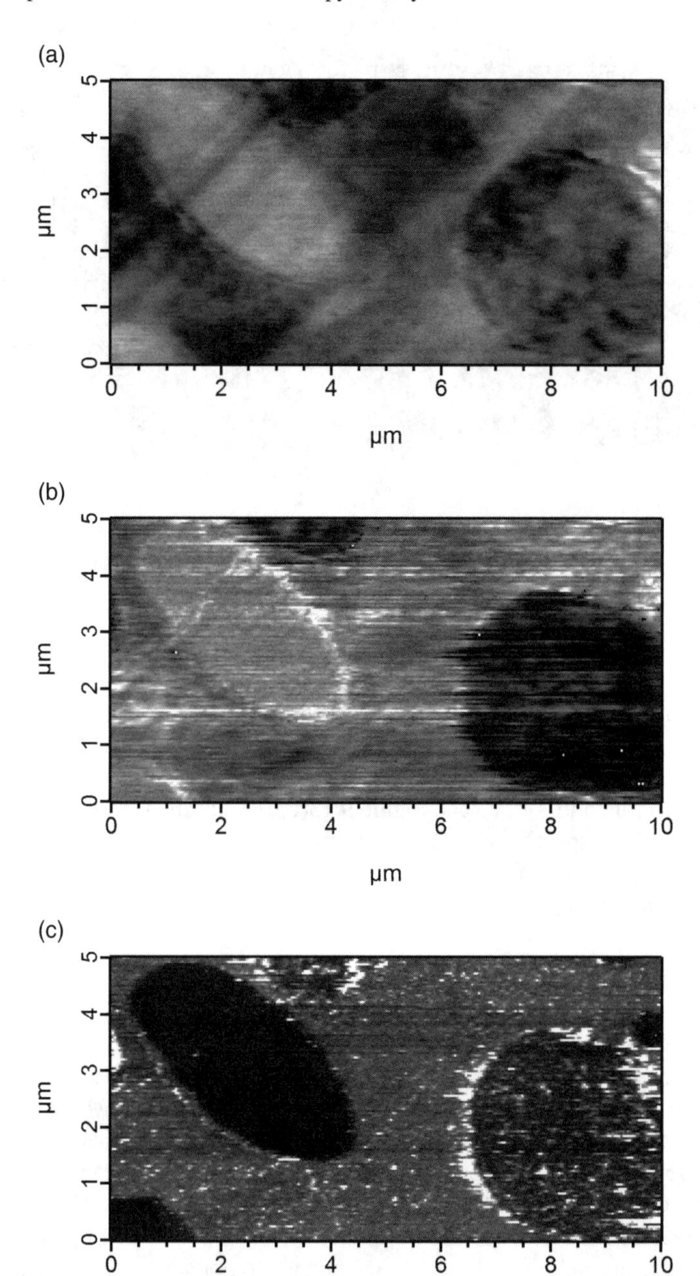

Figure 9.13 (a) Topography, (b) contact resonance storage modulus, and (c) contact resonance loss modulus of a ternary blend of PP, PE, and PS. Reprinted with permission from [96] Copyright 2012 American Chemical Society.

REFERENCES

1. Munk, P. *Introduction to Macromolecular Science*, Wiley: New York, 1989.
2. Rodrigeuz, F. *Principle of Polymer Systems*, 3rd ed., Hemisphere Publishing: New York, 1989.
3. Schultz, J. *Polymer Materials Science*, Prentice-Hall: New York, 1974.
4. Manson, J. A., and Sperling, L. H. *Polymer Blends and Composites*, Plenum: New York, 1976.
5. Binnig, G., Quate, C. F., and Gerber, C. *Physical Review Letters* **56** (1986): 930–933.
6. Sawyer, L. C., Grubb, D. T., and Meyers, G. F. *Polymer Microscopy*, 3rd ed., Springer: New York, 2010.
7. Drake, B., Prater, C. B., Weisenhorn, A. L., Gould, S. A., Albrecht, T. R., Quate, C. F., Cannell, D. S., Hansma, H. G., and Hansma, P. K. *Science* **243** (1988): 1586.
8. Ohnesorge, F., and Binnig, G. *Science* **260** (1988): 1451.
9. Stocker, W., Bar, G., Kunz, M., Moller, M., Magonov, S. N., and Cantow, H. J. *Polymer Bulletin* **26** (1991): 215.
10. Patil, R., Kim, S.-J., Smit, E., Reneker, D. H., and Weisenhorn, A. L. *Polymer Communications* **31** (1990): 455.
11. Reneker, D. H., Patil, R., Kim, S.-J., and Tsukruk, V. V. *Nato Adv Ser* **C405** (1993): 375.
12. Zhong, Q., Innis, D., Kjoller, K., and Elings, V. *Surface Science Letters* **290** (1993): L688.
13. Spatz, J. P., Sheiko, S., Moller, M., Moller, R. G., Winler, R. G., Reineker, P., and Marti, O. *Nanotechnology* **6** (1995): 40.
14. Tamayo, J., and Garcia, R. *Langmuir* **12** (1996): 4430.
15. Cleveland, J. P., Anczykowski, B., Schmid, A. E., and Elings, V. B. *Applied Physics Letters* **72** (1998): 2613–2615.
16. Tamayo, J., and Garcia, R. *Applied Physics Letters* **73** (1998): 2926.
17. Paulo, A. S., and Garcia, R. *Biophysics Journal* **78** (2000): 1559–1605.
18. Tsukruk, V. V. *Rubber Chemistry and Technology* **70** (1997): 430.
19. Tsukruk, V. V., and Spencer, N. D., Eds., *Advances in Scanning Probe Microscopy of Polymers*, Wiley-VCH: Weinheim, 2001.
20. Tsou, A. H., Lyon, M. K., Chapman, B. R., and Datta, S. *Journal of Applied Polymer Science* **107** (2008): 1362.
21. Galuska, A. A., Poulter, R. R., and McElrath, K. O. *Surface and Interfacial Science* **25** (1997): 418.
22. Tsou, A. H., and Waddell, W. H. *Kautschuk Gummi Kunststoffe* **55** (2002): 382.
23. Hugel, T., Strobl, G., and Thomann, R. *Acta Polym.* **107** (1999):
24. Hobbs, J. K., and Miles, M. J. *Macromolecules* **34** (2001): 353.
25. Hobbs, J. K., Farrance, O. E., and Kailas, L. *Polymer* **50** (2009): 4281.
26. Winesett, D. A., and Tsou, A. H. *Rubber Chemistry and Technology* **81** (2008): 26.
27. Anczykowski, B., Kruger, D., Babcock, K. L., and Fuch, H. **66** (1996): 251.
28. Tamayo, J., and Garcia, R. *Applied Physics Letters* **71** (1997): 2394.
29. Bar, G., Thomann, Y., and Wangbo, M.-H. *Langmuir* **14** (1998): 1219.
30. Wang, L. *Applied Physics Letters* **73** (1998): 3781.
31. Tsou, A. H., Zhang, G. E., and Boyce, M. C. *Annual Technical Conference—Society of Plastics Engineers (60th)* **3** (2002): 3197.
32. Tsou, A. H., Duvdevani, I., and McElrath, K. O. *Rubber Chemistry and Technology* **76** (2003): 318.
33. Tsou, A. H., Soeda, Y., and Hara, Y., U.S. Patent 7709575, 2010.
34. Klinov, D., and Magonov, S. *Applied Physics Letters* **84** (2004): 2697.
35. Mullin, N., and Hobbs, J. K. *Physical Review Letters* **107** (2011): 197807.
36. Paulo, A. S., and Garcia, R. *Physical Review B* **66** (2002): 041406/1–041406/4.
37. Garcia, R., Gomez, C. J., Martinez, N. F., Patil, S., Dietz, C., and Magerle, R. *Physical Review Letters* **97** (2006): 016103.
38. Melcher, J., Hu, S. Q., and Raman, A. *Review of Scientific Instruments* **79** (2008): 061301.
39. Kiracofe, D., Melcher, J., and Raman, A. *Review of Scientific Instruments* **83** (2012): 013702.
40. Ting, T. C. *Journal of Applied Mechanics* **33** (1966): 845.
41. Proksch, R., and Yablon, D. G. *Applied Physics Letters* **100** (2012): 073106.

42. Newey, D., Wilkins, M. A., and Pollock, H. M. *Journal of Physics E: Scientific INstrumentation* **15** (1982): 119.
43. Pethica, J., Hutchings, R., and Oliver, W. C. *Philos Mag* **A48** (1983): 593.
44. Doerner, M. F., and Nix, W. D. *J Mater Res* **1** (1986): 601.
45. Oliver, W. C., and Pharr, G. M. *J Mater Res* **7** (1992): 1564.
46. Oliver, W. C., and Pharr, G. M. *J Mater Res* **19** (2004): 3.
47. Tranchida, D., Piccarolo, S., Loos, J., and Alexeev, A. *Applied Physics Letters* **89** (2006): 171905.
48. Tsou, A. H., Westwood, A. D., Schulze, J. S., and Herbert, E. G. *Rubber Chemistry and Technology* **77** (2004): 678.
49. Clifford, C. A., and Seah, M. P. *Applied Surface Science* **252** (2005): 1915.
50. Tranchida, D., Piccarolo, S., and Soliman, M. *Macromolecules* **39** (2006): 4547.
51. Miyake, K., Satomi, N., and Sasaki, S. *Applied Physics Letters* **89** (2006): 031925.
52. Tranchida, D., Piccarolo, S., Loos, J., and Alexeev, A. *Macromolecules* **40** (2007): 1259.
53. Tang, B., Ngan, A. H., and Pethica, J. B. *Nanotechnology* **19** (2008): 495713.
54. Bedoui, F., Sansoz, F., and Murthy, N. S. *Acta Mater* **56** (2008): 2296.
55. Jee, A. Y., and Lee, M. *Polymer Testing* **29** (2010): 95.
56. Maivald, P., Butt, H. J., Gould, S. A. C., Prater, C. B., Drake, B., Gurley, J. A., Elings, V. B., and Hansma, P. K. *Nanotechnology* **2** (1991): 103.
57. Radmacher, M., Tillman, R. W., and Gaub, H. E. *Biophysical Journal* **64** (1993): 735–742.
58. Overney, R. M., Meyer, E., Frommer, J., Guntherodt, H. J., Fujihira, M., Takano, H., and Gotoh, Y. *Langmuir* **10** (1994): 1281.
59. Johnson, K. L. *Contact Mechanics*, Cambridge University Press: Cambridge, UK, 1985.
60. Derjaguin, B. V., Muller, V. M., and Toporov, Y. P. *Journal of Colloid and Interface Science* **53** (1975): 314.
61. Johnson, K. L., Kendall, K., and Roberts, A. D. *Proc R Soc London A* **324** (1971): 301.
62. Ogletree, D. F., Carpick, R. W., and Salmeron, M. *Rev. Sci. instrum* **67** (1996): 3298.
63. Soeda, Y., Tsou, A. H., Hashimura, Y., Kaidou, H., and Ball, J. W. U.S. Patent 7560514, 2009.
64. Rosa-Zeiser, A., Weilandt, E., Hild, S., and Marti, O. *Measurement Science and Technology* **8** (1997): 1333–1338.
65. Marti, O., Stifter, T., Waschipky, W. H., Quintus, M., and Hild, S. *Colloids Surface A* **154** (1999): 65.
66. Rezende, C. A., Lee, L.-T., and Galembeck, F. *Langmuir* **25** (2009): 9938.
67. Krotil, H. U., Stifter, T., Waschipky, H., Weiskaupt, K., Hild, S., and Marti, O. *Surface and Interface Analysis* **27** (1999): 336.
68. Meincken, M., Klash, A., Seboa, S., and Sanderson, R. D. *Applied Surface Science* **253** (2006): 805.
69. Lewtas, K., Bons, A. J., and Stuyver, J. U.S. Patent. 7765855, 2010.
70. Shonherr, H., Hruska, Z., and Vancso, G. J. *Macromolecules* **33** (2000): 4532.
71. Nagai, S., Fujinami, F., Nakajima, K., and Nishi, T. *Compos Interface* **16** (2009): 13.
72. Sun, Y., and Walker, G. C. *Langmuir* **20** (2004): 5837.
73. Wang, D., Fujinami, S., Nakajima, K., Inukai, S., Ueki, H., Magario, A., Noguchi, T., Endo, M., and Nishi, T. *Polymer* **51** (2010): 2455–2459.
74. Sahin, O., Magonov, S., Su, C., Quate, C. F., and Solgaard, O. *Nature Nanotechnology* **2** (2007): 501–514.
75. Sahin, O. *Review of Scientific Instruments* **78** (2007): 103707.
76. Mullins, N., Vasileve, C., Tucker, J. D., Hunter, C. N., Weber, C. H. M., and Hobbs, J. K. *Applied Physics Letters* **94** (2009): 173109.
77. Qu, M., Deng, F., Kalkhoran, S. M., Gouldstone, A., Robisson, A., VanVliet, K. J. *Soft Matter* **7** (2011): 1066–1077.
78. Schon, P., Bagdi, K., Molnar, K., Markus, P., Pukanszky, P., and Vancso, G. J. *European Polymer Journal* **47** (2011): 692.
79. Sahin, O., and Erina, N. *Nanotechnology* **19** (2008): 445717.

80. Minne, S. C., Hu, Y., Hu, S., Pittenger, B., and Su, C. *Microscopy and Micranalysis* **16** (2010): 464.
81. Young, T. J., Monclus, M. A., Burnett, T. L., Broughton, W. R., Ogin, S. L., and Smith, P. A. *Measurement Science and Technology* **22** (2011): 125703.
82. Stark, R. W., Drobek, T., and Heckl, W. M. *Applied Physics Letters* **74** (1999): 3296.
83. Sahin, O., and Atalar, A. *Applied Physics Letters* **79** (2001): 4455.
84. Crittenden, S., Raman, A., and Reifenberger, R. *Physical Review B* **72** (2005): 235422.
85. Hillenbrand, R., Stark, M., and Guckenberger, R. *Applied Physics Letters* **76** (2000): 3478.
86. Stark, R. W., and Heckl, W. M. *Review of Scientific Instruments* **74** (2003): 5111.
87. Rodriguez, T. R., and Garcia, R. *Applied Physics Letters* **84** (2004): 449.
88. Martinez, N. F., Patil, S., Lozano, J. R., and Garcia, R. *Applied Physics Letters* **89** (2006): 153115.
89. Proksch, R. *Applied Physics Letters* **89** (2006): 11312/1.
90. Dietz, C., Zerson, M., Riesch, C., Gigler, A. M., Stark, R., Rehse, N., and Magerle, R. *Applied Physics Letters* **92** (2008): 143107.
91. Martinez, N. F., Lozano, J. R., Herruzo, E. T., Garcia, F., Richter, C., Sulzbach, T., and Garcia, R. *Nanotechnology* **19** (2008): 384001.
92. Li, J., Cleveland, J., and Proksch, R. *Applied Physics Letters* **94** (2009): 163118.
93. Baumann, M., and Stark, R. *Ultramicroscopy* **110** (2010): 578.
94. Gigler, A., Dietz, C., Baumann, M., Martinez, N., Garcia, R., and Stark, R. *Beilstein Journal of Nanotechnology* **3** (2012): 456.
95. Yablon, D. G., Proksch, R., Gannepalli, A., and Tsou, A. H. *Rubber Chemistry and Technology* **85** (2012): 559.
96. Yablon, D. G., Grabowski, J., Killgore, J. P., Hurley, D. C., Proksch, R., and Tsou, A. H. *Macromolecules* **45** (2012): 4363–4370.

Chapter 10

Unraveling Links between Food Structure and Function with Probe Microscopy

A. Patrick Gunning and Victor J. Morris

Institute of Food Research, Norwich Research Park, Norwich, United Kingdom

10.1 INTRODUCTION

The birth of scanning probe microscopes in the early 1980s sparked one of the most actively researched areas of physical science in the new millennia: the exploration of the nanoscale world. It's hard to imagine that in the very early days of probe microscopy, Gerd Binnig and Heinrich Röhrer, working in their physics laboratory at IBM to develop methods to probe the variability of junctions in solid-state devices, had an inkling of the impact these methods would deliver, particularly in the biological sciences. However, the potential was soon recognized through the award of the Nobel Prize in Physics in 1986. The first instrument they developed, the scanning tunneling microscope (STM), provided the blueprint for a truly revolutionary technology for exploring the nanoworld that has blossomed and grown stronger with each passing year. Hot on the heels of the STM came, arguably, the most important of its progeny, the atomic force microscope (AFM). AFMs, which image by feeling or mechanical interaction rather than seeing their samples, have carved an incredibly diverse range of applications and have provided a new understanding of biological systems. Collectively, probe microscopes (such as the STM and AFM) have opened up nanoscale phenomena not only to visualization but also to manipulation. The term *nano*, previously understood by only a limited segment of the scientific community, has exploded into the public consciousness on a truly global scale, and to such an extent that *nano* has become a marketing tag for a myriad of industries and products.

Scanning Probe Microscopy in Industrial Applications: Nanomechanical Characterization, First Edition. Edited by Dalia G. Yablon.

How does nanoscience and nanotechnology impact the food industry? The "nanoworld" is inhabited by particulate nanomaterials with characteristic dimensions in the range of 1–100 nm. In foods this length scale embodies the molecular structures in raw materials, the molecular additives and ingredients introduced into foods to generate function, and the assembly and disassembly of these structures during the processing and storage of food products [1, 2]. Thus, although the general public are rightly wary of the term *nanofoods* in a large measure because of the wild and, in many cases, alarming claims that appear in the press on the subject of nanotechnology and food, the fact is that the main functionality and behavior of presently used food materials and processes are usually governed by components, structures, and events that do occur on the nanoscale. The major future concerns are about a relatively small proportion of applications that could lead to the inclusion or accidental contamination of foods with new types of nanoparticles that are not metabolized and will accumulate within the body. The safety of these nanomaterials is still ill defined, and, in some cases, new methods need to be developed to assess the risks and benefits associated with such materials [1].

There are a range of physical and chemical methods that have been and are being used for investigating the molecular nanostructures formed through processing or those that are naturally present in foods. The ability to visualize the molecular structure of foods began through the use of electron microscopy and has prospered through recent use of probe microscopy techniques such as AFM, which offers the prospect of imaging these structures under more natural conditions [3, 4]. Unlike electron microscopy, AFM can operate in a variety of modes that allow ultrahigh resolution to be obtained on samples in air or liquids at ambient temperatures and pressures. These fall into three categories: contact mode, noncontact mode, and a sort of halfway house known as tapping mode. In the first of these the AFM tip touches the surface throughout the scan, hence the name contact mode. In the second the tip hovers a few nanometers above it, and in the third the AFM tip touches the surface intermittently [4]. Whatever the sample environment or imaging mode, the common factor for successful high-resolution imaging of soft samples with the AFM is the requirement to control the interaction force between the AFM tip and the sample. This typically means operating at forces below about 300 pN. In liquids such as pure alcohols (as used in the examples in Figs. 10.1 and 10.2), this can be achieved relatively easily since the only significant contribution to the tip–sample force will be the deflection of the AFM cantilever, which can be precisely controlled by the instruments' feedback loop. This means that even in contact mode high resolution can be achieved by choosing a cantilever with a low spring constant and keeping its deflection minimal. In aqueous liquids things become more complicated because dissociation of surface groups on the AFM tip and sample gives rise to electrostatic forces that act independently of cantilever deflection. These can be repulsive or attractive and long or short range, depending upon the nature of the AFM tip, the sample surface, the ionic strength, and pH of the aqueous liquid. In order to image at high resolution in aqueous liquids all of the forces in the system need to be tuned to achieve an optimal balance where none of the forces are excessive [4]. This is achieved by adjusting not only the cantilever deflection but also the

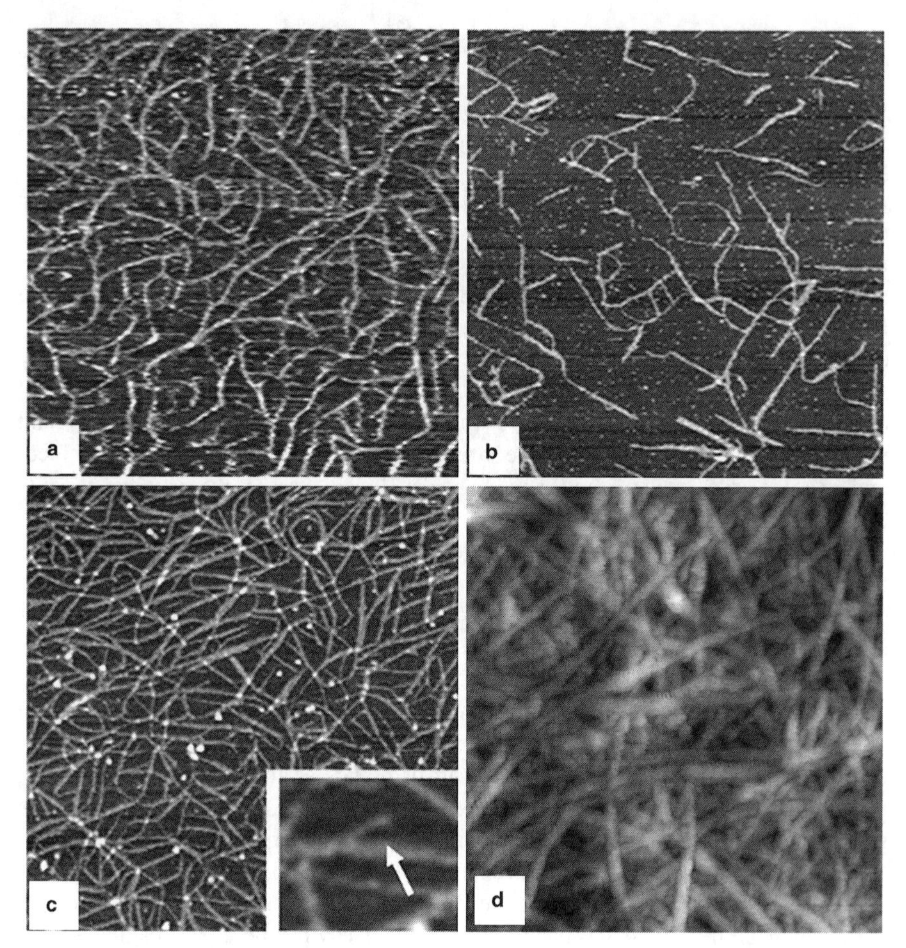

Figure 10.1 AFM topography images of the embryonic stages in gellan gum gelation: (a) in the presence of inhibitory counterions and (b) gel-promoting counterions. (c) As gelation proceeds further, network formation in the latter case becomes more extensive (inset magnified branch region). (d) The fibrous surface of a bulk gellan gel confirms the thin-film observations (d). Scan sizes: (a) 700 × 700 nm (b,c) 800 × 800 nm, (d) 1 × 1 μm. All images were acquired in contact mode under butanol.

ionic strength and pH of the buffer being used so that the forces exerted on the sample will neither destroy it nor allow it to float away during imaging (a common source of frustration in liquid AFM!). The alternative to imaging under liquids is to image the sample in air. In general, this is a much simpler option, but it has one drawback in terms of the tip–sample interaction: A capillary force due to a very thin condensed water layer is often present at ambient relative humidity (RH) (unless the sample is very hydrophobic), and this "glues" the AFM tip onto the sample surface with a large and uncontrollable force. This large capillary force damages soft samples, preventing high-resolution imaging in contact mode. A solution to this problem is to operate in noncontact or tapping mode. In both of these modes a relatively stiff

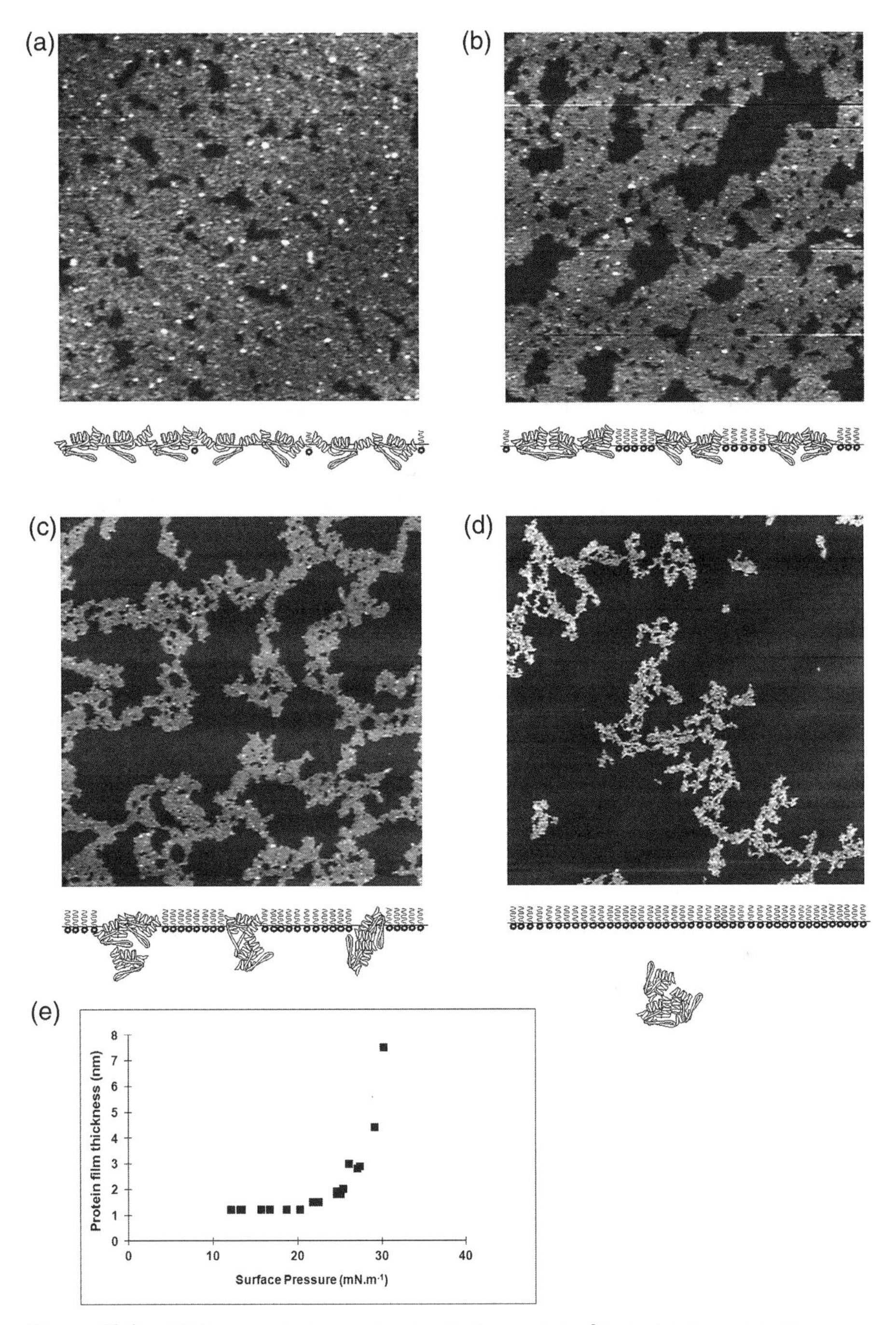

Figure 10.2 AFM topography images showing displacement of a β-lactoglobulin protein film (white) from an air–water interface by progressive addition of surfactant Tween 20 (black). Image sizes (a) 1.0 × 1.0 μm, $\pi = 22$ mN · m^{-1}; (b) 1.6 × 1.6 μm, $\pi = 25$ mN · m^{-1} ; (c) 3.6 × 3.6 μm, $\pi = 27$ mN · m^{-1} ; and (d) 6.4 × 6.4 μm, $\pi = 30$ mN · m^{-1}. (e) A graph of protein film thickness versus surface pressure. Note: the bright spots in (a) and (b) are small aggregates that were not included in the film thickness measurements. A schematic model of the displacement is illustrated below the images. All images were obtained in contact mode under butanol.

cantilever is oscillated close to its resonant frequency and scanned just above the sample, and for tapping mode the tip is allowed to strike the surface at the end of each cycle. With large enough oscillation amplitude, and a stiff enough cantilever, the tip will fly in and out of the water film as if it wasn't there. In this regime the force exerted on the sample is determined by the amplitude reduction during the collision of the tip with the surface, and this can be precisely controlled by the instrument. This regains instrument control over the force acting between the tip and sample, and the transient nature of the contact also provides the significant advantage that it eliminates the shear forces that cause the majority of sample damage in contact mode. Noncontact and tapping modes are not limited to operation in air; they can also be used in liquids to reduce the forces on soft samples and are particularly useful when working in aqueous media where a fully hydrated sample will be at its softest. Operation with resonating cantilevers is often referred to as AC (alternating current) mode imaging because it involves the processing of an AC response signal from the cantilever, whereas in contact mode the cantilever response signal undergoes a DC (direct current) level change, and hence this mode is sometimes referred to as DC mode [4]. Other advantages from the dynamic AC modes can be obtained by recording the energy transfer between the AFM tip and the sample surface to generate contrast that extends beyond simple topography and in addition the inherent signal-to-noise improvement provided by resonance. In addition to its use as a microscope, the AFM can be used to probe mechanical interactions between structures at both the molecular and colloidal level, a factor that adds a new dimension to the possibility for rational design of food structure to enhance quality. In this chapter the intention is not to list all the materials or structures investigated by AFM but rather to illustrate examples of where AFM has generated new insights into the understanding of food structure, where such understanding has led to new approaches to the design of food structure in order to enhance quality, and, finally, where new applications of probe microscopy techniques are likely to have a future impact in food science. The first half discusses the utility of AFM as an imaging tool to improve our understanding of the growth and associated mechanisms of networks of various food components. The second part of the chapter discusses the use of force spectroscopy (see Chapter 3) to probe various molecular and colloidal interactions important in food systems.

10.2 GELS AND THICKENERS: MOLECULAR NETWORKS

The texture of many foods is inextricably linked to a subset of the component molecules present. The most important class of functionally relevant molecules in this context are termed *hydrocolloids*. These are water-soluble polymers, which in food tend to be naturally occurring biopolymers such as polysaccharides and proteins. They are very complex in nature since each is made up of many components: covalently linked chains of sugars in the case of polysaccharides and amino acids in the case of proteins, and in some cases a mixture of both [5]. The mixture, ratio, and relative positions of these building blocks all have a profound effect upon the nature of the polymers, both in terms of their innate properties and in terms of how they interact with their environment. One of the more interesting properties of hydrocolloids is their propensity for emergent behavior (new properties that emerge as they interact with each other and their environment), and the subjects of polysaccharide gelation and interfacial stability provide two examples of how AFM can be used to tease out mechanisms that had resisted all previous attempts at characterization.

Thermoreversible gels are an important type of gelling agent used by the food industry in a hugely diverse range of applications to impart texture [6, 7]. Their gelling properties find direct application in foods such as chilled deserts, but the sensory attributes imparted by gelation means they are also used as fat substitutes, in products such as low-fat yogurt and mayonnaise. These types of gels are generally formed by charged helix-forming polysaccharides such as the carrageenans extracted from seaweed and new materials such as the bacterial polysaccharide gellan gum. A common feature of these materials is that gelation is triggered by helix formation on cooling the polysaccharide solution and the gels can be melted again by heating [6, 7]. The molecular changes that occur during polysaccharide gelation are similar and can be illustrated through studies on the model system gellan gum.

There is a substantial, but still controversial, literature on the physical chemical basis for gelation [6, 7]. However, it is agreed that gelation involves two discrete and separable steps: helix formation on cooling followed by helix aggregation. Because the polysaccharides are charged, an increase in ionic strength will screen the charge, promoting helix formation, stabilizing the helix, and raising its melting temperature. Certain types of *gel-promoting cations* have been found to enhance helix aggregation and gelation. In the absence of gel-promoting cations some molecular association can occur, but the network structures formed are weak, they are often not self-supporting, and they are easy to break at low deformation. The published physical chemical studies provide detailed information on the local, ordered structures formed within the gels, the so-called junction zones, but little information on the long-range network structure, which is what actually determines the mechanical properties of the gels. AFM has provided direct information on the nature of the molecular networks within gels [8]. Figure 10.1 illustrates the type of information that can be obtained by AFM.

When a dilute solution of gellan gum is deposited on mica and allowed to dry in air, the molecular concentration increases leading to aggregation. These aggregates or gel precursors formed in dilute solution provide clues to the modes of assembly at higher concentration. In the presence of tetramethyl ammonium (TMA) counterions, an example of a gel-inhibiting cation, the gellan gel precursors are thin branched fibrils [Fig. 10.1(a)] of uniform height around 1.5–2.0 nm, which is consistent with the helical diameter of gellan (feature height is a better measure than width because the latter suffers from probe broadening). Although the TMA cations do screen the charge, promoting helix formation, they are too bulky (radius 6.4 Å) to fit and bind into the ordered structures that would be formed on helix aggregation. The resulting fibrils represent the association formed in the weak gels in the absence of gel-promoting cations. The setting and melting temperatures of such gels show no hysteresis, suggesting that the aggregates are formed by elongation of the molecules in an end-to-end configuration, where disordered ends of neighboring gellan helices knit together to simply extend the innate double helical structure.

In the presence of potassium (radius 1.3 Å), an example of a gel-promoting cation, a different association mechanism comes into play. The fibrils can be seen to have associated into thicker branched fibrous structures [Fig. 10.1(b)], and the variable heights (2.0–6.0 nm) of the fibers is attributed in this case to a cation-induced side-by-side association of the fibrils. The height measurements suggest this typically involves two to three gellan helices. As well as the visual differences seen in the AFM images, this also leads to physical differences in the gels; namely a hysteresis in the setting and melting temperatures. The binding of cations within the aggregates stabilizes the structure, requiring more energy to disrupt the aggregates than that required to melt individual helices.

Images of gel precursors suggest the form of association likely to occur in bulk gels. Imaging the structures in hydrated bulk gels is difficult because the tip may either be sucked into the gel causing ripping and tearing of the network, or it will deform the surface of the gel during scanning, blurring the image. An intermediate model of a gel network is the structure formed in a monolayer hydrated film on the mica substrate, produced by drop deposition of gellan from very dilute solutions (typically 1–10 $\mu g \cdot mL^{-1}$). These networks are easier to image because the network is supported on a solid surface. The films are seen to consist of a continuous branched fibrous network of aggregated gellan fibrils (helices) [Fig. 10.1(c)]. This is illustrated clearly in the inset in Figure 10.1(c), which shows a closeup of a fibril emerging from a fiber. Fortunately, under acidic conditions gellan forms high-modulus gels (~10^4 Pa) that show negligible distortion on scanning, allowing collection of high-resolution AFM images [Fig. 10.1(d)] of the network structure present on the upper surface of the gel. The structure is equivalent to the branched fibrous structure observed in the thin hydrated films. Thus the AFM has provided new information on the long-range structure within the gels. It can be seen that the gels are fibrous networks in which the junction zones, the regions within which helices associate, are continuous throughout the fibrous bundles. This is apparent because the fibrils appear homogeneous and continuous in the AFM images of bulk gels—ends of molecules or nonfibrous regions were never seen. This is in contrast with previous well-accepted models for thermoreversible gels, which pictured them as rubberlike networks containing extended junction zones linked by disordered (i.e., nonhelical and thus nonfibrous) polysaccharide chains. The mechanism of energy storage on deformation in the two types of networks is very different, and the present fibrous description provides a better model for interpreting rheological and mechanical properties.

10.3 EMULSIONS AND FOAMS: PROTEIN–SURFACTANT COMPETITION

Proteins are useful emulsifiers with widespread application in the food industry for the stabilization of food foams and emulsions (e.g., creams, meringues, beer foam, mayonnaise, ice cream, smoothies). AFM has proved that proteins form two-dimensional networks at air–water and oil–water interfaces [9]. Adsorption at the interface drives some degree of structural rearrangement of proteins in order to accommodate the hydrophilic and hydrophobic environments at the interface. These changes expose structures that enable protein association at the interface [10]. The emulsifiers used in the food industry are usually not pure proteins, and hence the interfaces may be complex containing mixtures of various proteins.

Proteins are not the only type of surface-active molecules present in foods. Surfactants are lower-molecular-weight compounds that also stabilize interfaces. While proteins or surfactants alone can provide good stability, when both are present, they compete to occupy the interface, and the long-term stability offered by protein-stabilized interfaces is usually lost. Surfactants adsorb rapidly at the interface and rely on their high mobility at the interface to restore the local equilibrium when the interface is distorted in a process known as the Gibbs–Marangoni mechanism [10]. Such free diffusion at the interface is restricted by the presence of the protein network, but at sufficient concentrations the surfactants will generally displace the proteins. Given that the proteins are interconnected within a network, it is not obvious, despite the higher surface activity of the surfactants, how they displace the proteins. In addition to demonstrating the presence of elastic protein networks, AFM also led to the discovery of a new and unexpected displacement mechanism [9, 11, 12].

Model interfacial protein films were created at air–water and/or oil–water interfaces in a Langmuir trough. The protein networks and their displacement by surfactants were observed by transferring the interfacial films onto a solid substrate, using the Langmuir–Blodgett method in order to allow imaging by AFM. An illustrative set of AFM images is shown in Figure 10.2.

In the case of the neutral surfactant Tween 20, the images show that the surfactant occupies isolated regions at the interface that grow progressively in size. A potential explanation might be that growth of the surfactant domains arises through preferential loss of proteins at the expanding boundary between surfactant and protein. An advantage of the use of AFM is that it provides a three-dimensional image of the surface structure, allowing the height of the protein layer to be monitored as a function of the surface pressure (π) resulting from increased surfactant occupancy of the interface. These data showed that as the surfactant domains grew in area, the reduction in protein area was accompanied by an increase in the protein film thickness [Fig. 10.2(e)]. When these data were combined in the form of protein film volume (an indicative measure of the amount of protein in the film), the volume remained constant right up until the point where the surfactant domains merge forming a continuum at the interface. Close inspection of images of intact protein films revealed that the films contained holes or defects [13]. These holes presumably arise due to steric geometric constraints because the aggregation of the proteins during adsorption prevents perfect packing, and as the network develops less space is available for further adsorption until eventually the spaces that remain are insufficient to accommodate further proteins. These defects are the Achilles' heel of the structure. Surfactant can infiltrate into these defects, forming the nuclei for surfactant domains. As more surfactant is added, these domains grow, stretching the network until it breaks, ultimately freeing proteins or aggregates for displacement. The protein displacement is not continuous but occurs suddenly at a given surface pressure. The mechanism, known as orogenic displacement, was deduced from studies at air–water interfaces [11] but subsequently shown to apply also for oil–water interfaces [12]. Orogenic displacement has been found to be generic for all presently studied proteins and surfactants [9] and suggests that maintaining long-term stability of an emulsion or foam requires either strengthening the protein network or restricting access of the surfactant to the interface.

Food emulsions and foams contain a mixture of proteins, surfactants, and even on occasion some surface-active polysaccharide extracts. Protein samples used as commercial foam stabilizers or emulsifiers will also generally be isolates rather than pure proteins. Hence the interfaces are likely to contain mixtures of proteins as well as other surface-active components. It is important to assess whether the simple models of protein displacement described above can accommodate this level of structural complexity.

A difficulty in studying protein mixtures is that different proteins are hard to distinguish by AFM alone. However, by tagging proteins with different colored fluorescent labels and using a combination of AFM and fluorescence microscopy, surfactant displacement of binary mixtures of proteins can be followed [9, 14]. Such studies revealed that as protein mixtures adsorb at interfaces, they interact rapidly and become kinetically trapped in an immobile network, with little evidence for segregation of proteins into distinct domains. In mixed networks different proteins appear to interact with each to different extents, leading to weaknesses in the networks. Addition of surfactant leads to growth of surfactant domains, which appear to break the weakest links, preferentially displacing those proteins that, on their own, would form the weakest networks. An important but surprising observation is that the final failure of the mixed network is actually dominated by the protein components that, on their own, would form the strongest network—even when this protein is only a minor component of the mixture [9, 14–16]. Thus it is possible to predict the general behavior expected for networks formed by protein isolates, provided the composition is known [15, 16].

The use of studies on model interfaces as the basis for the design of food structures to deliver required functionality will only be possible if the conclusions drawn from them apply to real interfaces and also to the finite, curved surfaces present in real foams and emulsions. Neutral surfactant domains in the later stages of growth have been observed directly at air–water interfaces by Brewster angle microscopy [17], confirming that the displacement occurs at the interface and is not induced by the sampling methods. Liquid lamellae (e.g., soap films stretched across a wire loop) used as models for foam drainage show the presence of domains of different thickness when both protein and surfactant are present at the interface [18–20]. Diffusion measurements have revealed very different levels of molecular mobility in the thick "protein" and thin "surfactant" regions, suggesting that the mechanisms observed in model interfaces apply in real foams. As will be discussed later, force spectroscopy can be used to follow the breakdown of the interfacial protein network at the surface of an oil droplet on addition of surfactant [21]. For emulsions it is possible to monitor surfactant-induced displacement of proteins through a colorimetric method [21] or, in the case of charged proteins displaced by neutral surfactants the change in surface charge or potential during displacement [21]. In all cases an abrupt "orogenic-like" displacement process was observed. These studies all validate the use of model interfaces to predict behavior in discussions of emulsion stability [21].

The discovery of the orogenic model has suggested methods to be discussed later for the rational design of food structures to deliver health benefits. Although the orogenic displacement mechanism resulted from an interest in the interplay between proteins and surfactants in food systems, the general conclusions apply to any protein-stabilized emulsion or foam. Examples of nonfood applications based on the idea of orogenic displacement include the development of particulate antifoams for protein-stabilized foams [22], nanotechnological applications involving enhanced resolution in the printing of protein inks [23], and arrays [24] suggested clinical applications [25] in the attenuation of gas embolism-induced thrombin production that can lead to clot formation and strokes, plus a possible explanation for the water-induced de wetting of ultrathin polystyrene films prepared on hydrophilic surfaces [26].

10.4 INTERFACIAL STRUCTURE AND DIGESTION: DESIGNER INTERFACES

A reason for developing a more fundamental understanding of the molecular structure of foods is to enable design of food structures to promote health. This requires an understanding of how food structure is modified during digestion. A driver for such research is the aim of creating foods to reduce fat metabolism in order to address the growing problems associated with obesity and related health issues.

For processed food emulsions, one approach is to restrict access of the fat-digesting enzyme lipase to the oil phase. This will slow the rate of fat breakdown, and the lipolysis products will be released further down the gut. This generates the illusion that a high-fat meal has been consumed and the body should respond with a desire to reduce consumption—the so-called illeal-brake response [27]. In order to study by AFM the fundamentals of digestion, an interfacial protein film was created in a Langmuir trough and then subjected to the numerous conditions relevant to the passage of food through the stomach and small intestine, namely, changes in ionic strength, temperature, biosurfactant attack, and enzymatic attack. Since lipolysis of the fat contained within a protein-stabilized oil droplet is an inherently interfacial process, it cannot take place until the interfacial protein film is removed. Without this the lipolytic enzymes involved (lipase and colipase) are simply locked out of the game. Lipolysis occurs in the small intestine and for protein-stabilized emulsions it is enabled by bile salts. The bile salts are biosurfactants and displace the protein through an orogenic mechanism [28]. Bile salts are capable of fully displacing most presently used food protein emulsifiers. Furthermore, bile salts help to anchor the lipase–colipase complexes, which hydrolyse the lipids in the oil droplets. Clearly, if the coverage of the interface by bile salts (the total area of the bile salt domains) can be reduced by increasing the mechanical strength of the protein network, it should be possible to reduce the surface concentration of lipase and reduce the rate of lipolysis. Preliminary in vitro model studies on emulsions, under simulated duodenum conditions, suggest that enzymatic cross-linking of interfacial protein networks can reduce the rate of lipolysis, suggesting the plausibility of this process [29]. In order to apply this approach it is necessary that the modified networks survive transit through the stomach into the small intestine.

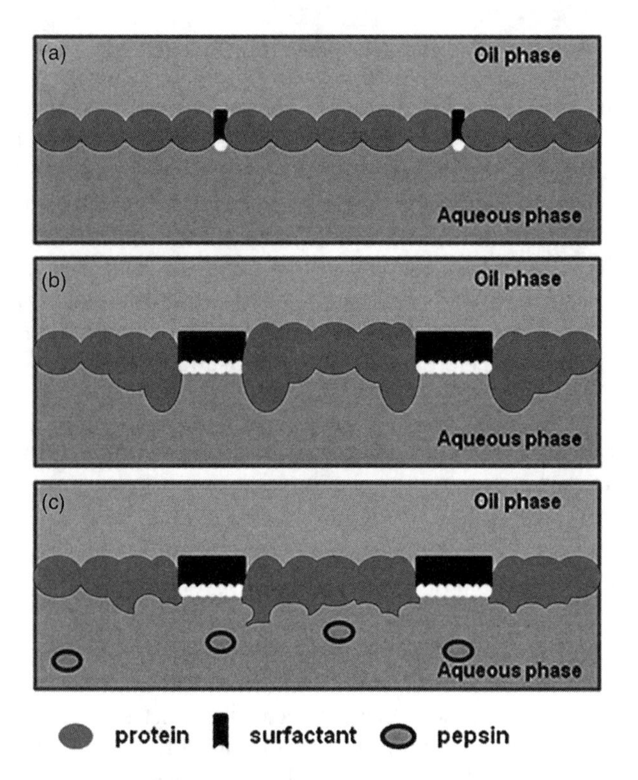

Figure 10.3 Schematic illustration of the synergy between surfactants and enzymes in the destruction of interfacial protein films during digestion. (a) Initial penetration of the protein film by surfactant leads to nucleation and domain growth (b) pushing out previously shielded regions of the proteins which become susceptible to pepsinolysis (c). For color details, please see color plate section.

Systematic studies using AFM to visualize the integrity of the interfacial networks, together with a range of interfacial techniques, have defined what happens during digestion [28, 30–33]. The results have led to unexpected findings. The changes in ionic strength, pH, and temperature encountered in the gastric environment can weaken interfacial protein networks, but the networks can remain intact [30, 31]. The nature of the oil phase is important and can stabilize the network [31, 33]. The nature of the oil determines the degree of unfolding of the protein at the interface, which in turn influences the strength of the network, but it also affects the structure of the proteins exposed in the aqueous phase and their susceptibility to enzymatic attack. In the absence of surfactants or biosurfactants, proteolysis (degradation of proteins by enzymatic attack) under gastric conditions leads to partial degradation of surface-adsorbed proteins, but the protein networks remain intact and strong. Partial displacement by surfactants force new regions of the proteins out into the aqueous phase, enhancing proteolysis, and can weaken and/or destroy the protein networks (Fig. 10.3). However, any approach used to strengthen the networks in order to reduce bile salt domain growth will also attenuate orogenic displacement in the stomach and thus inhibit the weakening effects

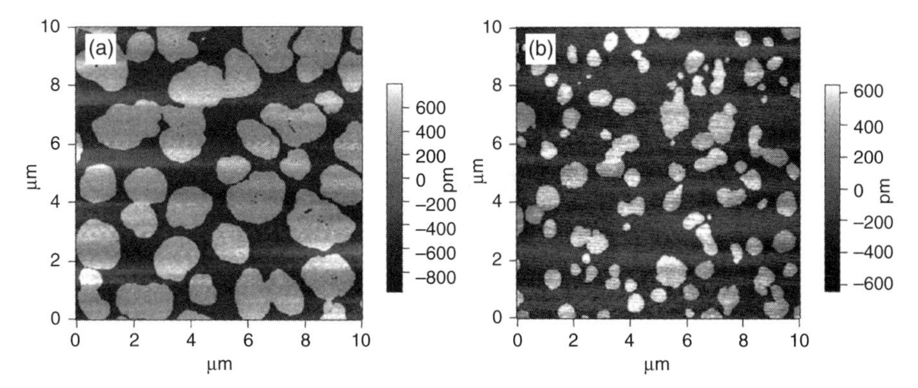

Figure 10.4 Effect of bile salts on the phase behavior of a mixed interfacial phospholipid film of dipalmitoylphosphatidylcholine (DPPC) and bile salt. Tapping mode topography images obtained in air of Langmuir–Blodgett films transferred at: (a) surface pressure $\pi = 18.7\ mN \cdot m^{-1}$, (b) $\pi = 24.2\ mN \cdot m^{-1}$. Bright regions, solid phase; dark regions, liquid phase.

of proteolysis. Thus the basic study of these interfacial networks suggests potential applications to moderate rates of lipolysis and fat uptake.

If protein-stabilized emulsions are broken during digestion, then the oils, like most fat and oils consumed in the diet, are emulsified through the adsorption of phospholipids. As in the previous study, the events that occur at the oil–water interface during passage through the upper stages of the gastrointestinal (GI) tract, can be recreated in the controlled environment of a Langmuir trough to allow investigation by AFM [34]. AFM is an excellent technique for characterizing the physical state (phase structure) of interfacial phospholipid films since various imaging modes are available that emphasize the orientation (topography), charge (scanning surface potential microscopy), and packing (frictional force microscopy) of the molecules. AFM imaging revealed the sensitivity of the packing of a phospholipid to the adsorption of the bile salts (Fig. 10.4). Within a certain range of surface pressures (which provides a measure of the interfacial packing), pure phospholipid monolayers exhibit coexisting phases—ordered "solid" phase and disordered "liquid" phase. As the surface pressure increases, the tighter packing of the molecules at the interface causes the solid-phase regions to grow at the expense of the more loosely packed liquid-phase regions. AFM topography imaging (where the grayscale depicts feature height) can discriminate the solid-phase regions from the surrounding liquid phase. Because they stand taller, they appear as bright islands within a darker background [Fig. 10.4(a)]. As bile salt invades the interfacial film, the surface pressure increases, which as explained would prompt growth of the ordered solid-phase islands for the pure phospholipid. However, the AFM image in Figure 10.4(b) clearly shows that in fact the opposite occurs and the solid-phase regions become smaller. The AFM images suggest that the bile salts disrupt the order in the film and, consequently, will be located in the liquid-phase regions. Since the bile salts mediate attachment of the lipase–colipase complexes to the interface, this suggests that another strategy for moderating lipolysis is to manipulate the phase behavior of the interfacial phospholipid layers. Recent visualization of the effects of galactolipids on the structure of mixed phospholipid–bile salt films by AFM is providing new insights into the potential roles of these compounds in the diet [34].

10.5 FORCE SPECTROSCOPY: MODEL EMULSIONS

Use of AFM as a microscope has generated new insights into the molecular structure of foods. AFMs image by measuring forces between the tip and the sample surface at the nanoscale, and this makes them unique when compared to other forms of microscopy [35]. The use of the AFM as a force transducer is termed force spectroscopy, and forces can be probed at the molecular or colloidal level. Force spectroscopy is discussed in detail in Chapter 3. In the latter case the AFM can be used to generate a model two-particle emulsion to probe underlying mechanisms of colloidal stability. Unlike other methods, the technique allows interactions between deformable droplets to be studied for the first time. Oil and water do not mix: Oil-in-water emulsions are created by providing energy to generate droplets of oil in water, which are stabilized by the adsorption of surface-active molecules such as proteins or surfactants.

As discussed earlier, AFM has proven useful in understanding the structures formed at the interface. The elastic skins formed by proteins yield long-term stability, whereas the high mobility of the surfactants permit rapid uptake but shorter lifetimes. The shelf life of an emulsion will depend on the rate at which collisions occur and the resultant flocculation and/or coalescence of the droplets. The suspended oil droplets of an emulsion are, of course, subjected to forces as the liquid they are in moves around. In food systems these can range from mild forces, such as gravitational driven buoyancy effects upon storage, to very high forces encountered during processing steps such as blending or heating. The integrity and adaptability of the interfacial films present on the droplets will determine to what extent the emulsion will be altered by such forces. This is an area ripe for study with AFM, which can push and pull small objects with extraordinary precision and quantify any resultant deformation, adhesion, or repulsion [35].

The first step in studying interacting oil droplets is to attach one to an AFM cantilever, a process that involves transferring weakly bound droplets from the surface of a glass slide under water to the end of an AFM cantilever [Fig. 10.5(a)]. The cantilever is composed of silicon or silicon nitride, which is slightly more hydrophobic than the glass and thus more attractive to the oil droplet [36]. More recently this method has been significantly refined with patches of defined surface chemistry being patterned onto the substrate and the end of AFM cantilevers to allow better control over the contact area and angle between the droplet and the cantilever [37]. The interfacial structure of the bare droplets attached to the substrate and the cantilever can be manipulated by adding surface-active species into the aqueous phase.

Once an interfacial structure has been established, then the forces generated between the droplets can be measured by pressing the captured droplet against a second droplet attached to the glass slide. This allows measurement of how the force

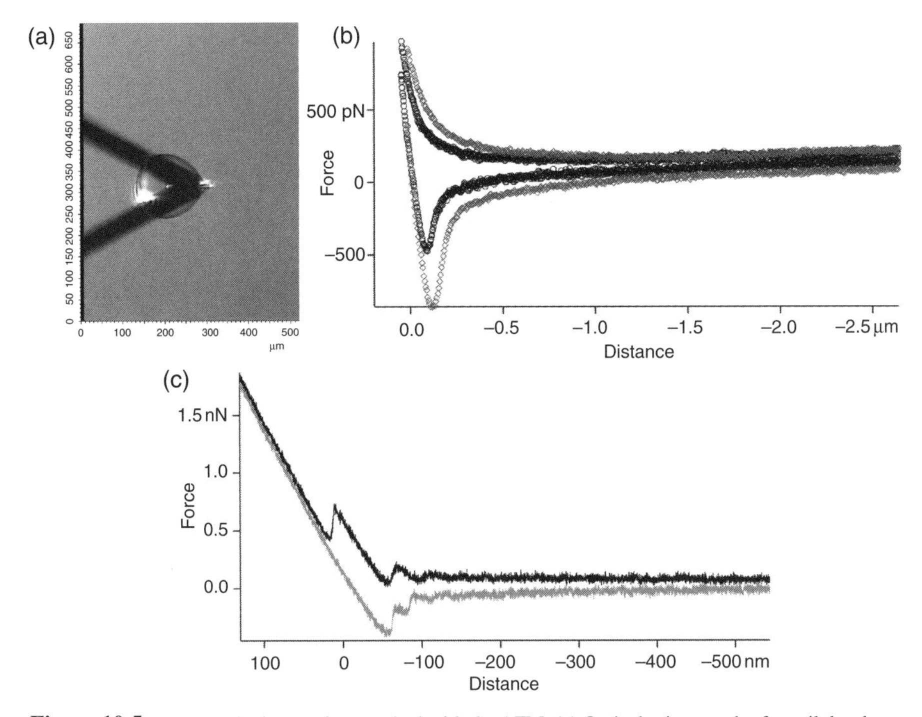

Figure 10.5 Oil droplet interactions probed with the AFM. (a) Optical micrograph of an oil droplet attached to an AFM cantilever. (b) Effect of droplet velocity on interdroplet hydrodynamic interaction forces; black curves $10\,\mu m \cdot s^{-1}$, gray curves $20\,\mu m \cdot s^{-1}$. (c) Oscillatory structural forces induced by the presence of nonadsorbing polymers in the aqueous phase; black curve, approach; gray curve, retract. Note: The distance axis roughly equates to droplet separation, however, the exact point of contact between deformable bodies is difficult to precisely determine.

of interaction between the two alters in response to various different parameters, such as approach and retract speed [Fig. 10.5(b)], ionic strength of the aqueous phase, for protein-stabilized interfaces the presence of competing surfactants (Fig. 10.6), and the influence of adsorbing or nonadsorbing polymers dissolved in the aqueous bathing solution in which the measurements are performed [Fig. 10.5(c)]. Such studies on deformable droplets are in their infancy [21, 36, 38–40], but they are already yielding new insights into behavior not seen when interfacial structures adsorbed onto hard nondeformable particles were used as model emulsions [41].

When droplets approach one another, the long-range repulsive interactions between them will lead to droplet deformation. At closer approach the force–distance curve will depend on the relative values of the approach speed and the rates of drainage of fluid between the droplets. At slow approach speeds the approach and retract curves are superimposable, but, when the approach speed and the rate of drainage become comparable, then the approach and retract curves show hysteresis effects as shown in Figure 10.5(b) (note the larger repulsion on close approach and the deeper adhesive minima upon retraction for the higher velocity data (gray curve). At slow approach speeds the force curves can be used to probe interfacial adsorption

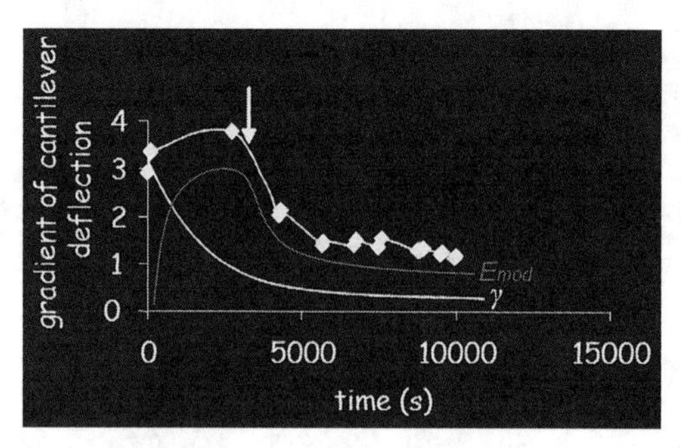

Figure 10.6 Effect of in situ exchange of the interfacial film on a pair of emulsion droplets. The yellow line links the AFM data points (arrow marks the time point for addition of surfactant). Red and white lines illustrate the overall trends for the interfacial elastic modulus and interfacial tension, respectively, as protein is exchanged for surfactant. Note, the Y axis values are depicted in arbitrary units. For color details, please see color plate section.

of surface-active species and the structures formed at the interface [21, 36]. Such studies provide the most realistic method for investigating the effect of interfacial structure on interactions between oil droplets in emulsions. The experimental system allows the correct unfolding of the proteins and their interaction to form networks at the interface. For example, Figure 10.6 shows how in situ exchange of an interfacial protein film for a surfactant film on a tetradecane droplet changes the deformability of the droplet. This was quantified by monitoring changes in the gradient of cantilever deflection (yellow line) as the droplets were forced together because a softer sample will exhibit a lower gradient for cantilever deflection than a harder sample. Adsorption of any species at the interface should decrease the interfacial tension (γ),as seen in the white curve in Figure 10.6, reducing the Laplace pressure within the droplet, making the droplets more deformable.

In practice the adsorption of protein leads to a hardening of the droplets, showing that droplet deformation is dominated by the elasticity of the protein skin covering the droplet [36]. The action of the surfactant in weakening the network is reflected in the increased deformability of the droplets, and this measure can be used to demonstrate orogenic displacement on finite-sized droplets in an aqueous medium [21, 36]. These studies validated the existence of orogenic displacement on droplets. This is important because it provides a basis for establishing whether changes found to strengthen networks in model studies will also achieve a similar effect in an emulsion, and the basis for examining the likely behavior of the resultant emulsions under different environmental conditions, or even the stability under simulated digestion regimes.

The absorption of the protein–polysaccharide complex, sugar beet pectin (which acts as an emulsion stabilizer) onto tetradecane droplets, generated a range of interfacial structures providing examples of normal electrostatic repulsion, steric repulsion, bridging, and depletion interactions [39]. The oscillatory shape of the force curves on close approach and separation of the droplets seen in Figure 10.5(c) illustrates that AFM force data on droplets can directly resolve a predicted effect from colloid theory, namely depletion interaction [42]. Depletion forces arise between colloidal particles when they are suspended in a liquid containing nonadsorbing polymers. As the particles approach one another, the polymer is squeezed out of the thinning liquid film that is caught between them, generating an osmotic gradient due to the localized depletion of polymer. The osmotic gradient results in a net adhesive force that pins the particles together, so that upon separation adhesion is observed [gray line Fig. 10.5(c)]. Before this occurs, however, the polymers must be pushed out of the closing gap between the particles and that requires additional work. In the oil droplet experiment as the final few layers of polymer are squeezed out, a modification to the force interaction between the droplets becomes resolvable as a series of blips on the approach curve [black line, Fig. 10.5(c)]. Such abrupt deviations are in stark contrast to the smooth trends seen in droplet force data where no polymers are present [Fig. 10.5(b)]. The origins of these type of force–distance curves have been successfully modeled in a recent ground-breaking study that demonstrated that deformable colloids such as oil droplets are particularly susceptible to such effects because they adapt their shapes in response to the changing forces caused by the nonadsorbed species (which can also include surfactant micelles) in the closing gap [43]. This opens up the potential for tailoring colloidal interaction within complex systems in a properly knowledge-based way for the first time.

10.6 FORCE SPECTROSCOPY: ORIGINS OF BIOACTIVITY

The health-promoting nature of carbohydrates in the diet has been recognized for many years, particularly in the area of dietary fiber and the beneficial effects this material has on gut health [44]. Dietary fiber includes nonstarch polysaccharides, which survive transit through the stomach and small intestine and exert an active role in the colon [45]. A primary role for dietary fiber is water retention and a bulking capacity of plant cell wall [46], and significant benefits are attributed to the fermentation of these polysaccharides by the colonic microflora of the carbohydrates to short-chain fatty acids [47]. Recently there has been interest in the potential for dietary polysaccharides to exert a more diverse range of bioactivities [48].

For example, there is considerable growing evidence that plant polysaccharides, such as the yeast and fungal-derived β-glucans [49] and modified forms of commercial citrus pectin may exhibit anticancer properties following oral consumption [50, 51]. In both cases the effects are attributed to the interaction of the carbohydrate material with mammalian lectins (proteins that have a structural affinity for specific carbohydrates). The important tumor-signaling molecule Galectin-3 (Gal-3) is a galactose binding lectin, and the hypothesis for the anticancer action of pectin (which is composed chiefly of galactose in one form or another) is that binding of the polysaccharide inhibits the biological activity of the Gal-3 [51] by preventing its association with natural receptors. This is potentially a rather neat deceptive tactic by an important dietary component since such interactions play decisive roles in several aspects of the spread of tumors [51]. Whether or not binding occurs between pectin and Gal-3, and the factors that affect it such as the role of molecular stereochemistry, is an eminently suitable subject for force spectroscopy.

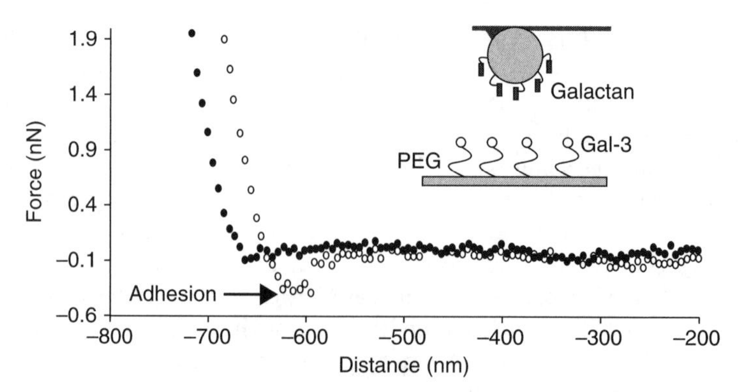

Figure 10.7 Force–distance curve demonstrating adhesion upon retraction (open circles) of a galactan-derivatised AFM tip from Gal-3 bound to a glass slide.

The modification of pectin can be mimicked by enzymatically generating the expected fragments and testing each of the pectin fragments for binding to Gal3 [52]. This was accomplished by attaching the pectin fragments to AFM tips using appropriate covalent linker chemistry [52–54]. The functionalized AFM tip was then pressed gently against an array of Gal-3 molecules tethered to the surface of a glass slide [52].

The interactions were monitored in an aqueous, buffered environment, and Figure 10.7 shows a typical force curve for one of the fragments, a pectin-derived galactan. Upon retraction of the functionalized cantilever from the Gal-3-coated surface of the slide, a notable adhesive event (labeled “adhesion” in Fig. 10.7) in the form of a negative peak is observed [52]. Lactose is a known specific inhibitor of Gal-3, and addition of this sugar to the liquid cell largely abolished the adhesive interactions, demonstrating that the interactions between the galactan and Gal-3 are specific and should inhibit the interaction of Gal-3 with natural receptors. The data is consistent with the hypothesis for the anticancer action of modified pectin and identifies pectin-derived galactans as the bioactive constituents. Such studies underpin research into whether such benefits can be obtained from consumption of fruits and vegetables, the cell walls of which contain pectin, and input into the possible development of new food supplements that could reduce the risk of the onset and progression of cancers.

10.7 CONCLUSIONS

In the last few decades AFM has matured from an interesting curiosity for picturing molecules to a sophisticated, although still not routine, tool for visualizing and understanding the behavior of complex biological systems. The wide variety of applications of AFM as an imaging technique have been reviewed elsewhere [4]. In this chapter a few selected case histories have been used to illustrate how the technique provides new insights into food structure and the impact of this in tackling previously intractable problems. While the use as microscope has become established and widely used, the potential applications as a force transducer to measure nanomechanical properties are emerging quite rapidly. The applications of force spectroscopy as a tool for probing molecular and colloidal interactions in food systems is likely to mirror the success of AFM as an imaging tool in the next few decades.

REFERENCES

1. Morris, V. J. *Trends Biotechnol* **29** (2011): 509–516.
2. Morris, V. J., Woodward, N. C., and Gunning, A. P. *J Sci Food & Agric* **91** (2011): 2117–2125.
3. Morris, V. J. In *Understanding and Controlling the Microstructure of Complex Foods*, McClements, D. J., Ed., Woodhead: Cambridge, UK, 2007, pp. 209–231.
4. Morris, V. J., Kirby, A. R., and Gunning, A. P. In *Atomic Force Microscopy for Biologists*, 2nd ed., Imperial College Press: London, 2010.
5. Phillips, G. O., and Williams, P. A. In *Handbook of Hydrocolloids*, 2nd ed., CRC Press: Boston, 2009.
6. Morris, V. J. In *Understanding and Controlling the Microstructure of Complex Foods*, McClements, D. J., Ed., Woodhead: Cambridge, UK, 2007, pp. 3–39.
7. Morris, V. J. In *The Chemical Physics of Food*; Belton, P. S, Ed., Blackwell: Oxford, UK, 2007, pp. 151–198.
8. Gunning, A. P., Kirby, A. R., Ridout, M. J., Brownsey, G. J., Morris, V. J. *Macromolecules*, 1996a **29**, 6791–6796.
9. Morris, V. J., and Gunning, A. P. *Soft Matter* **4** (2008): 943–951.
10. Dickinson, E., and Stainsby, G. *Colloids in Food, Applied Science Publishers*; London, York 1982.
11. Mackie, A. R., Gunning, A. P., Wilde, M. J., and Morris, V. J. *J Colloid Interface Sci* **210** (1999): 157–166.
12. Mackie, A. R., Gunning, A. P., Ridout, M. J., Wilde, P. J., and Morris, V. J. *Langmuir* **16** (2000): 2242–2247.
13. Gunning, A. P., Wilde, P. J., Clark, D. C., Morris, V. J., Parker, M. L., and Gunning, P. A. *J Colloid Interface Sci* **183**, (1996): 600–602.
14. Mackie, A. R., Gunning, A. P., Wide, P. J., and Morris, V. J. *Langmuir* **17** (2001): 6593–6598.
15. Woodward, N. C., Wilde, M. J., Mackie, A. R., Gunning, A. P., Gunning, P. A., Morris, V. J. *J Agric & Food Chem* **52** (2004): 1287–1292.
16. Woodward, N. C., Gunning, A. P., Mackie, A. R., Wilde, P. J., and Morris, V. J. *Langmuir* **25** (2009): 6739–6744.
17. Mackie, A. R., Gunning, A. P., Ridout, M. J., Wilde, P. J., and Patino, J. R. *Biomacromolecules* **2** (2001): 1001–1006.
18. Clark, D. C., Coke, M., Mackie, A. R., Pinder, A. C., and Wilson, D. R. *J Colloid Interface Sci* **38** (1990): 207.
19. Wilde, P. J., and Clark, D. C. *J Colloid Interface Sci* **155** (1993): 48.
20. Wilde, P. J., Mackie, A. R., Husband, F. A., Gunning, A. P., and Morris, V. J. *Adv Colloid & Interface Sci* **108–9** (2004): 63.
21. Woodward, N. C., Gunning, A. P., Maldonado-Valderrama, J., Wilde, P. J., Morris, V. J. *Langmuir* **26** (2010): 12560–12566.
22. Christiano, S. P., and Fey, K. C. *J Ind Microbiol Biotechnol* **30** (2003): 13.
23. Jung, H., Dalal, C. K., Kuntz, S., Shah, R., and Collier, C. P. *Nano Lett* **4** (2004): 2171.
24. Deng, Y., Zhu, X. Y., Kienien, T., and Guo, A. *J Am Chem Soc* **128** (2006): 2768.
25. Eckmann, D. M., and Diamond, S. L. *Anesthesiology* **100** (2004): 77.
26. Bonaccurso, E., Butt, H-J., Franz V., Graf, K., Kappi, M., Loi, S., Niesenhaus, B., Chemnitz, S., Böhm, M., Petrova, B., Janus, U., and Spiess, H. W. *Langmuir* **18** (2002): 8056.
27. Golding, M., and Wooster, T. J. *Curr Opin Colloid Interface Sci* **15** (2010): 90–101.
28. Maldonado-Valderrama, J., Woodward, N. C., Gunning, A. P., Ridout, M. J., Husband, F. A., Mackie, A. R., Morris, V. J., and Wilde, P. J. *Langmuir* **24** (2008): 6759–6767.

29. Woodward, N. C., Gunning, A. P., Wilde, P. J., Chu, B. S., and Morris, V. J. In *Gums and Stabilisers for the Food Industry 15*, Williams, P. A., and Phillips, G. O., Eds., Royal Society of Chemistry, Cambridge, UK, (2010): pp. 367–376.
30. Maldonado-Valderrama, J., Gunning, A. P., Ridout, M. J., Wilde, P. J., and Morris, V. J. *Eur Phys J E* **30** (2009): 165–174.
31. Maldonado-Valderrama, J., Miller, R., Fainerman, V. B., Wilde, P. J., and Morris, V. J. *Langmuir* **26** (2010):15901–15908.
32. Maldonado-Valderrama, J. Gunning, A. P., Wilde, P. J., and Morris, V. J. *Soft Matter* **6** (2010): 4908–4915.
33. Maldonado-Valderrama, J., Mulholland, F., Wilde, P. J., and Morris, V. J. *Soft Matter* **8**, (2012): 4402–4414.
34. Chu, B. S., Gunning, A. P., Rich, G. T., Ridout, M. J., Faulks, R. M., Wickham, M. S. J., Morris, V. J., and Wilde, P.J. *Langmuir* **26** (2010): 9782–9793.
35. Butt, H. J., Cappella, B., and Kappl, M. *Surface Sci Rep* **59** (2005): 1–152.
36. Gunning, A. P., Mackie, A. R., Wilde, P. J., and Morris, V. J. *Langmuir* **20** (2004):116–122.
37. Vakarelski, I. U., Manica, R., Tang, X., O'Shea, S. J., Stevens, G. W., Grieser, F., Dagastine, R. R., and Chan, D. Y. C. *Proc Natl Acad Sci USA* **107** (2010): 11177.
38. Gromer, A., Kirby, A. R., Gunning, A. P., and Morris, V. J. *Langmuir* **25** (2009): 8012–8018.
39. Gromer, A., Penfold, R., Gunning, A. P., Kirby, A. R., and Morris, V. J. *Soft Matter* **6** (2010): 3957–3969.
40. Dagastine, R. R., Stevens, G. W., Chan, D. Y. C., and Greiser, F. *J Colloid Interface Sci* **273** (2004): 339.
41. Tabor, R. F., Grieser, F., Dagastine, R. R., and Chan, D. Y. C. *J Colloid Interface Sci* **371** (2012): 1–14.
42. Asakura, S., and Oosawa, F. *J Chem Phys* **22** (1954): 1255–1256.
43. Tabor, R. F., Chan, D. Y. C., Grieser, F., and Dagastine, R. R. *J Phys Chem Lett* **2** (2011): 434–437.
44. Bingham, S. A., Day N. E., Luben, R., Ferrari, P., Slimani, N., Norat, T., Clavel-Chapelon, F., Kesse, E., Nieters, A, Boeing, H., Tjonneland, A., Overvad, K., Martinez, C., Dorronsoro, M., Gonzalez, C. A., Key, T. J., Trichopoulou, A., Naska, A., Vineis, P., Tumino, R., Krogh, V., Bueno-de-Mesquita, H. B., Peeters, P. H. M., Berglund, G., Hallmans, G., Lund, E., Skeie, G., Kaaks, R., and Riboli, E. *Lancet* **361** (2003): 1496–1501.
45. DeVries, J. W., Prosky, L., Li B., and Cho, S. *Cereal Foods World* **44** (1999): 367–369.
46. Armstrong, E., Eastwood, M. A., and Brydon, W. G. *Brit J Nutr* **69** (1993): 913–920.
47. Wong, J. M. W., de Souza, R., Kendall, C. W. C., Emam, A., and Jenkins, D. J. A. *J Clin Gastroenterol* **40** (2006): 235–243.
48. Keenan, J. M., Goulson, M., Shamliyan, T., Knutson, N., Kolberg, L., and Curry, L. *Brit J Nutr* **97** (2007): 1162–1168.
49. Chan, G. C., Chan, W. K., and Sze, D. M. *J Hematol Oncol* **2** (2009): article 25, doi:10.1186/1756-8722-2-25.
50. Nangia-Makker, P., Conklin, J., Hogan, V., and Raz, A. *Trends Mol Med* **8** (2002): 187–192.
51. Glinsky, V. V., and Raz, A. *Carbohyd Res* **344** (2009): 1788–1791.
52. Gunning, A. P., Bongaerts, R. J. M., and Morris, V. J. *FASEB J* **23** (2009): 415–424.
53. Bhatia, S. K., Shriver-Lake, L. C., Prior, K. J., Georger, J., Calvert, J. M., Bredehorst, R., and Ligler, F. *Anal Biochem* **178** (1989): 408–413.
54. Hinterdorfer, P., Gruber, H. J., Kienberger, F., Kada, G., Riener, C., Broken, C., and Schindler, H. *Colloids Surf B* **23** (2002): 115–123.

Chapter 11

Microcantilever Sensors for Petrochemical Applications

Alan M. Schilowitz

ExxonMobil Research and Engineering, Annandale, NJ

11.1 INTRODUCTION

Miniature sensor devices with high sensitivity and chemical specificity have been an object of significant research in recent years. Much of this activity was spurred by interest in medical devices and national security sensing applications where small sensor size, ultrahigh sensitivity, small sample size, and sensor redundancy can be vitally important. To a large extent developments in the micro- and nanoregimes were enabled by the broad availability and deployment of scanning probe microscopy (SPM) and, in particular, atomic force microscopy (AFM). This chapter discusses a microcantilever sensing platform that is based on AFM hardware but without the need for scanning or imaging. The equipment for assembling a microcantilever sensor is relatively straightforward [1]. An AFM microcantilever (MC) is used as a sensing probe, and ancillary hardware and software are needed to monitor the vertical up and down motion of the cantilever as it is exposed to analytes. By virtue of the cantilever's small size, this platform has ultrahigh sensitivity. Chemical specificity is imparted by modifying the surface properties of the cantilever. Detection accuracy and broad applicability can be imparted by using arrays of cantilever sensors that can add redundancy or can be uniquely treated for detection of different analytes [2–4].

Chemical and physical sensors based on AFM microcantilevers were initially advanced in the mid-1990s at IBM Zurich Laboratory [5, 6] and Oak Ridge National Laboratory [7–11] and soon thereafter applied to the study of self-assembled monolayers at the IBM Zurich Laboratory [12]. Subsequently, the field has expanded into sensors for many applications where small size and ultrahigh sensitivity are beneficial.

Scanning Probe Microscopy in Industrial Applications: Nanomechanical Characterization,
First Edition. Edited by Dalia G. Yablon.

Microcantilever utilization has been stimulated by wide commercial availability in numerous sizes and shapes and at low cost, which makes them essentially disposable.

Almost all published research on MC sensors has been conducted in liquid aqueous or gaseous systems [13] at relatively mild temperature and pressure because these are the environments of interest for medical/biological and national security applications. Medical applications such as blood analysis generally involve very small samples that can easily be accommodated by microcantilevers. Both biological [14–19] and national security samples (e.g., explosive detection [20–25]) may require very high sensitivity to ultralow concentration of analyte. These are also relatively clean environments with little interference to complicate measurements. Microcantilever sensors are ideal for these applications because they are very sensitive and work best in clean environments.

This chapter will deal with applications in the petroleum industry where numerous uses potentially exist for small, highly sensitive MC sensors. Most relevant are applications for hydrocarbon analysis where sensitivity, speed, and accuracy may be important. Limitations of such MC sensors for hydrocarbon analysis will also be discussed. Hydrocarbons are often viscous and complex mixtures that may present difficulties for tracking MC motion—if the MC moves at all. Some common hydrocarbon mixtures contain significant quantities of surface-active molecules that can interfere with any detection scheme dependent on surface activity. Hydrocarbons also have higher refractive indices than air and water, which imposes challenging geometrical boundaries on MC optical detection schemes. Nevertheless, due to their small size, high sensitivity, and relative simplicity, MC sensors present useful sensing capabilities in hydrocarbon liquids that can be applied in industrial applications in general and the petroleum industry in particular. Examples of measurement applications important to the petroleum industry that will be described are viscosity, temperature, acidity, mercaptan detection, and quantification of corrosivity. Microcantilever sensors are especially amenable to rapid high-throughput analysis of small samples which can be of high value in the petroleum laboratory.

Since the hardware used for MC sensing is essentially identical to that used for AFM measurements-described in earlier chapters (see Chapter 1), it will be left primarily to others to describe the state of current hardware and recent hardware developments. The focus here will be on applications. A brief hardware description will be given focusing on the essential sensing components. It is important to note that most references are to laboratory measurements and research applications and not field implementations of miniature sensing devices. In fact, much MC sensor development research has been done using commercial AFM equipment, although some work has been done using purpose-built sensing hardware.

11.2 BACKGROUND

Microcantilever sensors operate in two modes: dynamic and static. In dynamic mode the sensor functions as a small oscillating mass balance and detects mass changes due to surface adsorption where the surface may have been modified to adsorb chemical species of interest. Similarly, the vibrating sensor can be used to measure viscosity where the surrounding environment imposes a viscous drag resistance on the cantilever.

11.2.1 Hardware Setup

In all sensing applications it is necessary to expose the microcantilever to the liquid or gas analyte in such a way that motion of the microcantilever can be tracked. This is essentially the function of an AFM head, and most commonly an AFM will be used for making such sensing measurements in the laboratory. In dedicated sensing applications, a purpose-built apparatus can be advantageous. Figure 11.1 shows a typical experimental setup for tracking the motion of a microcantilever using a laser diode and position-sensitive diode (PSD). In the case of liquid analytes an AFM liquid cell can be used. One small complication arises in the case of hydrocarbon systems. Due to the relatively high refractive index of hydrocarbons relative to air and water and the restricted optical path of some AFMs, it may not be possible to achieve optical alignment within the confines of the AFM head [26]. In such cases it is desirable to use a dedicated sensing system for hydrocarbons with an open optical path to achieve optical alignment. All data shown in this chapter were generated using the optical detection method. An example of a purpose-built flow cell for making measurements is shown in Figure 11.2. The cell has a glass lid to allow optical access to the cantilever while enclosing fluid inside. Problems associated with optical alignment can be mitigated by using active microcantilevers such as piezoresistive cantilevers [27–31]. Self-sensing integrated piezoresistive microcantilevers eliminate the need for optical alignment and can simplify the measurement. This is especially valuable where a self-contained sensor system with integrated

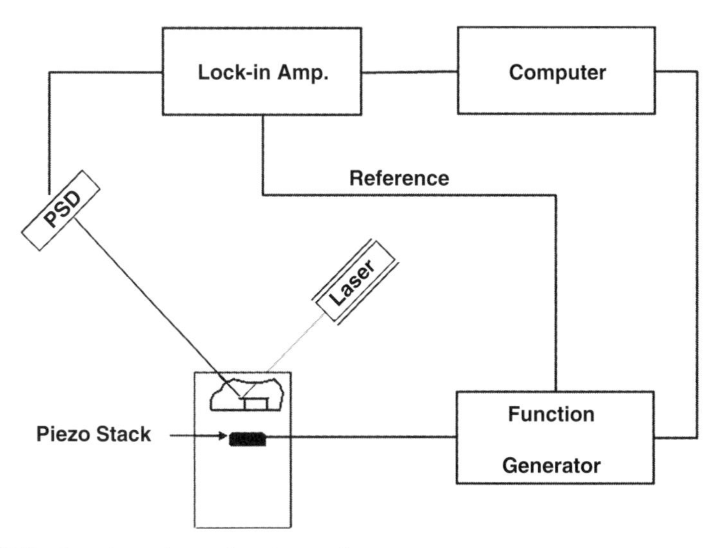

Figure 11.1 Experimental setup for microcantilever sensor.

Figure 11.2 Liquid flow cell with transparent lid for making microcantilever sensor. Inlet is to left and outlet is to right. Wires in foreground are connected to embedded piezoelectric stack.

readout is desired. However, such microcantilevers are expensive and have inherently high noise levels and corresponding lower sensitivity. The optical detection method affords greater flexibility and sensitivity especially in static MC applications that will be described below.

Any system that is dependent on optical tracking of MC motion is also limited by optical density of the fluid medium. The sensor will not work if the fluid absorbs light emitted by the tracking laser. Alternatively, particulates will scatter laser light, making the sensor inoperable. We have found that air bubbles, especially in a flow cell, can also be problematic. Careful cell design may be necessary to minimize bubble interference.

11.2.2 MC Sensing in Dynamic Mode

If the cantilever is to be operated in dynamic mode, as noted below, than it is also necessary to mount a piezoelectric actuator underneath the MC to activate the MC into resonance. It is possible to rely on thermal excitation of the cantilever if it is mounted in air. However, this requires very high sensitivity to cantilever motion [32]. Cantilever oscillation can be captured by exciting the cantilever into resonance using a limited number of excitation pulses to the piezoelectric stack at or near the microcantilever's resonance frequency and monitoring the microcantilever's decaying oscillatory motion on the PSD. This will result in a ringdown curve as shown in Figure 11.3(a). The rapid oscillation is indicative of the resonance frequency. The relatively slow decay of the oscillation amplitude is indicative of Q, the quality factor, a measure of energy dissipated by the cantilever. A longer timescale for the ringdown is associated with a larger Q value.

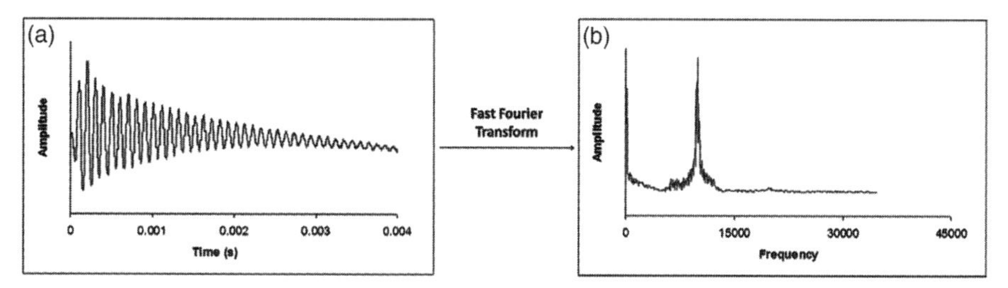

Figure 11.3 (a) Microcantilever ring down in air after excitation is turned off. (b) Resonance spectrum of microcantilever after Fourier transform.

Frequency can be directly derived from the ringdown data by taking the Fourier transform as shown in Figure 11.3(b).

Alternatively, a function generator can be used to provide a sinusoidal excitation pulse to vibrate the cantilever while analyzing the oscillating signal on the PSD with a lock-in amplifier and using the function generator pulse as the lock-in reference signal. Frequency is continuously swept from a value below the resonance frequency to a value above the resonance frequency. In this way amplitude versus frequency of the cantilever's motion can be obtained along with phase information, which can also be useful.

In its simplest embodiment the cantilever's oscillatory motion can be modeled as a damped harmonic oscillator using the following equation:

$$m\frac{d^2x}{dt^2}+c\frac{dx}{dt}=-kx \tag{11.1}$$

where m is mass, c is the damping coefficient, x is the cantilever displacement, and k is the spring constant. The damping coefficient can be related to the quality factor, Q, as a measure of energy dissipation per cycle:

$$Q=\frac{\sqrt{mk}}{c} \tag{11.2}$$

where Q can be an effective surrogate for measuring viscosity. Figure 11.4 shows a typical resonance curve for an oscillating microcantilever in air acquired by sweeping frequency and analyzing PSD output with a lock-in amplifier. A fit of the data to the harmonic oscillator function from Eq. (11.1) is also shown. Using a lock-in amplifier, it is also possible to monitor the phase angle between excitation and MC motion. Phase can also be used to locate the phase resonance frequency, which occurs at the frequency at which phase undergoes the steepest change.

More accurate models of MC motion have been derived by modeling the cantilever as an oscillating flexible bar fixed at one end [33]. Models for flexible bars, which include internal and external damping, such as Eq. (11.3) [26, 34] yield more accurate descriptions of MC motion:

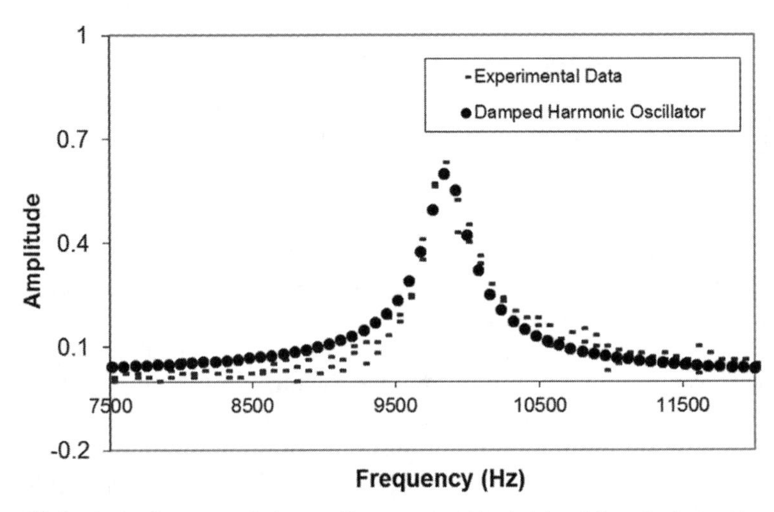

Figure 11.4 Resonance curve of microcantilever generated in air (–) and fit to the damped harmonic oscillator function. Reprinted from [26] with permission.

$$\frac{EI}{A}\frac{\partial^4 u(x,t)}{\partial x^4}+\rho\frac{\partial^2 u(x,t)}{\partial t^2}+\beta\frac{\partial u(x,t)}{\partial t}+\gamma\frac{\partial^4 u(x,t)}{\partial x^3 \partial t}=0 \tag{11.3}$$

where $u(x, t)$ is the vertical deflection at position x and at time t, E is Young's modulus, I is the cross-sectional moment of inertia, A is the cross-sectional area of the cantilever, ρ is the effective density, β is the external damping coefficient, and γ is the internal frictional damping coefficient of the cantilever. In a perfect vacuum, energy would still be dissipated by an oscillating cantilever due to the internal friction in the silicon crystal lattice. Vibrational characteristics of oscillating MC beams exposed to viscous fluids have been worked out in detail by Sader and collaborators [35, 36].

11.2.3 MC Sensing in Static Mode

The second MC sensing mode is known as static mode. In static mode the cantilever is still vibrating due to thermal excitation; however, this oscillatory motion is ignored, and the average or equilibrium position of the cantilever is tracked as the cantilever deflects due to stress induced by adsorption on the MC surface. The stress is generated by coating the cantilever on one side with a surface-active material. For example, a native silicon cantilever might be coated on one side with gold. In this way the MC will bend due to stress induced by a chemical imbalance according to Stoney's formula [37, 38], which quantifies the degree of cantilever bending imposed by an imposed stress:

$$\sigma=\frac{\Delta z\, Et^2}{4L^2(1-\nu)} \tag{11.4}$$

where σ is stress, z is deflection, E is Young's modulus, t is cantilever thickness, L is cantilever length, and ν is Poisson's ratio.

Application of an activating surface coating to one side of the cantilever will induce a stress imbalance to the cantilever, causing it to bend even before any chemical sensing has been attempted. This initial deflection point is considered the neutral or zero reference position. When analyte adsorbs to the surface-activated layer, the cantilever will deflect again with respect to its new reference point. Typically, relatively floppy or soft microcantilevers with spring constants less than 0.5 N/m are used for static mode sensing. Care must be taken to ensure that an opposing stress is not induced on the "bottom" or underside of the cantilever due to adsorption on that side of the microcantilever. This might cause a nulling effect resulting in spurious results. The underside of the cantilever must be made inert to adsorption in the sensing environment. If stress causes the cantilever to bend downward, away from the activated surface, the stress is referred to as compressive in that it causes compression of the underlying cantilever. If stress causes the cantilever to bend upward toward the activated surface, than the stress is referred to as tensile.

The electronics needed for a static mode sensor are relatively simple in that only the slow average direct current (DC) output signal of the PSD needs to be collected. This generally does not require much filtering or averaging. It is often not necessary to get an absolute quantitative measure of deflection if a single microcantilever is used for sensing. If necessary, it is possible to calculate an absolute measure of deflection [39]. For best sensitivity it is suggested that a reference cantilever be used to correct for thermal effects. It has been estimated that under carefully controlled conditions 10^{-16} grams of adsorbate can be detected under ideal conditions.

11.3 APPLICATIONS

11.3.1 Physical Property Measurements

Temperature and Heat

In its simplest embodiment the MC sensor is a thermostat. If the native cantilever is coated on one side with a dissimilar material (e.g., gold coated onto a silicon cantilever), then the cantilever will bend due to the dissimilar coefficients of thermal expansion. While this may be a useful platform for measuring temperature, the primary manifestation of this phenomenon is as an interference in most other measurements in static mode and requires significant attention to ensure valid and reproducible measurements by careful temperature control or by using a compensating reference microcantilever. Even in dynamic mode, temperature sensitivity of the MC sensor might be possible, if the temperature changes are relatively large, due to changes in Young's modulus that change resonant frequency. Microcantilever sensitivity to external conditions such as temperature is both a strength and a weakness of using such ultrasensitive devices in uncontrolled environments outside the laboratory.

As noted above, a bimaterial cantilever will bend when temperature changes, and MC sensors have been shown to be ultrasensitive thermometers and calorimeters [40]. The ultimate sensitivity to heat has been estimated to be about 10 pW. Due to low mass and relatively fast response, energy sensitivity has been estimated to be on the order of 20 fJ [41]. Similarly, it has been demonstrated that by treating the cantilever with an optically absorbing surface coating it can be used as a photothermal sensor, which could be used for imaging or possibly as a detector on an infrared spectrometer [10].

Viscosity and Density

Viscosity and density are important physical properties routinely measured on hydrocarbon fluids. Some petroleum products, for example, lubricants, are categorized by viscosity grade, and their performance may depend on viscosity, among other things. Other refinery streams are characterized by density. While many types of viscometers and densitometers are available, microcantilever-based devices enable measurements on tiny samples of less than a milliliter.

Vibrating microcantilevers have been applied in several embodiments for measuring fluid viscosity ranging from gases to relatively heavy liquids. Oden et al. [42] demonstrated the effectiveness of MC viscometers for measuring viscosity of water and glycerol mixtures up to a viscosity of 2 P. Microcantilever response was modeled by assuming that the fluid occupied a sphere around the microcantilever, and frequency change was due to an additional effective mass of fluid adsorbed to the MC surface. Reasonable agreement between experiment and model was achieved at low viscosities. Both Q and frequency could be used for measuring viscosity.

Thundat et al. [43] describes a slightly different type of viscometer where a microcantilever is excited into stable resonance and then switched off. Vibration amplitude is measured as a function of time giving the ring down time, which is the rate of energy dissipation from which Q can be directly calculated. Resonant frequency is also measured. These quantities can then be used to calculate fluid viscosity and density. This method requires calibration in vacuum or fluids of known properties.

A different embodiment of an MC viscometer patented by Oden [44] describes a flexible beam moved through a viscous fluid at constant velocity. The measured static beam deflection is an indication of fluid viscosity. As with other viscosity sensors this one also requires calibration to determine the relationship between viscosity and deflection. However, correcting for density differences is not anticipated in the method.

Blood analysis is one application for which microsensors would be ideal due to small sample availability. Grzegorzewski [45] has patented a disposable biosensor for measuring blood coagulation time based on a small vibrating piezoelectric viscometer, which could also be accomplished with a microcantilever sensor.

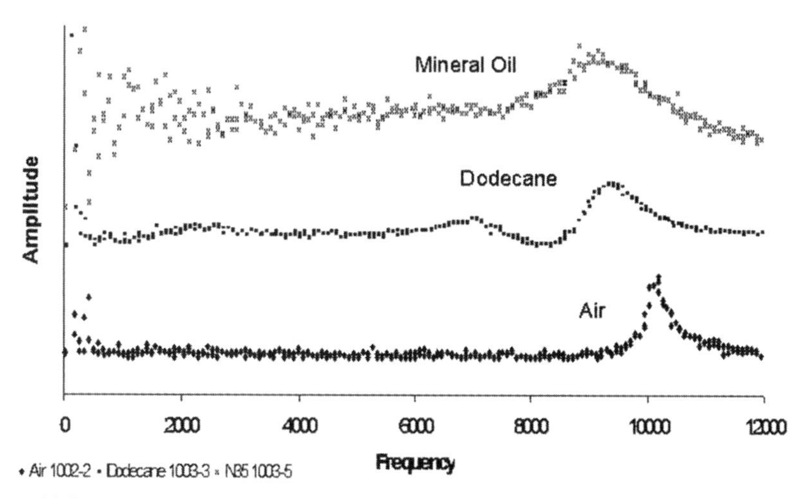

Figure 11.5 Resonance curves of microcantilever generated in air, dodecane, and mineral oil.

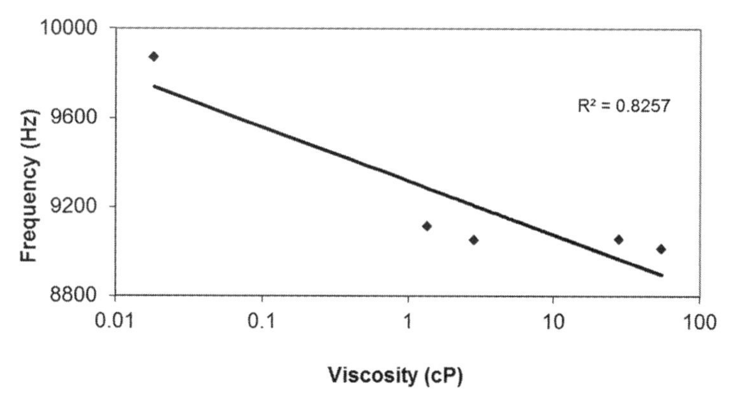

Figure 11.6 Plot of microcantilever resonant frequency vs. viscosity for air, dodecane, hexadecane, and two heavier mineral oils. (The measurements in Figure 11.5 and 11.6 were made with similar but different microcantilevers.)

An example of the effect of viscosity on the resonance curve of a microcantilever is shown in Figure 11.5 for air, dodecane, and a hydrocarbon mineral oil basestock. As viscosity increases, resonance frequency decreases, and the resonance peak broadens due to increased energy dissipation. A plot of the change in microcantilever resonance frequency with viscosity is shown in Figure 11.6 for air, dodecane, hexadecane, and two heavier hydrocarbon mineral oil basestocks. In this case fluids were introduced into the flow cell depicted in Figure 11.2. An immersed microcantilever was excited with a piezoelectric stack mounted in the cell directly underneath the base of the microcantilever using the experimental setup shown in Figure 11.1.

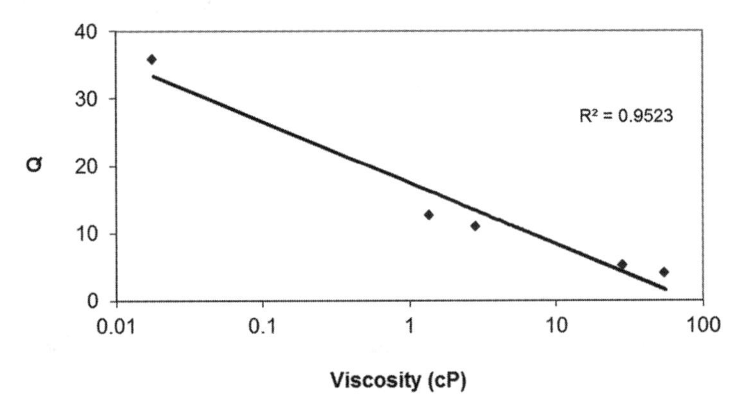

Figure 11.7 Plot of microcantilever quality factor (Q) vs. viscosity for air, dodecane, hexadecane, and two heavier mineral oils.

Figure 11.7 shows the effect of viscosity on the microcantilever quality factor, Q. While viscosity has a measureable effect on frequency, the effect on Q is also pronounced and appears to be at least as good an indicator of viscosity as frequency.

11.3.2 Chemical Measurements

A significant benefit of microcantilever sensors and the focus of much research has been on adaptation to chemical sensing. In dynamic mode the mass sensitivity of a microcantilever mass balance is estimated to be 10^{-16} grams, which is more sensitive than a quartz crystal microbalance. If the physical dimensions of the cantilever are reduced further to a length of less than 5 μm and a thickness on the order of hundreds of nanometers, it might be possible to detect a single strand of DNA (deoxyribonucleic acid) [46]. It is this high sensitivity combined with small size that has led to interest in MC chemical sensors. As noted above, almost all applications have been demonstrated for gas analysis and for aqueous liquids.

Since the development of MC sensors in the early to mid-1990s, many chemical sensing applications have been anticipated and developed. Early applications were for detection of mercury vapors [9], vapor-phase mercaptans [47], trinitrotoluene vapor [48], humidity [48], calcium ions [49, 50], hydrogen [51], mercury ions [52], aromatics [53], and various gases [6]. As noted above, MC sensors are especially amenable to biological systems, and biochemical sensing applications include sensors for detecting lipoproteins [54], DNA hybridization [55], glucose [56], and prostate specific antigen [57]. Essentially, all applications depend on developing an effective surface treatment for adsorbing the analyte of interest. Different surface treatments aside from pure metals (e.g., gold) have been used. For example, hydrogels [58] and coordination polymers such as metal organic and zeolitic frameworks have been tried [59]. Biological systems typically require specific biological agents for binding analytes. The high sensitivity of surface-treated microcantilevers can also be a disadvantage because poor selectivity can be associated with high sensitivity.

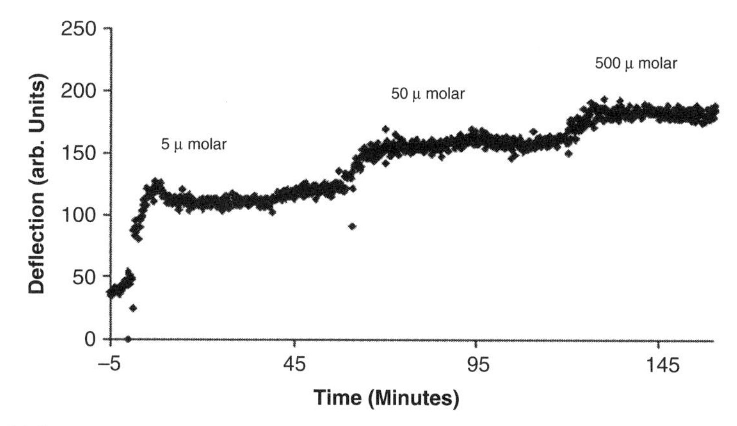

Figure 11.8 Plot of microcantilever bending with time upon exposure of a silicon microcantilever coated with gold on one side to nonanethiol dissolved in heptane at three different concentrations—5, 50, and 500 μM. Deflection units are arbitrary.

It may be hard to develop a surface treatment that is sensitive yet specific enough to avoid interferences. For example, it may be difficult to develop a highly sensitive binding agent specific to trinitro toluene (TNT) but not sensitive to nonexplosive aromatic compounds containing oxygen and nitrogen.

One of the first applications of chemical sensing with a microcantilever sensor was the detection of mercaptan vapor with gold-coated silicon cantilevers in static mode [12]. Analysis for sulfur-containing molecules is also of interest in the petrochemical business since sulfur is a common element in crude oil. In this experiment a silicon microcantilever was coated on one side with a thin layer of gold and exposed to alkanethiol vapor with differing alkane chain lengths. Cantilever bending occurred due to a compressive stress applied by formation of an alkanethiol self-assembled monolayer (SAM) on the gold surface. Stress increase could be observed in real time as adsorption occurred. In fact, different degrees of bending (i.e. stress) occurred depending on alkane chain length. This was attributed to the degree of packing, which could be established in the two-dimensional SAM crystal phase. A more ordered and dense monolayer resulted in larger compressive stress and more deflection.

Figure 11.8 shows results from a similar experiment where a single gold-coated silicon microcantilever was exposed consecutively to increasing concentrations of nonanethiol dissolved in liquid heptane. The results in Figure 11.8 show that the cantilever bends more with increasing concentration. Presumably, increased bending is due to higher surface coverage of the gold with adsorbed thiol. This is demonstrated in Figure 11.9, which shows a plot of nonanethiol concentration against the amount of bending at equilibrium achieved with exposure to increasing concentration of nonanethiol. Using the assumption that amount of bending is directly proportional to surface coverage and applying the standard assumptions made in Langmuir adsorption theory, the data in Figure 11.9 can be fit to the Langmuir adsorption isotherm equation by replacing surface coverage with degree of bending:

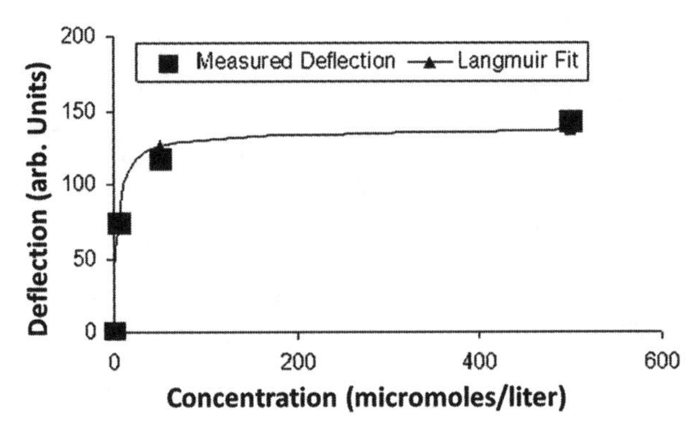

Figure 11.9 Plot of the equilibrium deflection of a silicon microcantilever coated on one side with gold at different concentrations of nonanethiol dissolved in heptane. Data are fit with the Langmuir adsorption isotherm equation. Deflection units are arbitrary. Reprinted from [60] with permission.

$$\text{Deflection} \approx \theta = \frac{CK}{1+CK} \tag{11.5}$$

where θ is the fractional coverage of the surface, K is the equilibrium constant for adsorption, and C is a constant. Fitting microcantilever deflection data to this equation assumes that deflection is proportional to the fractional coverage of adsorbate on the microcantilever surface, θ. Furthermore, since

$$\Delta G_{\text{adsorption}} = -RT \ln K \tag{11.6}$$

the Gibbs free energy of adsorption can also be measured directly from the deflection data. For nonanethiol in n-heptane it was found that $\Delta G_{\text{adsorption}} = -7.2\,\text{kcal/mol}$ [60].

The goodness of fit of the deflection data to the Langmuir isotherm shown in Figure 11.9 demonstrates that for this MC system the standard Langmuir adsorption assumptions hold true, and relative cantilever deflection can be used as a surrogate for quantifying analyte concentration. Obviously, each application will require an independent confirmation of the quantitative relationship between deflection and concentration. This is no different than would be required for any new analytical chemistry test method application.

One further complication regarding the relationship between MC deflection and analyte concentration is the nature of the adsorbate structure on the MC surface. As shown in Ref. [12] deflection is not a matter of concentration alone. Ultimately, deflection is caused by stress created by intermolecular forces among adsorbate molecules. In the case of SAM formation, it is known [61] that monolayer formation evolves with time as the monolayer structure evolves, eventually forming a dense crystalline phase. Similar evolving surface stress development could lead to misleading MC sensor results by assuming a simple relationship between surface stress and concentration. Nevertheless, Langmuir adsorption behavior accurately describes many adsorption processes. It might also be safer to use a final equilibrium value of cantilever

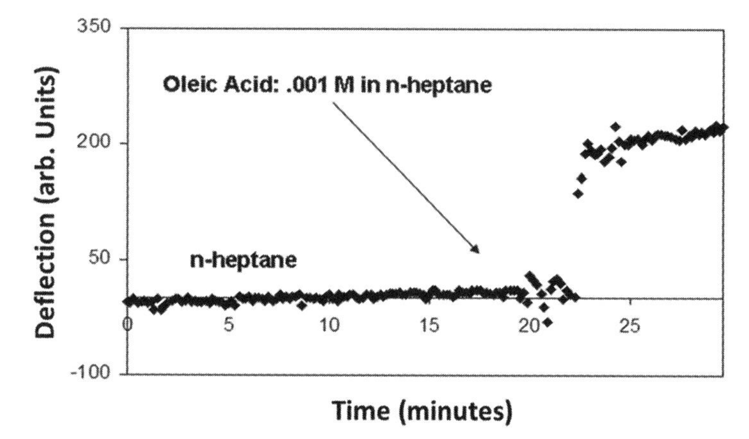

Figure 11.10 Deflection time trace of microcantilever coated on one side with gold that has been stabilized in *n*-heptane. A solution of 0.001 M oleic acid was injected as indicated by the arrow. Note that increasing deflection in these figures represents a compressive deflection of the cantilever. Deflection units are arbitrary. Reprinted from [60] with permission.

deflection after any packing readjustments have had time to occur. Figure 11.9 is based on using an equilibrated value for the MC deflection and therefore relative deflection can be used to measure concentration. However, one needs to be careful not to compare deflections representing different levels of adsorbate film stress development. In such a case deflection may not be representative of analyte concentration.

In addition to measuring chemical concentration that is the typical objective of any chemical sensor, MC sensors used in static mode afford the ability to monitor concentration with high temporal fidelity—given the caveat of the last paragraph—with relative ease as shown in Figure 11.8 for nonanethiol and Figure 11.10 for oleic acid.

As noted above, a key question is the relationship between stress-induced deflection and concentration. While the Langmuir fit to the data in Figure 11.9 demonstrates that equilibrium deflection is directly related to surface coverage for the system studied, it does not demonstrate that this is true throughout the adsorption process. Is deflection at every point in the adsorption process directly related to surface coverage? It has been suggested [62] that stress development on the microcantilever can be effected by SAM rearrangement during the adsorption process. Accordingly, MC deflection at two different times in the adsorption process may not indicate the relative surface coverage at those two times. This will depend on the nature of the system. In multilayer adsorption the analysis may be more complicated. As always, careful calibration will be necessary in all cases.

Figure 11.11 shows a series of experiments demonstrating the impact of competitive adsorption on deflection. Initially, a microcantilever coated with gold on one side was immersed in *n*-heptane and left until a stable equilibration deflection was established. After about 20 minutes a solution of oleic acid (0.001 M) in *n*-heptane was injected and the microcantilever immediately deflected due to adsorption. The cantilever system was left unperturbed for 10 minutes and then the cell was flushed

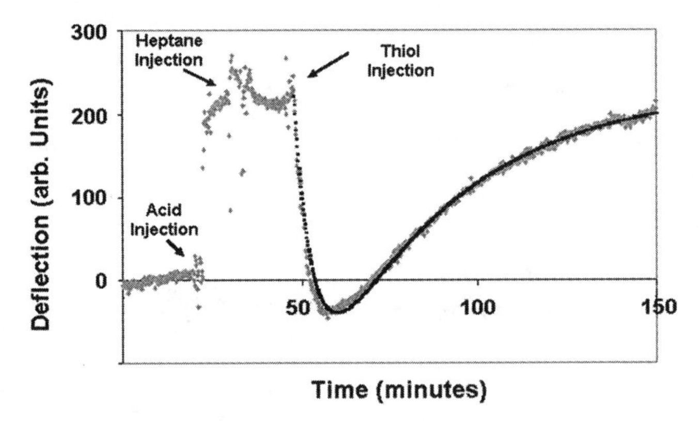

Figure 11.11 Deflection time trace of microcantilever coated on one side with gold that has been stabilized in *n*-heptane. Oleic acid (0.001 M), *n*-heptane, and 1-nonanethiol (0.003 M) solutions were injected at various times as indicated by arrows. Solid black line is double exponential fit to data points. Deflection units are arbitrary. Reprinted from [60] with permission.

with *n*-heptane. Finally, a solution of *n*-nonanethiol (0.003 M) was injected at about 50 minutes into the experiment. Initially, this resulted in desorption of oleic acid indicated by the microcantilever deflecting back toward its original starting position. However, after desorption of the acid, the microcantilever deflected back as the oleic acid was displaced by thiol. Figure 11.11 also shows a fit of the desorption and readsorption processes to a double exponential. The agreement between the experimental deflection data with a double exponential indicates that this part of the process follows Langmuir behavior and that the adsorption process (i.e., surface coverage) can be monitored in real time with temporal fidelity.

This experiment also demonstrates a more practical result. By showing that desorption and adsorption can be simultaneously monitored with microcantilever deflection, it also suggests that the selectivity of MC sensors toward specific analytes can be controlled by pretreating microcantilever sensors with specific binding agents, which desorb in the presence of the analyte of interest [63]. A selective cantilever array might consist of cantilevers pretreated with binding agents of different surface affinities.

Microcantilever sensors have also been used to monitor pH in aqueous liquids [58, 64, 65]. A sensor for monitoring acid and base in hydrocarbons may also be useful in applications where hydrocarbons oxidize into acids. This might occur, for example, in high-temperature applications such as lubricants. Figure 11.10 shows the direct adsorption of oleic acid to a gold-coated cantilever, showing that even without further surface treatment the cantilever is responsive to acid. Similar results were observed with stearic acid.

The ability to reversibly detect acid and base was further demonstrated by coating a silicon microcantilever on one side with a thin layer of gold. This cantilever was then soaked in a solution of the sodium salt of mercaptoacetic acid so that the sulfur end of the molecule adsorbed to the gold surface while exposing the sodium carboxylate salt end of the molecule to the liquid environment while anchored to the MC.

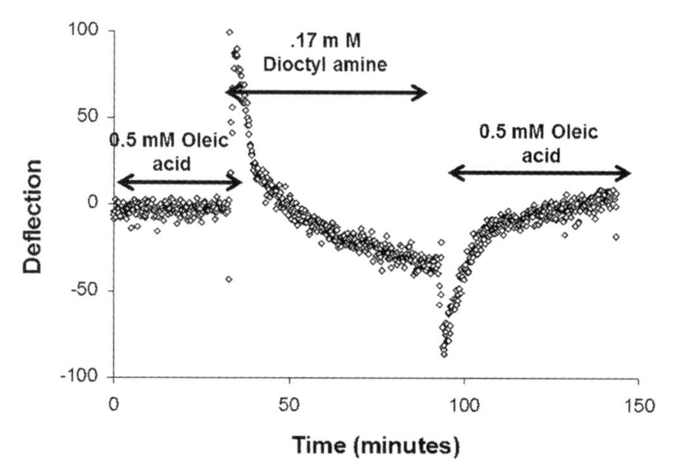

Figure 11.12 Silicon microcantilever coated on one side with gold and treated with sodium salt of mercaptoacetic acid. MC is initially exposed to a 0.5-mM solution of oleic acid in *n*-heptane. At 30 minutes a solution of dioctyl amine (0.17 mM) in *n*-heptane is injected causing rapid deflection. At 90 minutes a 0.5-mM solution of oleic acid is reinjected causing rapid deflection in the opposite directly.

The surface-modified MC was than exposed to a 0.5-mM solution of oleic acid in *n*-heptane to establish a constant baseline as shown in Figure 11.12. After about 30 minutes a 0.17-mM solution of dioctyl amine in *n*-heptane was injected, causing the cantilever to rapidly deflect. The cause of this stress-induced deflection is unclear, but it is likely due to an electrostatic interaction on the MC surface or between it and the liquid environment causing a detectable MC deflection. Following introduction of the amine, the MC slowly re-equilibrated toward its initial position. After about 90 minutes a 0.5-mM solution of oleic acid in *n*-heptane was injected, and the MC rapidly deflected in the opposite direction followed again by a slow equilibration back to its original position at the start of the experiment. While this experiment was not conducted in a quantitative manner, it does indicate that in hydrocarbon liquids a microcantilever sensor can respond reversibly to acid–base interactions with a response that is similar to that found for MC pH sensors in aqueous liquids.

Corrosion

High-throughput experimentation, which involves rapid screening of a large number of very small laboratory samples is one of the most useful and implementable applications of MC sensors [66, 67] for industrial and petrochemical applications. High-throughput samples may be robotically generated and require rapid screening using an automated measurement protocol. Some of the methods described above can be employed in a high-throughput mode and many others can be envisioned. One application that has been considered is microcantilever-based screening of corrosivity. This is a specific example of a chemical reaction that can be observed in real time with an MC sensor. Observing and characterizing chemical reactions in real time using microresonators was demonstrated as early as 1994 for the reaction of hydrogen

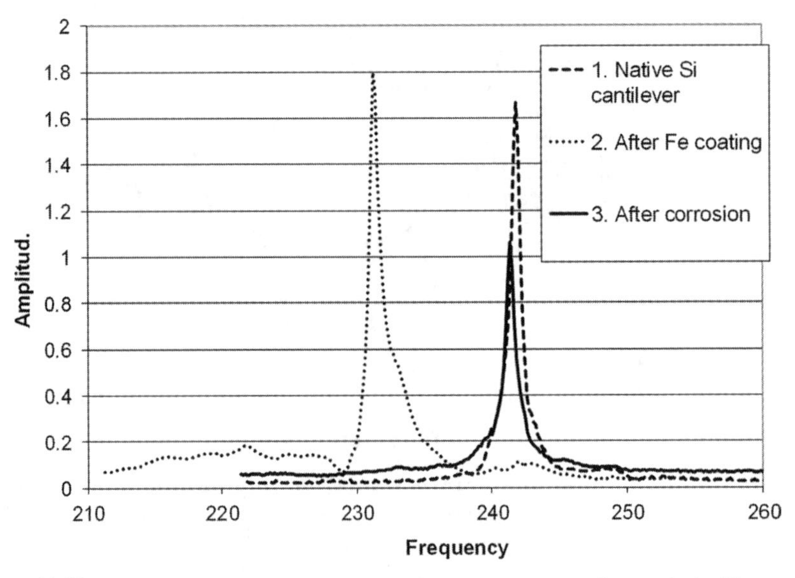

Figure 11.13 Resonance spectra for three consecutive measurements made on a single silicon microcantilever: (1) before coating with iron, (2) after coating with approximately 20 nm of iron, and (3) after corrosion in 2-M acid solution.

and oxygen to form water. In that case the reaction was followed by using the cantilever as a calorimeter and monitoring the reaction exotherm. Corrosion is an example of a reaction with particular relevance to industry.

Typically, the corrosivity of a fluid or a metal is measured using relatively large samples and qualitatively measuring corrosion using a standard long-term test method [68]. Materials testing can be tedious and time consuming, and the qualitative visual data can be highly variable. The process can be simplified by using an MC platform for measuring corrosion.

A silicon microcantilever was coated with a thin iron coating, and its resonance spectrum was measured before and after coating. After soaking the cantilever in acid solution for 15 minutes, the resonance spectrum of the microcantilever was measured again and compared to the prior measurements. The three resonance spectra are depicted in Figure 11.13 demonstrating that corrosion can be measured rapidly on very small samples. This process is readily adaptable to an automated robotic laboratory unit. In general, high-throughput materials testing is an ideal application for microcantilever sensors since it minimizes sample size, reduces test time, and can be adapted to automation.

11.4 CONCLUSION

Microcantilever sensors have evolved considerably over the last 15–20 years from laboratory curiosities to useful research tools and are approaching adaptation as useful sensor systems. Microcantilever sensors have many important benefits with potential to impact a variety of industrial applications. These benefits include low cost, low-volume sampling, high sensitivity, and potential for high-throughput experimentation. The applications in the petroleum industry outlined here include the ability to sense physical properties such as temperature and viscosity and chemical properties such as various chemical species and corrosion, all in real-world hydrocarbon environments.

The easiest implementations will be those where few interferences are present. Optical tracking of the MC is convenient yet requires that the environment be optically transparent and free of particles and bubbles. Otherwise it is necessary to use active (self-actuated) devices such as commercially available piezoresistive microcantilevers. Laboratory detectors and sensors for high-throughput experiments where the environment is relatively stable and controlled make ideal applications.

As with all new analytical techniques, it is especially important to develop a careful and well-understood calibration of any microcantilever sensor, especially when the microcantilever is used in static mode. The details of surface stress development are not fully understood, and cantilever deflection may not always be consistently related to adsorbate surface coverage. This is especially true when applied to new chemical environments. On the other hand, when used in dynamic mode microcantilevers are essentially sensitive mass balances and require similar calibration methods to other mass balances.

Most importantly, microcantilever sensors can be relatively easy to implement. The microcantilevers are readily available in numerous varieties and the associated hardware is available in many laboratories, and so the entrance barrier to experiments is low.

REFERENCES

1. Lang, H. P., and Gerber, Ch. *Topics in Current Chemistry*, Springer: Berlin, 2008, p. 1.
2. Battison, F. M., Ramseyer, J. P., Lang, H. P., Baller, M. K., Gerber, Ch., Gimzewski, J. K., Meyer, E. and Guntherodt, H. J. *Sensors and Actuators B* **77** (2001): 122.
3. Lang, H. P., Hegner, M., and Gerber, Ch. *Materials Today* **8**(4) (2005): 30.
4. Boisen, A., Dohn, S., Keller, S. S., Schmid, S., and Tenje, M. *Rep Prog Phy* **74**(3) (2011): 36101.
5. Gimzewski, J. K., Gerber, Ch., Meyer, E., and Schlittler, R. R. *A Chemical Physics Letters* **217** (5, 6) (1994): 589.
6. Lang, H. P., Berger, R., Battiston, F., Ramseyer, J. P., Meyer, E., Andreoli, C., Brugger, J., Vettiger, P., Despont, M., Mezzacasa, T., Scandella, L., Guntherodt., H. J., Gerber, Ch., and Gimzewski, J. K. *Applied Physics A* **66** (1998): S61.
7. Thundat, T., Warmack, R. J., Chen, G. Y., and Allison, D. P. *Applied Physics Letters* **64** (1994): 2894.
8. Thundat, T., Oden, P. I., and Warmack, R. J. *Micro Thermophys Eng* **1** (1997): 185.
9. Wachter, E. A., and Thundat, T. *Rev Sci Instrum* **66** (1995): 3662.
10. Thundat, T., Oden, P. I., and Warmack, R. J., *Electromechemical Society Proceedings* **97**(5), (1997): 179.
11. Lavrik, N. V., Sepaniak, M. J., and Datskos, P. G., *Review of Scientific Instruments*, **75**(7), (2004): 2229.
12. Berger, R., Delamarche, E., Lang, H. P., Gerber, Ch., Gimzewski, J. K., Meyer, E., and Guntherodt, H. J. *Science* **276** (1997): 2021.
13. Lang, H. P. *Solid State Gas Sensing*, Comini, E. et. al., Eds., Springer Science + Business Media: New York, 2009, 305.
14. Ziegler, C. *Analytical and Bioanalytical Chemistry* **379**(7–8), (2004): 946.

15. McKendry, R., Zhang, J., Arntz, Y., Strunz, T., Hegner, M., Lang, H. P., Baller, M. K., Certa, U., Meyer, E., Guntherodt, H.-J., and Gerber, Ch. *Proc Natl Acad Sci USA* **99** (2002): 9783.
16. Arntz, Y., Seelig, J. D., Lang, H. P., Zhang, J., Hunziker, P., Ramseyer, J., Meyer, E., Hegner, M., and Gerber, Ch. *Nanotechnology* **14** (2003): 86.
17. Fritz, J., Baller, M. K., Lang, H. P., Rothuizen, H., Vettiger, P., Meyer, E., Guntherodt, H.-J., Gerber, C., and Gimzewski, J. K. *Science* **288** (2000): 316.
18. Weeks, B. L., Camarero, J., Noy, A., Miller, A. E., Stanker, L., and De Yoreo, J. J. *Scanning* **25** (2003): 297.
19. Kaur, G., Sawhney, R. S., and Vohra, R. *International Journal of Advanced Research in Computer Engineering and Technology* **1**(5) (2012): 285
20. Thundat, T., Pinnaduwage, L., and Lareau, R. *Electronic Noses and Sensors for the Detection of Explosives*, Gardner, J. W., and Yinon, J. Eds., Kluwer Academic: Dordrecht, 2004, p. 249.
21. Pinnaduwage, L. A., Thundat, T., Gehl, A., Wilson, S. D., Hedden, D. L., and Lareau, R. T., *Ultramicroscopy* **100**(3–4), (2004): 211.
22. Muralidharan, G., Wig, A., Pinnaduwage, L. A., Hedden, D., Thundat, T., and Lareau, R. T., *Ultramicroscopy* **97**(1–4), (2003): 433.
23. Zhu, W., Park, J. S., Sessler, J. L., and Gaitas, A. *Appl Phys Lett* **98** (2011): 123501.
24. Pinnaduwage, L. A., Gehl, A., Hedden, D. L., Muralidharan, G., Thundat, T., Lareau, R. T. Sulchek, T., Manning, L., Rogers, B., Jones, M., and Adams, J. D. *Nature (London)* **425** (2003): 474.
25. Pinnaduwage, L. A., Wig, A., Hedden, D. L., Gehl, A., Thundat, D. Y. T., and Lareau, R. T. *J Appl Phys* **95** (2004): 5871.
26. Schilowitz, A. M., Yablon, D. G., Lansey, E., and Zypman, F. R., *Measurement* **41**(10), (2008): 1169.
27. Itoh, T., and Suga, T. *Appl Phys Lett* **64** (1994): 37.
28. Su, Y., Evans, A. G. R., Brunnschweiler, A., and Ensell, G. *J Micromechanic Microengineer* **12**(6) (2002): 780.
29. Gaitas, A., and French, P. *Sens Actuators A: Phys* **186** (2012): 125.
30. Li, M., Tang, H. X., and Roukes, M. L. *Nature Nano-technology* **2** (2007): 114.
31. Calleja, M., Rasmussen, P. A., Johansson, A., and Boisen, A. *Proceedings of SPIE* **5116** (2003): 314.
32. Muralidharan, G., Mehta, A., Cherian, S., and Thundat, T. *Journal of Applied Physics* **89**(8), (2001): 4587.
33. Chon, J. W. M., Mulvaney, P., and Sader, J. E., *J. Appl Phys* **87** (2000): 3978.
34. Schilowitz, A. M., Yablon, D. G., and Zypman, F. *MRS Proceedings* (2005): 838–O10.17.
35. Sader, J. E. *J Appl Phys* **84**(1), (1998): 64.
36. Van Eysden, C. A., and Sader, J. E. *J Appl Phys* **101**(4), (2007): 044908.
37. Stoney, G. G. *Proc Royal Soc London Ser A* **82** (1909): 172.
38. Godin, M., Tabard-Cossa, V., and Grutter, P. *Appl Phys Lett* **79**(4), (2001): 551.
39. Beaulieu, L. Y., Godin, M., Laroche, O., Tabard-Cossa, V., and Grutter, P. *Appl Phys Lett* **88** (2006): 083108.
40. Gimzewski, J. K., Gerber, Ch., Meyer, E., and Schlitter, R. R. *Chem Phys Lett* **217**(5,6), (1994): 589.
41. Stephenson, R. J., Woodburn, C. N., O'Shea, S. J., Welland, M. E., Rayment, T., Gimzewski, J. K., and Gerber, Ch. *Review of Scientific Instruments* **65**(12), (1994): 3793.
42. Oden, P. I., Chen, G. Y., Steele, R. A., Warmack, R. J., and Thundat, T. *Appl Phys Lett* **68**(26), (1996): 3814.
43. Thundat, T. G., Oden, P. I., Warmack, R. J., and Finot, E. L. U.S. Patent 6311549, Nov. 6, 2001.
44. Oden, P. I., U.S. Patent 6269685, August 7, 2001.
45. Grzegorzewski, A. U.S. Patent 5494639, Feb. 27, 1996.
46. Datskos, P. G., Thundat, T. and Lavrik, N.V. *Encylopedia of Nanoscience and Nanotechnology*, Nalwa, H. S., ed., Volume X, American Scientific Publishers; Valencia, CA, 2004, p. 1.
47. Datskos, P. G., and Sauers, I. *Sensors and Actuators B: Chemical* **61** (1999): 75.

48. Pinnaduwage, L. A., Hawk, J. E., Boiadjiev, V., Yi, D., and Thundat, T. *Langmuir* **19** (2003): 7841.
49. Cherian, S., Mehta, A., and Thundat, T. *Langmuir* **18** (2002): 6935.
50. Ji, H. F., and Thundat, T. *Biosensors and Bioelectronics* **17** (2002): 337.
51. Hu, Z., Thundat, T., and Warmack, R. J. *Journal of Applied Physics* **90** (1), (2001): 427.
52. Xu, X., Thundat, T., Brown, G. M., and Ji, H. F. *Analytical Chemistry* **74** (2002): 3611.
53. Riley, A. E., Schilowitz, A. M., Yablon, D. G. and Disko, M. M. *MRS Proceedings*, 0915-R05-08 (2006).
54. Moulin, A. M., O'Shea, S. J., and Welland, M. E. *Ultramicroscopy* **82** (2000): 23.
55. Alvarez, M., Carrascosa, L. G., Moreno, M., Calle, A., Zaballos, A., Lechuga, L. M., Martinez-A., C., and Tamayo., J. *Langmuir* **20** (2004): 9663.
56. Subramanian, A., Oden, P. I., Kennel, S. J., Jacobson, K. B., Warmack, R. J., Thundat, T., and Doktycz, M. J. *Appl Phys Lett* **81** (2002): 385.
57. Guanghua, W., Datar, R. H., Hansen, K. M., Thundat, T., Cote, R. J., and Majumdar, A. *Nature Biotechnology* **19** (2001): 856.
58. Zhang, Y., Ji, H. F., Snow, D., Sterling, R., and Brown, G. M. *Instrumentation Science and Technology* **32** (4) (2004): 361.
59. Allendorf, M. D., and Hesketh, P. J. U.S. Patent 8065904, Nov. 29, 2011.
60. Schilowitz, A. M., and Yablon, D. G. *MRS Proceedings* **1318** (2011): 85.
61. Poirier, G. E. *Langmuir* **15** (1999): 1167.
62. Godin, M., Williams, P. J., Tabard-Cossa, V., Laroche, O., Beaulieu, L. Y., Lennox, R. B., and Grutter, P. *Langmuir* **20** (2004): 7090.
63. Schilowitz, A. M., and Yablon, D. G. U.S. Patent 7765854, Aug. 3, 2010.
64. Ji, H. F., Hansen, K. M., Hu, Z., and Thundat, T. *Sensors and Actuators B* **72** (2001): 233.
65. Bashir, R., Hilt, J. Z., Elibol, O., Gupta, A., and Peppas, N. A. *Applied Physics Letters* **81**(16), (2002): 3091.
66. Carrascosa, L. G., Moreno, M., Alvarez, M., and Lechuga, L. M. *Trends in Analytical Chemistry* **25**(3), (2006): 196.
67. Yablon, D. G., and Schilowitz, A. M. *Sensors, 2010 IEEE* Vol. no. 1–4, Nov. 2010: 373–377.
68. ASTM Standard D130–10.

Chapter 12

Applications of Scanning Probe Methods in Cosmetic Science

Gustavo S. Luengo and Anthony Galliano

L'Oreal Research and Innovation, Aulnay sous Bois, France

12.1 INTRODUCTION

A product applied onto the various external parts of the human body for cleaning, perfuming, changing appearance and/or correcting body odors, and/or keeping them in good condition is considered to be a cosmetic according to European Union (EU) regulation. Part of the performance of a cosmetic product, apart from its necessary and obvious harmlessness, lies in its interaction with hair, skin, and nails to fulfill various needs, protection, adorning, enhancing appearance, and modifying hair shape and color. The development of efficient and innovative cosmetic products implies an acute knowledge of the substrates onto which they are applied and their various needs with age, ethnic specificities, exposure, and the like.

As an example, the hair surface is known to be hydrophobic but turns hydrophilic in the presence of water. Behind such macroscopic evidence, many mechanisms still need to be elucidated. Scanning probe microscopy (SPM) methods, which provide local information on surface structures and their related changes, can be of great help for developing cosmetics products, as well as understanding and evaluating their mode of action.

Scanning Probe Microscopy in Industrial Applications: Nanomechanical Characterization, First Edition. Edited by Dalia G. Yablon.

12.2 SUBSTRATES OF COSMETICS

12.2.1 Hair

Hair is roughly a cylindrical fiber ~50–100 μm in diameter. The total number of hairs covering a normal human head amounts to 120,000–150,000, making a large developed surface area (typically ~6 m^2 for ~20-cm-long hairs). Changes in the surface properties will, therefore, play an important role in the self-appraisal of its physical and aesthetic features. Hair is a good example of a biocomposite material composed of two distinct morphological components [1]: the outer protective layers known as the cuticle and the central core known as the cortex, both representing about 10 and 90% of total volume, respectively. A third component is known as the medulla, a central, porous, irregular component in the center of hair, is always seen in animal furs or feathers but is not always present in humans as shown in the scanning electron microscope (SEM) micrographs in Figure 12.1.

(a)

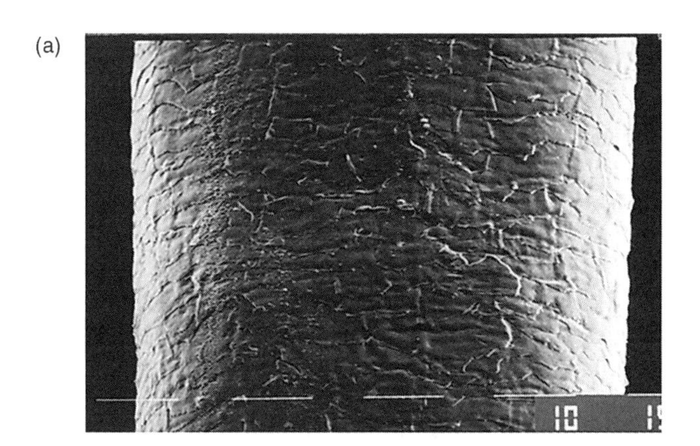

(b)

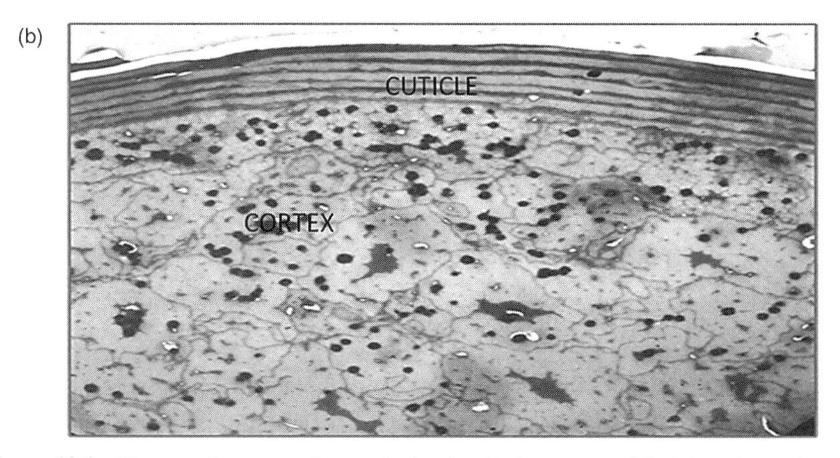

Figure 12.1 Electron microscopy micrographs showing (a) the structure of the hair surface and (b) transversal section where the cortex and the cuticle can be distinguished.

The cortex is the major component of the hair. It is constituted of elongated cortical cells (~100 μm in length and ~1–6 μm in diameter) aligned along the axis of the fiber and filled with partially crystallized keratin.

The keratin biochemical family has the ability to crystallize in the form of an α-helical structure, a basic element of a four-strand rope arranged in a coiled-coil configuration [1, 2]. Serine, glycine, *N*-acetyl serine, and cystine are the main amino acids, and the chains are stabilized by many inter- and intramolecular interactions: electrostatic, van der Waals, hydrogen, and covalent (disulfide bonds). α-Keratin units form hexagonal crystals called microfibrils (~7.5 nm in diameter) or intermediate filaments (IF). Moreover, IF are embedded in an amorphous protein (matrix), as a semicrystalline substructure of the cells called macrofibrils (~100–400 nm in diameter) [3, 4].

These intricate crystalline structures, strongly anisotropic, confer on the hair true bionanocomposite features with unique mechanical properties such as high rupture stress and elongation, high elastic modulus, and rapid and total recovery even after high deformation [5].

The Cuticle

The cuticle forms on the outer surface of hair. It has an overlapping scalelike (approximately 50 μm across, 0.5 μm thick) structure protecting the cortex and oriented from the root to the tip of the fiber producing a series of edges on the outer surface of each hair. A normal hair has about six–eight overlapping scales in close contact. The final structure of the outermost layer of the cuticle (epicuticle) is of special interest: the surface properties of hair depend on the physico-chemical properties of this layer.

The most recent findings have shown that the epicuticle membrane of hair fibers contain highly cross-linked protein (~75%) and fatty acids (~25%). Among adsorbed fatty acids, 18-methyl eicosanoic acid (18-MEA) is the most abundant (50% w/w) [6] and appears to be covalently grafted on the outer surface (constituting the outer β-layer) via covalent thio-ester linkages to the protein [7, 8]. It is considered to play an important role in the physico-chemical and tribological properties of hair.

Based on these observations, a model has emerged to explain the fine molecular structure of the hair surface [9], depicting a highly ordered monolayer of 18-MEA grafted to the proteins of the epicuticle that would explain the excellent tribological properties of natural hair in ambient conditions [10, 11].

12.2.2 Skin

The human body is covered by skin, whose surface area can reach ~1.6 m^2 in adults and whose main role is protection. The structure of skin includes three compartments: epidermis, dermis, and subcutaneous tissue. The epidermis structure is made up of several cell layers: basal, spinous, granular, and horny. The latter, otherwise called the stratum corneum (SC) is the outermost and exfoliative layer of the skin. From the basal layer, epidermal cells (*keratinocytes*) migrate upward, progressively

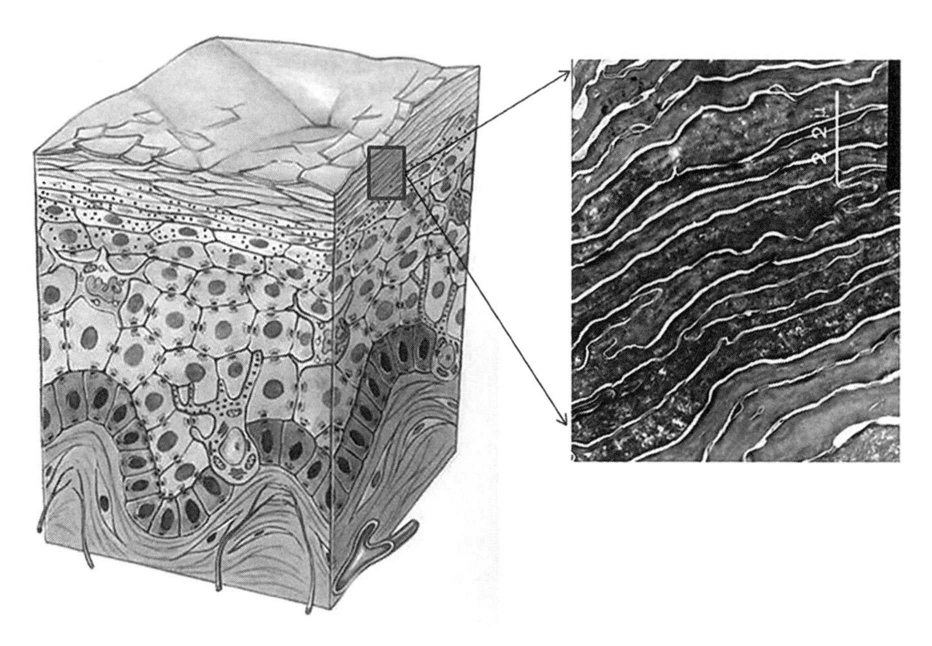

Figure 12.2 Diagram showing the main structures of skin and the organization of the stratum corneum in the zoomed inset.

flatten, lose their nuclei, and become filled with keratin filaments (*keratinisation process*) embedded within the remnants of cell proteins. The anucleated cells of the horny layer (*corneocytes*) (see Fig. 12.2) appear roughly hexagonal in shape, overlapping at the edges with neighboring cells [12].

The 10- to 20-μm thick layer of the SC [see inset in Fig. (12.2)] is able to protect the body against water loss and physical and chemical aggression from a variety of external assaults. The barrier function and the cohesiveness of the SC are closely related to the highly structured organization of the corneocytes and the intercellular matrix. According to the simplest structural model widely accepted [13], the matrix is mainly composed of a lamellar phase of various lipids organized in bilayers, and the cells (corneocytes) are attached among each other by protein adhesion structures called corneodesmosomes, providing a barrier effect.

The lipids are mainly composed of ceramides, cholesterol, fatty acids, and cholesterol esters. Ceramides, that make up ~40% of all SC lipids, have a sphingosine-based polar head (hydroxyl terminal groups) and two nonpolar saturated hydrocarbon (fatty) chains. They form highly ordered and impermeable membranes acting as an additional link to keep membranes tightly associated.

Adhesion protein cell structures (*desmosomes*) bridge the space between adjacent cells at specific places in the membrane. They are still present in the SC (*corneodesmosomes*) and play an important role in the mechanical integrity of this layer, ensuring intercellular adhesion.

12.3 MECHANICAL PROPERTIES AND MODIFICATIONS BY COSMETIC PRODUCTS

Much work has been accomplished in improving our knowledge of the structure of hair and skin as *biocomposites* mostly using optical and electron microscopy techniques. In particular, scanning and transmission electron microscopy (SEM and TEM) are classical techniques used to describe these substrates at the submicron range.

The complex structures of hair and skin and their related properties can be impaired by various factors. Hair fibers, for instance, although considered to have exceptional resistance are affected by continuous exposure to sunlight, weathering their physical properties. It is still unclear how and to what extent the cuticle and its particular scaley structure contribute to the strong mechanical properties (e.g., when pulled), mostly attributed to the structure of the internal cortex. The softness and elasticity of skin are other examples, where the different skin layers and fiber networks (collagen, elastin) of the dermis undergo progressive and deep changes resulting in dramatic manifestation of aging signs (wrinkles, etc.). The SC has to be mechanically strong to protect our body from the harsh environmental factors and mechanical stress, while at the same time maintaining softness and its ability to distort.

The overall influence of the SC layer on the mechanical properties of skin is recognized [14, 15]. Human SC can be isolated from skin, enabling its tensile properties to be studied using common mechanical traction methods such as stress relaxation testing [16, 17]. A typical stress–strain (load–elongation) curve of human SC conditioned in water at 25°C for one hour reveals three distinct regions separated by inflections at approximately 25 and 125% of elongation. During extension of the SC, structural changes are observed: The lipidic layers progressively become disorganized soon after 5% extension, unlike desmosomes that resist until just before tissue breakage [18]. Other experiments describe the resistance of the lipid intercellular organization up to 60% extension of the SC [19], although the detachment of lipid layers from one of adjacent corneocytes was generally observed.

12.3.1 Influence of Cosmetic Products

Many cosmetic products aim at improving or restoring the intrinsic properties of skin and hair. These effects are in some cases reflected in enhanced mechanical and tribological behavior.

Traction experiments have been used to study the effect of cosmetic ingredients and products. For example, the softening effect of emollients on the SC is clearly seen on stress–strain curves [17]. Takahashi et al. [20] assessed the change in skin softening effect brought on by different moisturizing formulas with time allowing the influence of some ingredients to be distinguished.

The results also contribute to better define the mechanisms by which different moisturizers do act: hydrating effect of glycerol, plasticizing effect of urea associated with a strong interaction with protein components, and the like.

12.4 SCANNING PROBE TECHNOLOGIES ADAPTED TO COSMETIC SCIENCE

The concept of using a physical mechanical probe to explore the properties of a surface offers a considerable asset. In fact, the term "probe" can also be applied to the use of radiation (i.e. X-rays) or ion beams in the case of secondary ion mass spectrometry (SIMS). This chapter will restrict the scope to a physical mechanical probe that guarantees the interaction between two bodies despite the presence of an interfacing medium (i.e., water). The principles of the physics and mechanics of two spherical bodies in contact can be applied as a first approach. Practically, the use of a physical probe allows the tissue to remain connected (as is the case when exploring natural sensorial properties such as touch).

- *Instrumented nanoindentation* (as a stand-alone technique) is a type of indentation hardness test that uses mechanical probes to penetrate small volumes of material. It is described briefly in Chapter 1 and more fully in Chapter 7. The load L applied by the indenter (of a specific geometry) and the penetration h into the substrate are continuously monitored. It is possible to extract information about the hardness and Young's modulus of the material. A three-sided diamond Berkovich tip of curvature radius of ~0.1 μm is generally used. This technique has been successfully applied to the characterization of relatively hard solids (i.e., metals). A dynamic version of this measurement, the *continuous stiffness measurement* (CSM), is an additional resource in which stiffness is measured continuously during the loading of the indenter by imposing a small dynamical oscillation of the force and measuring the in-phase and out-of-phase components. Apart from providing continuing results, its application to softer materials in which dissipation is important (polymers, biomaterials) is possible.
- "Nanoindentation" (force curves or load versus depth measurements) using atomic force microscopy (AFM) enables easier access to the nanoscale world as the size of the probe is significantly reduced (~5–10 nm). It is also possible to visualize micro- and nanostructures by scanning the tip over the surface, therefore obtaining accurate topographical images. Also, AFM can make measurements in many different environmental conditions. This is an important advantage as in general the use of vacuum techniques (i.e., SEM or TEM) to image surface samples prevents in vivo or ex vivo analysis, since biological materials generally cannot withstand prolonged vacuum (nowadays environmental electron microscopy makes it possible to overcome such limitations, expanding the use of these classical techniques). While in contact, single-point measurements in the form of force curves (described in detail in Chapter 3) with an ultrastiff cantilever and a diamond tip can also be conducted with the possibility of extracting hardness and Young's modulus information in a very localized region. Here, these AFM-based nanoindentation measurements will be referred as "force curve measurements." Unfortunately, the lack of complete knowledge of the tip geometry makes it more prone to a lower accuracy than *instrumented* nanoindentation. Another problem is the higher degree of freedom of motion of the cantilever that may twist during the test, a problem not present in instrumented indentation.

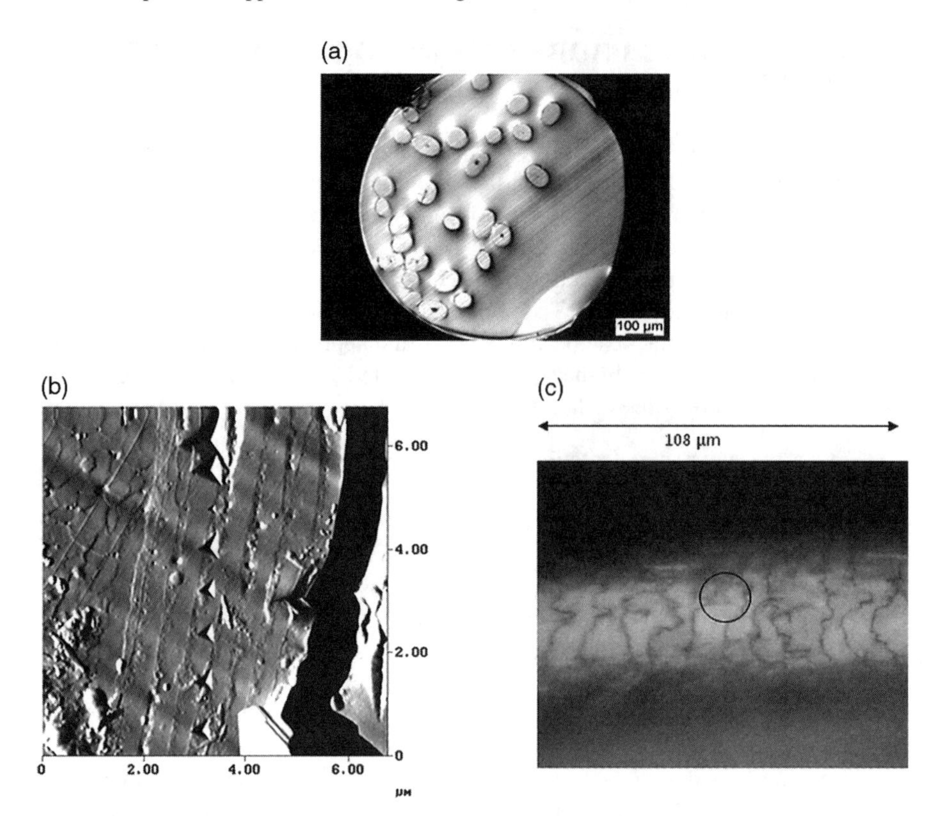

Figure 12.3 Examples of types of indentation of hairs. (a) Microtomed hair cuts embedded in a resin. (b) Example of precise indents on the cuticle of hair performed using an AFM-based indentation. (c) Examples of indents done at the surface of hair fibers using instrumented nanoindentation. For color details, please see color plate section.

- Finally, to expand the capabilities of *instrumented* nanoindenters, other hybrid approaches have been used. The recent Hysitron instruments with improved resolution introduce the localization and imaging capabilities that the first instrumented nanoindenters lacked. Used as a self-contained instrument or by taking advantage of AFM instrumentation, this technologic approach may be considered as a good, serious advance in nanomechanics.

In the cosmetic domain, these techniques are used for measuring some particular physical properties of:

- A specific component within the overall substrate structure (localized measurement)
- A previously isolated structural component (i.e., an epithelial skin cell)
- The characterization of external ingredients in contact with the substrate

The choice of the most suitable technique (instrumented nanoindenter versus AFM) in cosmetics will highly depend on the type of sample and the type of information needed. Figure 12.3 presents some illustrative examples of indents on different areas

Table 12.1 Comparison of the Different Indentation Methods and their Experimental Parameters

Methods	Nanoindentation by AFM	Instrumented Indentation	Combining AFM and Instrumented Nanoindentation
Type of instrumental setup	AFM (Bruker)	Nanotest XP (MTS)	Triboscope (Hysitron)
Accuracy of results	No	Yes	No
Achieving fast results	No	Yes	Yes
Price in favor of system purchase	N/A	Yes	Yes
adaptability to test on different materials	Yes	No	No
Position accuracy	Yes	Yes	Yes
Automation	No	Yes	Yes
potential to differentiate different parts of the hair	Yes	No	No

of hair, showing how the choice of the technique and tip depends on the type of structure to be reached.

Our studies showed that the ideal solution to assess the local mechanical properties of a hair fiber is to combine the two techniques: AFM, as a microscopic technology at a nanoscale level, and instrumented nanoindenter, for both its accurate measurements and its easy automation (Table 12.1).

12.4.1 Instrumented Nanoindentation Approaches

The examples below illustrate various themes where the instrumented nanoindentation technique has been used for hair care applications:

- Differences in structures/properties of hair from various ethnic origins
- Impact of chemical treatments
- Environmental effects [relative humidity and ultraviolet (UV) exposure] and different ways to validate formulas

Nanoindenters have been readily applied to measuring properties of hair using a standard Berkovich diamond tip and maximum load fixed at 10 mN (1 g). A dynamic stress (connected with small discharge/load diagrams) makes it possible to quantitatively measure the modulus of elasticity and hardness according to the indentation depth. The dynamic measurement is carried out at a frequency of 32 Hz with an oscillation amplitude of 3 nm.

First tests were carried out on the cross section of natural hair or hair after conventional bleaching or bleaching followed by treatment with proprietary product (Intra-cylane). Results showed that the latter increased the modulus and local hardness of the cortex by about 30%, whereas bleaching only increased these values by about 14%. The same measurements were also carried out perpendicular to the surface of the hair (the hair fiber being positioned longitudinally). No difference between samples could be found despite noticeable stiffer contact at the root than at the tip of the hair fiber after bleaching (about 50–60%). This difference is removed when applying the proprietary product.

Mechanical Properties of Hair of Different Ethnic Origin

Studies have been performed to compare the mechanical properties of Caucasian and African hair. These types of hair are differentiated by the amount of melanin granules (~0.5 μm) present within the cortex, the pigment also present in skin that confers hair its color (see granules in Fig. 12.1). Little is known about the mechanical constituents of hair, and access to more local information requires better positioning techniques. Data was obtained using the Hysitron nanoindenter head on an AFM.

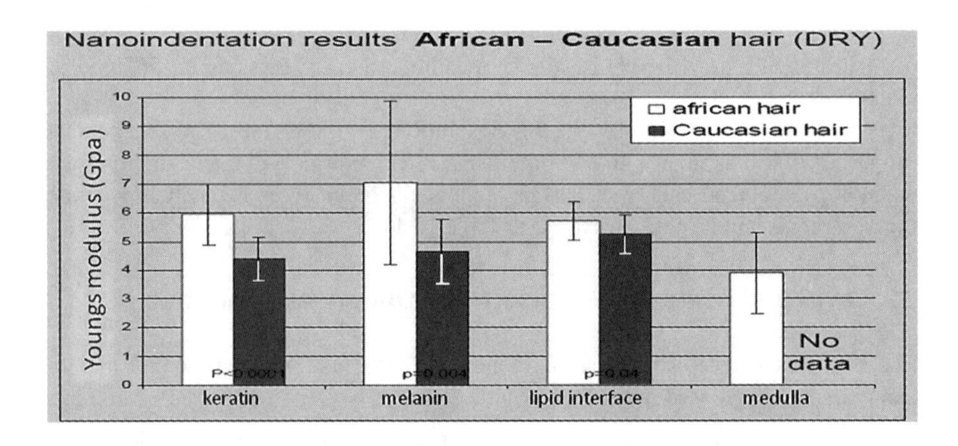

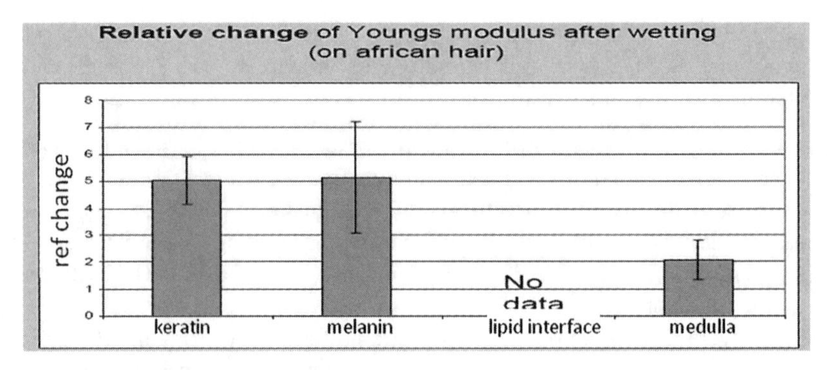

Figure 12.4 Indentation measurements carried out with the Hysitron indenter on Caucasian and African hair under wet/dry conditions.

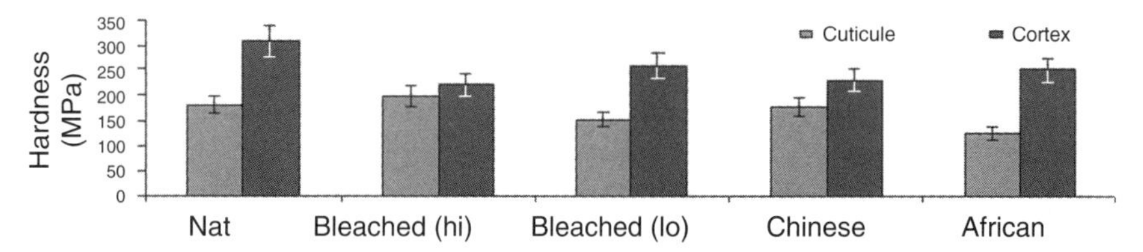

Figure 12.5 Median values of the hardness of cortex and cuticle obtained on various hair types.

Figure 12.4 shows that the mechanical properties of melanin and keratins differ between the two groups. Moreover, the elasticity of these components is changed by a factor of 5 after immersing in water. It also appears that the elastic modulus of the cortical cells is close to that of hair keratin.

Mechanical Properties of Cuticle and Cortex

Figure 12.5 shows the values of the hardness of cuticle (standard measurement on the surface of the hair) and cortex (measurement on transverse sections) of natural Caucasian hair, the same hair after medium and high bleaching, Chinese hair and African naturally curly hair as measured by an instrumented nanoindenter. Three hair fibers of each type were studied. Seven indents were carried out on each fiber. Tests were carried out at room temperature (25°C) and at ambient relative humidity (40%RH, except for the study under variable HR conditions).

Results clearly show differences between the properties of cortex and cuticle. The hardness of the cuticle appears systematically lower than that of cortex, depending on the nature of the hair. This approach can discriminate structural specificities of a hair according to its ethnic origin and, consequently, evaluate whether the impact of a given treatment on the fiber differs with its origin. This data is not in agreement with that presented in Table 12.1, in particular for Chinese hair. However, the tested parameters are different since cortex and cuticle were probed perpendicular to the transverse axis of the hair (and not on cross section). There is likely anisotropy in the physical properties of hair depending on whether the fiber is probed perpendicularly or transversely. This anisotropy could be related to the orientation and the organization of the fibers. A deeper knowledge of these properties is necessary to better understand the differences between various hair types.

Effect of Bleaching: Properties of Natural and Bleached Ethnic Hair Types

Figure 12.6 presents some values of cuticle hardness obtained from bleached hair using various techniques compared to the natural hair of the same origin.

Results show that bleaching increases cuticle hardness. The interest of this study lies in the selective analysis of the hair components when probing the alterations brought to the hair surface. Such a measurement complements the characterization of mechanical properties previously provided by the "traditional" measurements, which lead to a more macroscopic evaluation of the hair fiber.

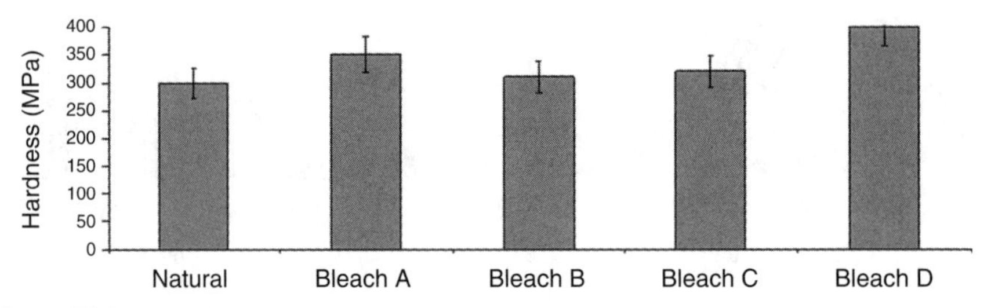

Figure 12.6 Median values of cuticle hardness of natural hair and hair bleached using various products.

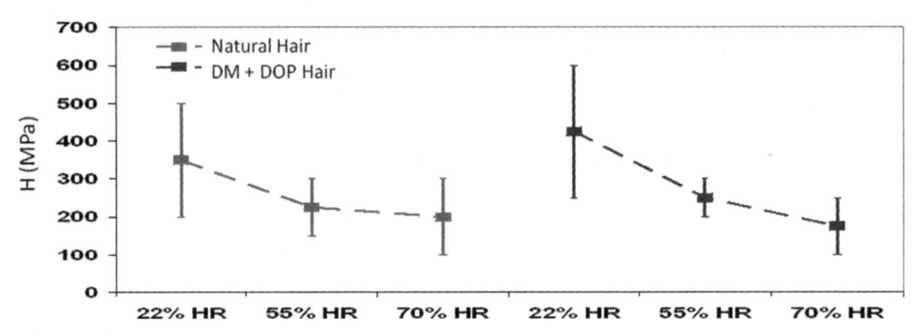

Figure 12.7 Median values of cuticle hardness of natural and medium bleached hair.

Impact of Environmental Conditions (RH and UV)

Studies were carried out to evaluate the impact of relative humidity on the local mechanical properties of the cuticle, in both natural and bleached hair using instrumented nanoindentation. Figure 12.7 shows the values of cuticle hardness obtained on natural and medium bleached hair under three different RH conditions: 22, –55, and 70% RH.

Results highlight a decrease in cuticle hardness with increasing RH for all hair types studied. On the first hand, bleached and natural hair (treated with a standard shampoo) show identical surface mechanical properties at medium and high RH (55 and 70%). On the other hand, the values are slightly higher at low RH (22%) for bleached hair, making the decrease of cuticle hardness with RH more noticeable.

Nanoindentation can be used to evaluate changes in mechanical properties of hair with exposure to solar radiation, that is, to follow and record the photodegradation process. Figure 12.8 illustrates the changes in mechanical properties of both natural and bleached hair exposed to different exposure times of UV radiation.

The curves show the effects of increased UV doses on the behavior of hair. Little change in the mechanical properties of hair is observed between T_0 and T_{10} hours of exposure, for both natural and bleached hair. However, the hardness of the cuticle largely increases after 40 hours of exposure. This increase is more noticeable for natural hair than bleached hair. The surface of natural hair thus appears more sensitive to UV than that of bleached hair (cuticle becoming more brittle).

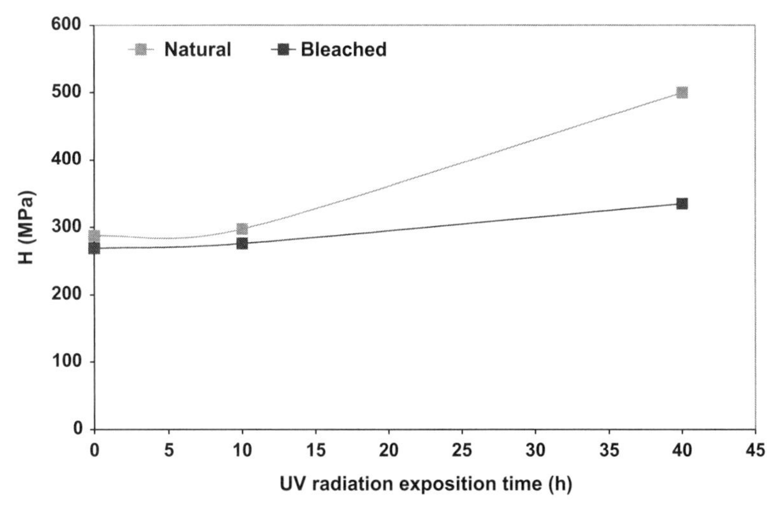

Figure 12.8 Average values for cuticle hardness at an indentation depth of 10 nm obtained for natural hair and bleached hair (Platifiz formula) following exposure to UV for increasing time.

These observations confirm the degradation of cuticle following exposure to UV for a total of 40 hours. The impact of the irradiation is more marked on the scale surface and on the overall cuticle, in particular on natural hair.

The effectiveness of a proprietary photoprotective product (not shown) can be measured by its ability to preserve the mechanical properties of hair. The nanoindentation technique, for a certain number of targeted studies, can be an efficient approach to develop protective formulations, in particular when a given ingredient plays a significant role on the surface properties at a scale and cuticular level.

12.4.2 A Closer Look at Atomic Force Microscopy

The characterization of local mechanical properties, in particular hardness and friction, are often measured using AFM [21–24]. AFM gives access to more localized properties (depending, among other things, on the size of the tip), since the amount of material under scope is relatively small as compared to instrumented nanoindentation.

Using this method, Bhushan [25] specifically studied the nanomechanical properties of the surface and transverse sections of hair [26]. Nanomechanical properties such as hardness, elastic modulus, and scratch resistance of the surface of hair were compared before and after application of various hair care products (conditioners). Some interesting conclusions indicate that only the first three to four layers of scales are affected and that the modulus of the surface of hair increases with increased RH. The protocols used in these studies were similar to those developed for materials with similar mechanical and geometrical properties, namely glass fibers [27] and keratin fibers [28].

Hardness of cuticular components of hair have also been measured on transversal cuts of hair of different ethnic origin (i.e., Chinese and Caucasian) using AFM and the analysis of force curves [29]. These measurements revealed that the hardness of the endocuticle of Asian hair was slightly higher than that of Caucasian hair. More recently, comparable measurements were carried out by another group [30] at a more fundamental level, affording applications to a broad range of composite materials and biomaterials [31]. Transverse and perpendicular measurements were performed, in particular on hair and glass fiber [32] to propose a mechanical model in accordance with findings from literature [33, 34].

In our laboratories, AFM was initially used for studying the mechanical properties of hair with the aim of obtaining information at a more local level, that is, that of constituents of hair (cortical cells, melanin, etc.), and measuring local variations resulting from chemical treatments

- A thorough study of residual prints after indentation [cf. Fig. 12.3(c)]. For example, the cuticle was tested on a transverse section of hair showing values of hardness similar to those found on wool (~0.9 GPa).
- A thorough analysis of the approaching AFM curves and fitting to models of different contact theories (Hertz, Sneddon, JKR, etc; discussed in Chapter 2). The calculated modulus of elasticity was significantly different between cuticle and cortex.

Effect of Thin Films at the Surface of Hair

Atomic force microscopy enabled the exploration of properties of the outermost layers of hair and, in particular, the effect of a particular lipid, 18-methyl eicosanoic acid (18-MEA), on the overall surface properties of hair. 18-MEA is known to play an important role in the properties of hair, manageability being the major lipid present at the hair surface. This lipid is a remnant of the "old" cell membrane constituents transformed during cell eratinization process in the hair bulb. It is replaced by straight-chain fatty acids on hairs of subjects affected with maple syrup urine disease (MSUD) [35]. Availability of hairs from twins (one twin carrying the disease, the other being unaffected) allowed us to better understand the role of 18-MEA. In parallel, both types of hair were treated by standard bleaching protocols and were used to compare the effects of well-known lipid removal from hair.

Accordingly, AFM adhesion studies were performed using two different instruments: a Discover TMX2000 scanning probe microscope (SPM) (Veeco Instruments, Cambridge, UK) and an AutoProbe CP Research SPM (Veeco Instruments, Sunnyvale, CA). Adhesion was extracted from force curve data acquired using V-shaped, silicon nitride (Si-N) cantilevers of nominal force constant 0.03 and 0.37 N $m^{-1.}$ In addition, friction traces were obtained in lateral force mode while increasing the applied load during each scan. A linear 'Amonton's law' fit was found between friction and applied load. The slope of the fit allowed us to calculate the friction coefficient. This procedure was repeated on three areas of five sections along the length of a hair and on two fibers from each sample type.

Hairs were cleaned using an established procedure: sonication in a 1% aqueous sodium dodecylsulfate solution, followed by thorough rinsing with twice distilled

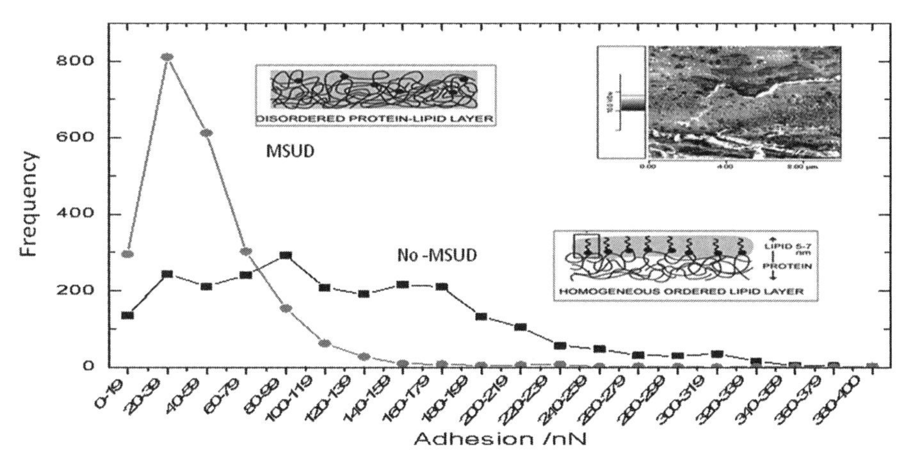

Figure 12.9 Adhesion between an AFM tip and hair of a patient lacking 18-MEA lipid or hair of a healthy individual. For color details, please see color plate section.

water and careful drying under nitrogen [36]. Hairs were bleached by a hydrogen peroxide solution. Two different methods were used for removing 18-MEA from nonbleached hair surface: (i) a mild treatment with anhydrous hydroxylamine, following the method of Evans and Lanczki [37] and (ii) a coarser treatment in methanolic potassium hydroxide, reported to completely remove 18-MEA [38].

As illustrated in Figure 12.9, a significantly lower adhesion between a Si–N tip and the hair of a patient with MSUD was observed as compared to that of the hair of a twin unaffected with MSUD. An adhesion study also revealed that the 18-MEA layer was partly removed by treatment with hydroxylamine (see Fig. 12.9, inset) and was not uniform on gray hairs. The inset of Figure 12.9 shows a friction image of the hair fiber after removal of the 18-MEA layer. The brighter zones are areas where this lipid has been fully removed. This effect is not homogeneous. The native hair (nontreated, not shown) shows a homogeneous distribution of friction (lipid layer fully present). The friction coefficient of bleached and hydroxylamine-treated hairs were found to be greater compared to virgin hair.

All these studies suggested that the presence of this lipid layer can be easily detected using AFM–force curve measurements, likely due to the capillary force produced on the tip by the lipid layer, increasing the adhesion value during pull-off.

Effect of Hydration in Cells and Tissue (Stratum Corneum)

Cell research is another application where the AFM–force curve approach is better suited than an instrumented indentation. This can be illustrated by studies on the surface layers of skin and in particular the stratum corneum.

Some studies use isolated dry and wet SC. Yuan and Verma [39] used AFM together with a Triboscope nanoindenter and a nano-dynamic mechanical analysis (DMA) (Hysitron, Minneapolis, MN) to measure viscoelastic moduli (E' and E'') at varying penetration depths. Elastic moduli values obtained with a pure elastic model were of the order of 100 and 10 MPa for dry and wet SC, respectively. Tan δ increased from approximately 0.1 to 0.25. An apparent modulus variation with indentation depth was noticed.

Other studies focused upon isolated individual components of the stratum corneum (lipids, cells, etc.). The cell (corneocyte) is likely to be the most studied component of SC. Although several studies have looked into the structure of corneocytes using optical and electron microscopes, few give insights into mechanical properties. Much work done in this field used "indirect" observations based on morphological changes of isolated corneocytes as a function of RH. Richter et al. [40] investigated and quantified the swelling of corneocytes in water using AFM. They mainly found changes in thickness without a significant lateral alteration. A more direct approach was reported by Lévêque et al. [41]. The authors measured the force needed to elongate isolated corneocytes immersed in water using a microhandling technique. The calculated Young's elastic modulus was $E \sim 4.5 \times 10^8$ Pa, although this modulus was considered by the authors as being underestimated due to technical difficulties.

Our experiments used the force curve capabilities of AFM to get more detailed information on the mechanical properties of corneocytes and eventually help to understand the effect of common cosmetic products. A silicon wafer was used as a reference for an infinitely hard surface. Surface mechanical indentations were then performed on this reference substrate prior to the corneocyte experiments, which were carried out in ambient conditions (approximately 45% RH).

Measured deflection curves are analyzed using Hertz and Sneddon elasticity models to derive elastic modulus of the material. The Hertzian model for a spherical tip and a flat surface is given by [42]

$$F = \frac{3}{4} E \sqrt{R} \delta^{3/2}$$

where E denotes elastic modulus, δ indentation depth, and R the radius of the sphere. This model works well at low indentation depths. At higher loads and with soft materials, the Sneddon model or even more sophisticated ones such as the JKR (Johnson, Kendall, Roberts) [43] might be more appropriate. Our studies proved that the Hertz model worked reasonably well for characterizing the surface elasticity of isolated corneocytes as long as the indentation depth is small enough (ca. 10 nm) relative to corneocytes thickness.

As for corneocytes under ambient conditions, several surface indentations were performed at different positions in a square array. Apart from helping to average the values obtained, another advantage of this approach is that information can be obtained on specific areas of corneocyte surface by superimposing the indents on top of the topographic image. At low loads, no indent was noticed at the surface confirming that the experimental parameters adequately fitted the level needed for elastic distortion of corneocyte membrane. The force curves were then analyzed following the previously described procedure. The curve fits were satisfying and Young's modulus could be calculated. Values fell within about 4.0×10^9 Pa and, as expected, higher than those obtained in water by Lévêque et al. [41]. Further experiments are still needed to understand the effect of water using this direct method.

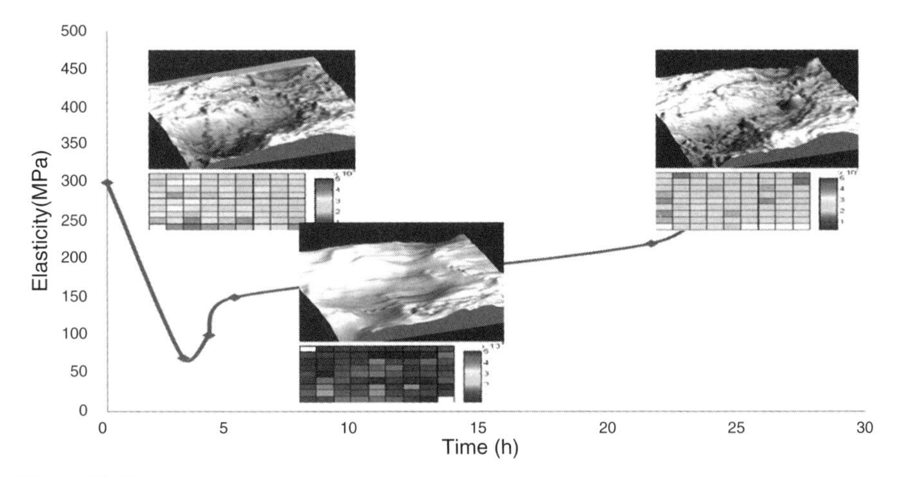

Figure 12.10 Changes in mechanical properties of the stratum corneum after applying a layer of glycerol onto its surface (AFM nanoindentation measurements). For color details, please see color plate section.

Finally, AFM can also be applied to the surface of the stratum corneum (Fig. 12.10), bringing the advantage of monitoring the surface mechanical elasticity while applying a moisturizer such as glycerol (7%) applied onto the stratum corneum. The extreme surface elasticity is then measured. At the surface of SC, glycerol forms a slowly penetrating, smooth layer. The initial surface topography is recovered after 27 hours. AFM force curves have been performed at different times during this process. SC initial elastic modulus (ca. 2.5×10^8 Pa) first decreases (to ca. 1.0285×10^8 Pa) following glycerol application then slowly returns to initial value as glycerol penetrates through SC.

12.5 CONCLUSIONS

Nanoindentation techniques described here have obviously to be carefully chosen to determine the most appropriate for a given study. Their combination can bring information at both the micro- and submacroscale of hair and skin substrates. This information has to be compared with more currently used macroscale tests (i.e., traction tests) to provide a better knowledge of the effect of the products on structures of hair and skin with a goal of improving our understanding of the ultimate sensorial perception of consumers.

REFERENCES

1. Mitsui, T. In *New Cosmetic Science*, Elsevier: New York, 1997.
2. Crick, F. H. C. *Acta Cryst*, **6** (1953): 689.
3. Fraser, R. D. B., Macrae, T. P., and Suzuki, E. *J Mol Biol*, **108** (1976): 435.
4. Parry, D. A. D. *Int J Biol Macromol*, **19** (1996): 45.
5. Robbins, C. R., Ed. In *Chemical and Physical Behavior of Human Hair*, 4th ed., Springer: New York, 2002, pp. 211–226.
6. Yorimoto, N., and Naito, S. *Proc Int Symp Fiber Sci Technol Yokohama*, (1994): 215.

7. WERTZ, P. T., and DOWNINGS, D. T. *Lipids*, **23** (1988): 878–881.
8. EVANS, D. J., and LANCZKI M. *Text Res J*, **67** (1997): 435–444.
9. NEGRI, A., RANKIN, D. A., NELSON, W. G., and RIVETT, D. E. *Text Res J*, **66** (1996): 491.
10. BREAKSPEAR, S., SMITH, J. R., and LUENGO, G. *J Struct Biol*, **149** (2005): 235–242.
11. HUSON, M., EVANS, D., CHURCH, J., HUTCHINSON, S., MAXWELL, J., and CORINO, G. *J Struct Biol*, **163** (2008): 127–136.
12. GARSON, J. C., DOUCET, J., LEVÊQUE, J. L., and TSOUCARIS, G. *J Invest Dermatol*, **96** (1991): 43.
13. WERTZ, P. W., MADISON, K. C., and DOWNING, D. T. *J Invest Dermatol*, **92** (1989): 109.
14. AGACHE P., and VARCHON, D. Skin Mechanical Function, in *Measuring the Skin*, AGACHE, P., and HUMBERT, P. Eds., Springer: New York, 2004, pp. 429–445.
15. BATISSE, D., BAZIN, R., BALDEWECK, T., QUERLEUX, B., and LÉVÊQUE, JL. *Skin Res Technol*, **8** (2002): 148.
16. WILDNAUER, R. H., BOTHWELL, J. W., and DOUGLASS, A. B. *J Invest Dermatol*, **56** (1970): 72–78.
17. ROCHEFORT, A., DRUOT, P., and AGACHE, P. *Int J Cosmet Sci*, **8** (1986): 27–36.
18. RAWLINGS, A. V., WATKINSON, A., HARDING, C. R., ACKERMAN, C., BANKS, J., HOPE, J., and SCOTT, I.R. *J Soc Cosmet Chem*, **46** (1995): 141–151.
19. LÉVÊQUE, J. L., HALLEGOT, P., DOUCET, J., and PIERARD, G. *Dermatology*, **205** (2002): 353–357.
20. TAKAHASHI, M., YAMADA, M., and MACHIDA, Y. *J Soc Cosmet Chem*, **35** (1984): 171–181.
21. PARBHU, A. N., ALMQVIST, N., BRYSON, W. G., and LAL, R. *Mol Biol Cell*, **9** (1998): 105A.
22. BLACH, J., LOUGHLIN, W., WATSON, G., and MYHRA, S. *Int J Cosmet Sci*, **23** (2001): 165.
23. SMITH, J. R., and SWIFT J. A. *J Microsc*, **206** (2002): 182.
24. SUNDARARAJAN, S., BHUSHAN, B., NAMAZU, T., and ISONO, Y. *Ultramicroscopy*, **91** (2002): 111.
25. BHUSHAN, B. Progress in Materials, *Science*, **53** (2008): 585–710.
26. WEI, G., and BHUSHAN, B. *Ultramicroscopy*, **106** (2006): 742–754.
27. LI, X., BHUSHAN, B., and MCGINNIS, P. B. *Mater Lett*, **29** (1996): 215–220.
28. PARBHU, A. N., BRYSON, W. G., and LAL, R. *Biochemistry*, **38** (1999): 11755–11761.
29. TAKAHASHI, T., HAYASHI, R., OKAMOTO, M., and INOUE, S. *J Cosmet Sci*, **57** (2006): 327.
30. CAO, G., CHEN, X., XU, Z., and LI, X. *Composites Part B: Engineering*, **41** (2010): 33–41.
31. ERIK, B., HAVITCIOGLU, H., AKTAN, S., and KARAKUS, N. *Skin Res Technol*, **14** (2008): 147–151.
32. TUREK, J., and YANG, D. Nanoindentation of a Three-Layered Optical Fiber. Available from: www.hysitron.com.
33. XIAO, Q., SCHIRER, J., TSUCHIYA, F., YANG, D. Nanotensile Study of Single Human Hair Fiber. Available from: htt://www.hysitron.com.
34. AKKERMANS, R. L. C., and WARREN, P. B. *Phil Trans R Soc Lond A*, **362** (2004): 1783.
35. CHUANG, D. T. *J Pediatr*, **132** (1998): S17–S23.
36. SWIFT, J. A. *Int J Cosmet Sci*, **13** (1991): 143–159.
37. EVANS, D. J., and LANCZKI, M. *Text Res J*, **67** (1997): 435.
38. NEGRI, A. P., CORNELL, H. J., and RIVETT, D. E. *Text Res J*, **63** (1993): 109–115.
39. YUAN, Y., and VERMA, R., *Colloids and Surfaces B: Biointerfaces*, **48** (2006): 6–12.
40. RICHTER, T., MULLER, J. H., SCHWARZ, U. D., WEPF, R., and WIESENDANGER, R. *Appl Phys A*, **72** (2001): S125–S128.
41. LÉVÊQUE, J. L., POELMAN, M. C., DE RIGAL, J., and KLIGMAN, A. M. *Dermatologica*, **176** (1988): 65–69.
42. ROUND, A. N., YAN, B., DANG, S., ESTEPHAN, R., STARK, R. E., and BATTEAS, J. D. *Biophys J*, **79** (2000): 2761–2767.
43. ISRAELACHVILI, J. N., PEREZ, E., and TANDON, R. K. *J Coll Inter Sci*, **78** (1980): 260.

Chapter 13

Applications of Scanning Probe Microscopy and Nanomechanical Analysis in Pharmaceutical Development

Matthew S. Lamm

Merck Research Laboratories, Merck & Co., Summit, NJ

13.1 INTRODUCTION

Scanning probe microscopy (SPM) and nanomechanical analysis has been, until relatively recently, a technique generally underutilized in the pharmaceutical field. The primary reason is because traditional pharmaceutical formulation and processing involves handling of materials in powder form and SPM is not immediately suited to characterizing powders directly. However, as pharmaceutical scientists began exploring the impact of single-particle properties on the bulk behavior of the powders with respect to flow, blend uniformity, and drug product performance, SPM became an increasingly attractive tool. The other recent trend in the pharmaceutical field is the development of new types of formulations beyond the traditional powder blend in a tablet or capsule. Self-emulsifying systems, amorphous solid dispersions, liposomes, cyclodextrins, controlled-release formulations, and inhaled formulations, for example, rely on micro- and nanoscale phenomena and are thus ideally suited for characterization by SPM. In addition to imaging, nanomechanical methods such as nanoindentation have also been of increasing interest in the pharmaceutical field. For most formulations, the drug compound, referred to as the active pharmaceutical ingredient (API) is typically isolated from the final chemical synthesis step via crystallization. The mechanical and surface properties of these organic crystals can affect further downstream processes

Scanning Probe Microscopy in Industrial Applications: Nanomechanical Characterization, First Edition. Edited by Dalia G. Yablon.

such as particle size reduction, blending, compaction, powder flow, and tableting. Single crystals of many APIs can generally not be grown larger than a few hundred microns and thus bulk-scale mechanical tests are not usually possible. However, nanomechanical analysis provides a unique opportunity to probe the mechanical properties of these particles and predict how they will perform in various processes. In this chapter, the applications of SPM and nanomechanical analysis to pharmaceutical development will be explored. Although there are examples of application of these techniques in the areas of drug discovery and medicinal biology, they are not in the scope of the current discussion. Here the focus is on how SPM and nanomechanical characterization enable the drug development process from API isolation and processing to formulation and final dosage form preparation.

To provide insight into how SPM can be applied to pharmaceutical development, a brief background of the process may be useful. First, the active pharmaceutical ingredient is prepared via chemical synthesis and is then isolated from solution, typically via crystallization, and then dried. From here, the powder may be subjected to milling in order to reduce the particle size to the desired range. Particle size is important in that it can affect blend uniformity, flow properties, and dissolution rate. Milling can be of various types and intensity ranging from a gentle powder delumping to jet-milling where the particles are reduced to the micron size regime. At this point, the API is blended with other components of the formulation known as excipients that function as binders, disintegrants, flow aids, lubricants, and chemical stabilizers. The blends may undergo processing such as roller compaction or granulation prior to pressing into a tablet or filling in a capsule for simple oral formulations. Tablets may be coated for aesthetic or functional purposes such as slowing disintegration for controlled release. It is the surface and mechanical properties of the API and excipients that determine how the materials will behave in each of these processes, thus SPM is a powerful tool available to the pharmaceutical scientist.

13.2 APPLICATIONS OF SPM IMAGING

13.2.1 Surface Properties of Formulation Ingredients

Surface properties of the particles in a formulation including the API and other excipients such as binders, disintegrants, and flow aids affect how the blends will behave with respect to flow, blending, and compaction. As a surface technique, SPM allows investigation into particle surface attributes such as roughness and morphology. The effect of processing such as crystallization or milling on the surfaces of API crystals can be imaged [1–6]. In one study, the authors evaluated the relationship between crystal size and defect density of an API, flufenamic acid (FFA) [7]. By etching the surface of the single crystals briefly with a solvent, material at defect sites was preferentially dissolved and then imaged with SPM (Fig. 13.1). It was found that the defect density could be correlated to crystal size as well as the relative rate of transformation to a more thermodynamically stable crystalline phase. These polymorphic transformations are very

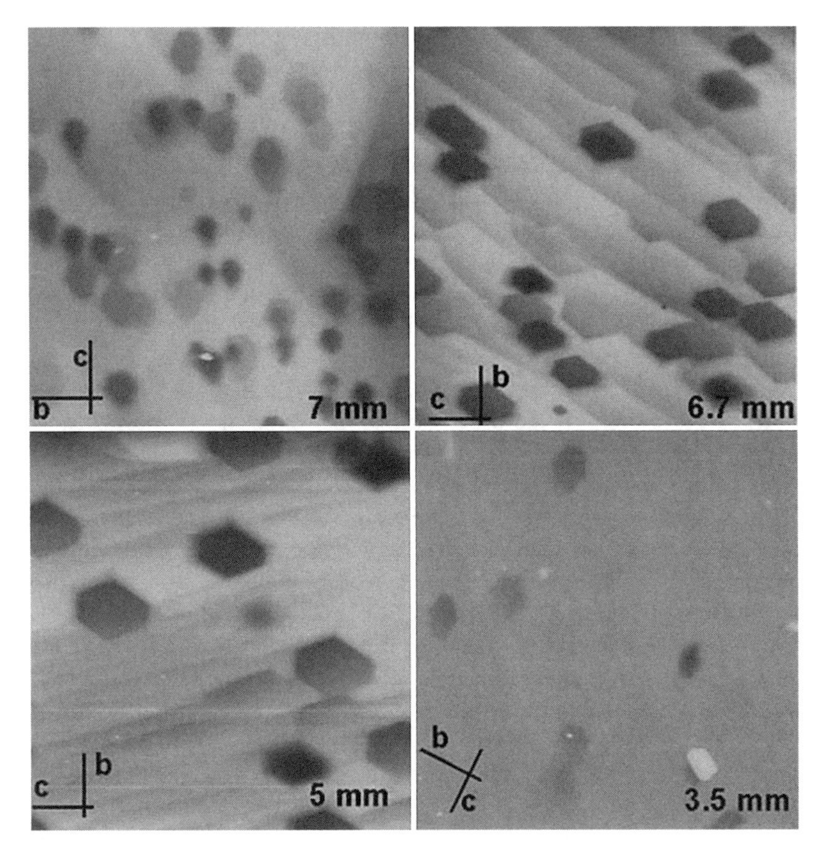

Figure 13.1 AFM height images collected in contact mode of different sized crystals of FFA Form I (1 0 0) face etched with *n*-pentane, with dark or red representing lower topography and bright or yellow representing higher topography. AFM scan size ~100 μm. Crystal size (lower right) and crystallographic orientation (lower left) are labeled on each image. Reprinted with permission from Ref. [7]. For color details, please see color plate section.

important in the pharmaceutical field as different crystalline and amorphous phases will have different solubility, which can impact bioperformance. SPM surface imaging techniques have been used to look at the effect of changes in environmental conditions, such as relative humidity, on the surface properties of particles as they undergo polymorphic changes such as hydrate formation or amorphous to crystalline transformations [8–14].

13.2.2 Characterization of Formulation Intermediates

Amorphous Solid Dispersions

One of the most promising applications of SPM and nanomechanical analysis in current pharmaceutical development is for the characterization of formulation intermediates known as amorphous solid dispersions. A common issue with many compounds developed as drugs is the low solubility of the molecule in biorelevant media. As discussed above, the API is typically isolated as a crystalline phase, with the most thermodynamically stable crystalline phase generally preferred to avoid uncontrolled phase transformations during downstream processing or storage. Much effort in screening for polymorphs is typically undertaken early in development to map the phase diagram for a typical compound and ultimately select the phase for formulation development.

The disadvantage of utilizing the most thermodynamically stable phase for development is that it will have the lowest solubility. In fact, solubility measurements can be used to identify which polymorph is the most thermodynamically stable at a given temperature (and pressure). Low solubility can sometimes lead to lower absorption and thus less of the compound entering the bloodstream, reducing the effectiveness of the drug. If the drug could be delivered in an amorphous state, a higher solubility as compared to the crystalline phase could be achieved that could possibly translate into greater bioavailability. The risk with delivering the drug in an amorphous state is the potential for the material to crystallize during formulation, shipping, storage, or on the shelf at a pharmacy or in a consumer's medicine cabinet. The decrease in solubility due to crystallization of the molecule could have a negative effect on the bioavailability of the compound and potentially affect the efficacy of the drug. To mitigate the risk of crystallization yet enable the use of the amorphous phase to take advantage of the greater solubility, a strategy commonly employed is to solubilize the drug molecule in a water-soluble polymer [15–18]. These polymers often have glass transition temperatures greater than drug compound and thus provide an antiplasticizing effect that reduces molecular mobility and can inhibit the kinetics of crystallization.

Two of the most commonly employed techniques for preparing amorphous solid dispersions are spray drying and hot melt extrusion. In spray drying, the polymer, API, and sometimes surfactants are dissolved in a common solvent and then sprayed at high pressure out of a nozzle, oftentimes with a heated gas. During the spray drying process, the solvent in the atomized droplets rapidly evaporate, yielding microsized particles comprised of the API, polymer, and surfactant, which are then collected and used in downstream processes. In hot melt extrusion, the polymer and API powders are blended and then melt-mixed at elevated temperatures, typically in a twin-screw extruder. Liquid surfactants or plasticizers may also be injected to aid in processing or dissolution behavior. The material is then extruded through a die where it is rapidly quench-cooled to room temperature. The extrudate is typically milled to reduce particle size, blended with other excipients, and filled in a capsule or pressed in a tablet. It is generally desirable for these solid dispersions to be isolated as a single amorphous phase that is then expected to remain as a single phase as the material is formulated, packaged, and ultimately dosed by the patient.

To assess the physical state of the material, techniques commonly employed are powder X-ray diffraction (PXRD), Fourier transform–Raman and Fourier transform–infrared (FT-IR) spectroscopy, and differential scanning calorimetry (DSC). Scanning probe microscopy is an additional tool that can provide insight into the phase behavior of these dispersions and has particular advantage over the other techniques, especially in cases where the drug compound phase separates as an amorphous domain. For these cases, PXRD can be of limited utility if the drug compound phase separates but does not crystallize. And, if the polymer and API have similar T_g's, DSC may not be able to resolve the T_g for each domain should it phase separate. With SPM, changes in topography and phase contrast, as well as newer techniques of nanomechanical mapping, can be used to visualize phase-separated domains with spatial resolution exceeding optical or laser-based techniques. It is relatively straightforward to directly image the surface of material coming from a hot-melt extrusion process with SPM in order to understand the effect of process variables such as barrel temperature, screw speed, or feed rate on the phase behavior of the system. If the surface is very rough or if a milled particle is to be imaged, the material can be embedded in epoxy and sectioned with a microtome.

13.3 SPM AS A SCREENING TOOL

In addition to characterizing solid dispersions from bulk-scale processes, SPM is an excellent tool to use in the screening phase or preformulation phase of development. Because SPM is a surface analytical technique and it scans relatively small areas, it requires very little material. Oftentimes in early development, the amount of API available for formulation activities may be limited to only a few grams, yet many experiments need to be conducted to identify optimal formulation compositions. If it is decided that a drug compound may be a good candidate for solid dispersions, experiments need to be performed to identify a polymer that the drug will be miscible with as well as any additional surfactants or plasticizers that may be included. Furthermore, the relative concentrations of each component need to be selected.

With limited API quantities it may be unreasonable to prepare by spray drying or hot-melt extrusion all the permutations of polymer, API, surfactant, and composition options. A solution for these early phase screening experiments involves dissolving the polymer, API, and surfactant in a suitable solvent and casting a thin film onto a flat surface such as a glass slide. This process can be easily automated in a high-throughput fashion with solvent-handling robots to prepare, for example, in a 96-well plate format. And, because SPM is a surface-imaging technique with maximum scan sizes in the range of 100 μm, a few microliters may be all that is required to cast a film, and thus material requirements can be minimal. The plate can be placed in a vacuum oven to rapidly evaporate the solvent and if desired, heated to elevated temperature to dry and anneal the films. Although this method is not an exact replication of the spray-drying or hot-melt extrusion process, it provides a starting point to assess miscibility over a wide range of compositions. Each film can then be imaged with SPM to assess the phase behavior of the system and rapidly identify compositions with poor miscibility. And, with recent advances in high-speed imaging and automation, hundreds of compositions could theoretically be prepared and analyzed. As an extension, the plates can then be placed in chambers for accelerated stability testing where the temperature and/or humidity is elevated to, for example, 40°C/75% relative humidity (RH), and after a set time removed from the chamber and again imaged with SPM. This testing can be especially important as many of the polymers used can be hygroscopic, and, although the API may be soluble in the polymer in the dry state, it may phase separate or crystallize as the polymer absorbs moisture due to its low aqueous solubility. Figure 13.2 shows data collected using HarmoniX atomic force microscopy (AFM), an advanced nanomechanical imaging mode using specialized cantilevers [19], on a solid dispersion consisting of an API, a surfactant, and a polymer that has been stressed at 40°C/75%RH. In addition to variations in topography, the technique provides maps of nanomechanical properties extracted from force curves collected in real time at each pixel. The modulus map, in particular, shows what appears to be three distinct phases resulting from humidity-induced phase separation.

In one study, Lauer et al. used AFM tapping mode phase mapping and image analysis in conjunction with Raman microscopy to assess the surface potential for demixing of two APIs prepared as solid dispersions with five different polymers [20]. Phase separation in these systems under stressed conditions was observed to occur via nucleation and growth or spinodal decomposition, depending on the API. The researchers in this study prefer to evaluate fractured surfaces of the dispersions to avoid preferential enrichment artifacts during processing of a given phase at either the glass or air interfaces for poorly miscible systems.

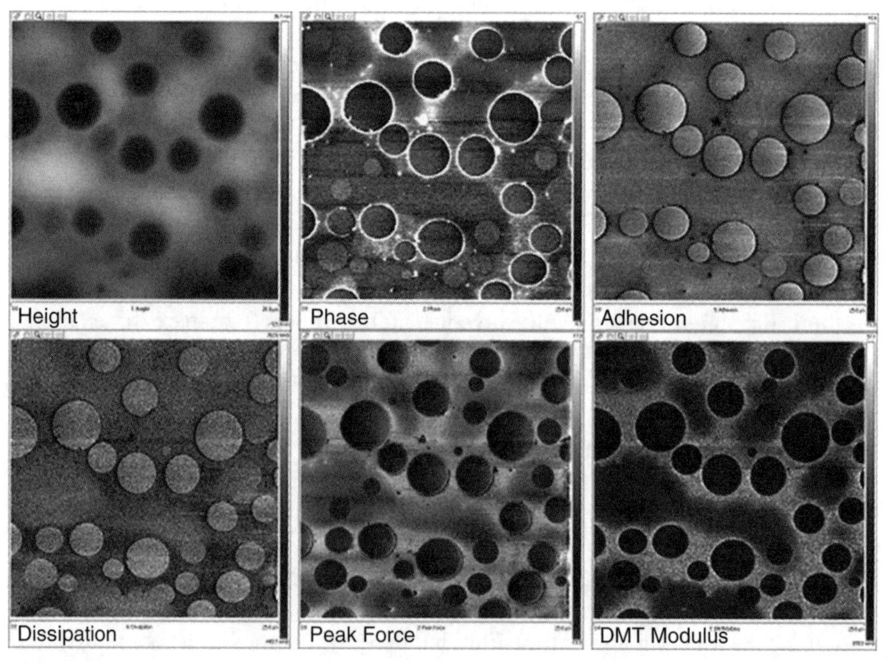

Figure 13.2 AFM HarmoniX image of humidity-induced phase separation of an amorphous solid dispersion. Scan size is 25 μm. For color details, please see color plate section.

13.4 APPLICATIONS OF NANOINDENTATION

13.4.1 Probing Particle Nanomechanical Properties

The mechanical properties of API and excipient particles will affect how they perform in various downstream processes such as particle size reduction, roller compaction, powder flow, and tableting. Because of the small size of these particles, nanomechanical analysis via a scanning probe microscope or with an instrument specifically designed for nanoindentation is ideal. In the pharmaceutical field, there is interest in predicting how API particles will perform in a mill using measurements made with nanoindentation [21–24]. For example, it has been shown that the breakage probability of a particle in a mill with a given input energy will depend upon the intrinsic mechanical properties such as hardness, elastic modulus, and fracture toughness, which can be measured on single particles using nanoindentation [25]. A recent study showed that the selection of crystallization conditions of a drug compound impacted the mechanical properties of resulting crystals as well as their resulting milling performance in a micronizer [26]. Other studies involve using nanoindentation to understand topics such as crystallographic slip planes and compaction for organic crystals [27–32].

The effect of polymorphic phase changes on particle mechanical properties can also be assessed using nanoindentation. For example, Figure 13.3 shows the hardness measured with a nanoindenting AFM cantilever (stainless steel with diamond tip) on the same crystal as it was transformed from a hydrate to a crystallographically similar anhydrous phase by cycling the humidity in the AFM chamber. Interestingly, the anhydrous phase always had a greater hardness than the hydrate phase. Increased

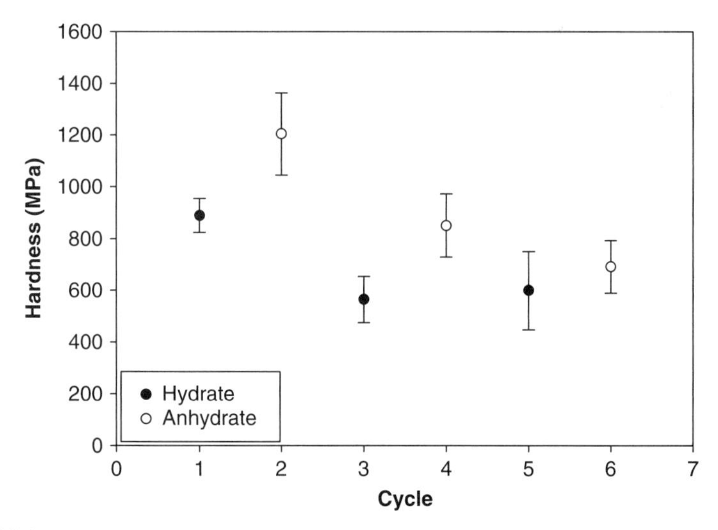

Figure 13.3 Hardness measured by nanoindentation for the same crystal cycled through hydrate and anhydrous phases.

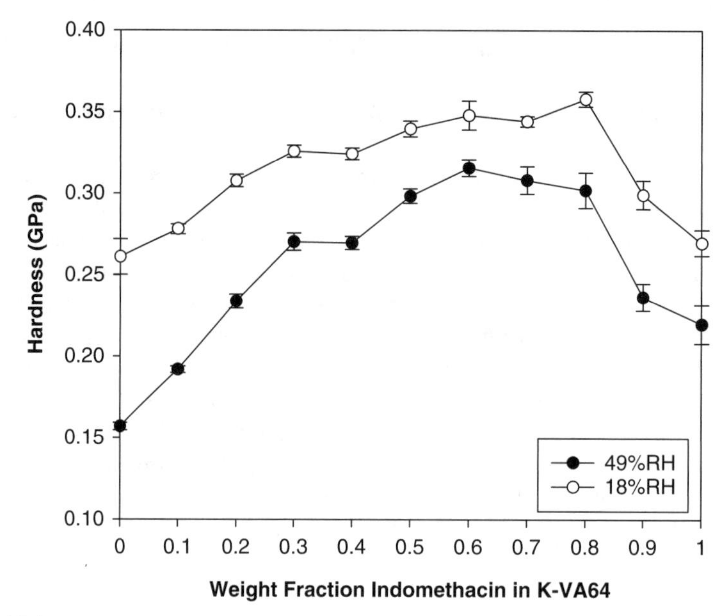

Figure 13.4 Hardness of amorphous solid-dispersion films as function of indomethacin content measured at 18% (○) and 49% (●) relative humidity.

cycling between phases caused the overall hardness to decrease, presumably as the amount of crystal defects built up during phase changes.

13.4.2 Probing Mechanical Properties of Amorphous Solid Dispersions

In addition to crystalline materials, the mechanical properties of amorphous solid dispersions described above can be probed with nanoindentation. For example, the effect of drug loading on the mechanical properties of solid dispersions can be assessed by casting films from solution. Also, environmental conditions such as temperature and humidity on the mechanical properties of the solid dispersions can be assessed. Figure 13.4 shows how the hardness changes as measured by nanoindentation of films composed of incremental amounts of indomethacin, a nonsteroidal anti-inflammatory drug (NSAID), in a water-soluble copolymer (Kollidon VA 64) at 18 and 49% relative humidity. Interestingly, the hardness increases with drug loading up to 60% by weight, where further addition of the drug results in a decrease in hardness. This result is unexpected because the glass transition temperature of indomethacin is lower than the polymer, and the T_g of these mixtures decreases with increased drug loading as expected, despite the fact that the hardness increases. The antiplasticizing effect of the drug molecule on the polymer may be the result of strong intermolecular interactions, in this case hydrogen bonding. The effect of increased humidity on lowering the overall hardness of the samples is consistent with the fact that water is a known plasticizer for the polymer.

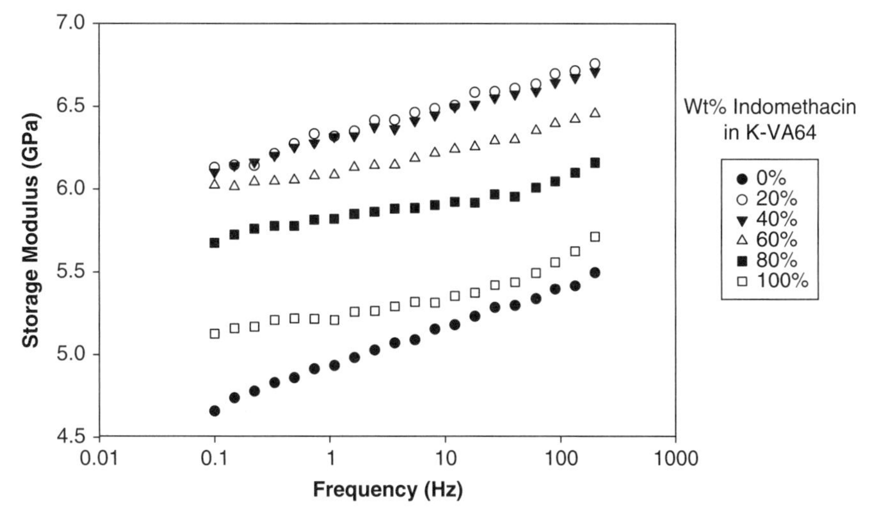

Figure 13.5 NanoDMA storage modulus as a function of frequency for amorphous solid dispersions comprised of indomethacin in Kollidon VA 64.

In addition to quasi-static nanoindentation methods, nanoscale dynamic mechanical analysis (DMA) can also be performed that is especially useful for viscoelastic materials such as polymers. Figure 13.5 shows the effect of drug loading on the storage modulus of the same solid dispersions of indomethacin in Kollidon VA 64 discussed above. A trend similar to the hardness data is observed where the modulus increases up to approximately 50% drug loading and then decreases with further addition of indomethacin. Also, there is a frequency dependence of the storage modulus typical of many viscoelastic materials as evident by the positive slope in the frequency sweep data in Figure 13.5.

13.4.3 Adhesion Measurements

The ability of AFM to quantify adhesion forces in the nanonewton regime has been well documented and is an area of special interest within the pharmaceutical field ranging from adhesion of API to carrier particles for pulmonary delivery, to measuring the adhesion of API to processing equipment, and to predict sticking in a tablet press. Techniques to attach particles to the ends of cantilevers are relatively straightforward, especially if the instrumental setup has an optical microscope integrated with the sample stage. A simple method involves spreading a thin layer of epoxy on a glass slide with a razor blade and engaging the surface using the AFM feedback control. The scan size is set to zero so the tip is not moved in the *x*-*y* directions. As soon as the tip touches the surface, the cantilever should be retracted. Now that the tip is coated in epoxy, the cantilever is moved over the particle to be attached, and the tip is again engaged with a scan size of zero. The tip is left in contact with the particle until the epoxy cures. The adhesion between the particle and various surfaces can be measured with force spectroscopy techniques where the particle is put in contact with the surface and then pulled off. With the measured cantilever deflection during pull-off and knowledge of the cantilever spring constant, a force of adhesion can be calculated.

Accurate force measurements are complicated by particle and surface roughness issues that prevent knowledge of actual contact area as well as spring constant variability with respect to where on the longitudinal axis of the cantilever the particle is actually attached. The simplest way to avoid these complications is to use the same particle-functionalized cantilever to perform adhesion measurements on all of the surfaces of interest. Assuming that the particle surface is not damaged or changed during the measurements, the relative adhesion among all the surfaces can be compared. Still, large variability in adhesion values on a given surface are typically obtained due to surface roughness affecting contact area. To reduce this effect, many researchers perform these measurements on faces of single crystals where the surface roughness can be subnanometer. Also, special attention should be paid to tribocharging where repeated adhesion experiments result in electrostatic charge buildup between the particle and surface, which can affect the measured values of adhesion. A steady growth of adhesion force versus time can be a tell-tale sign of tribocharging. Adhesion force frequencies can typically be fit to either Gaussian or log-normal distributions with the geometric mean and standard deviation reported.

Adhesion of API to Carrier Particles for Inhalers

For drug compounds indicated for respiratory diseases and formulated as dry-powder inhalers, aerodynamic particle size is important in determining where in the lung the API will be delivered and thus also affects the potency. One formulation strategy involves attaching the micronized API particles to larger carrier particles, typically sugars such as lactose. The adhesion of the micronized API to the lactose particles must be balanced such that there is sufficient attractive force to keep the particles attached but not too much as to prevent the API from detaching and being delivered to the deep portion of the lung. Price et al. have developed a method for these types of investigations where the cohesive–adhesive balance (CAB) is probed by attaching an API particle to a cantilever and measuring the relative adhesion against carrier particle surfaces of interest as well as the cohesion to API surfaces [33]. The process is repeated with different cantilevers prepared using the same API but different particles. A plot of cohesive and adhesive adhesion forces for each cantilever is prepared and a linear fit applied through the data. In one study, the authors correlated the slope of the CAB plot and the fine particle fraction measured using an Andersen Cascade Impactor for API blended with various sugar carrier particles and formulated into a dry powder inhaler (Fig. 13.6) [34]. They found that for strongly interactive blends where adhesion forces are greater cohesive forces, the fine particle fraction trended with CAB slope, that is, stronger adhesive interactions between the API and the carrier particle resulted in less deaggregation of the particles and thus less fine particles being delivered.

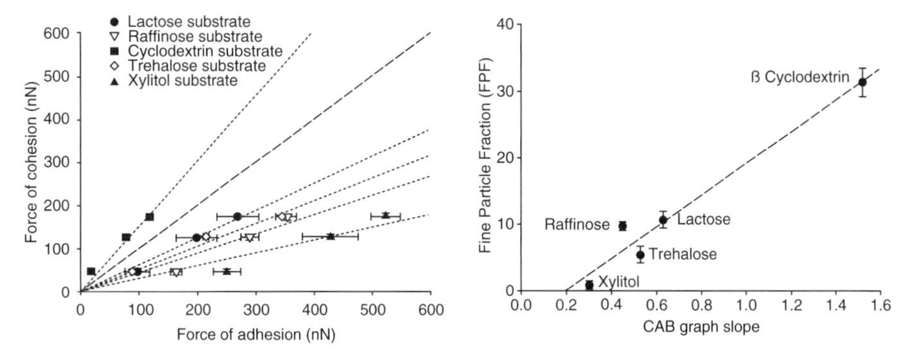

Figure 13.6 (*Left*) CAB plots of the interfacial interaction between salbutamol sulfate probes and various sugars. (*Right*) Relationship between fine particle fraction (FPF) measurements of salbutamol sulfate from various excipients and their respective slopes from the CAB plots. Reprinted with permission from Ref. [34].

Adhesion of API in pMDI

A second type of device for respiratory delivery of drugs is known as a pressurized metered-dose inhaler, or pMDI. In this case, the API is suspended in a propellant, typically a hydrofluoroalkane (HFA) along with any additional cosolvents, stabilizers, or bulking agents. The propellant pressurizes the device and when actuated, aerosolizes the API, which is then delivered to the deep lung. One potential issue with pMDIs is the adhesion of API particles to components within the device or the canister lining. This adhesion can lead to caking or material hang-up within the device and ultimately result in inconsistent emitted-dose uniformity. To probe the potential for API caking or identify potentially problematic device component materials, API adhesion studies can be performed as described above. Figure 13.7 shows the results for a study where the relative adhesion of an API crystal was tested against various components of a device. The adhesion in a matrix of 10×10 force curves was collected in three different locations for four different device components made of different materials. Despite the large variation in measurements due to surface and particle roughness, significantly greater adhesion was observed to the polybutylphthalate (PBT) component versus the other three. These measurements were conducted in ambient conditions, although they would ideally be conducted in the HFA propellant used in the device to avoid the effects of humidity on the measured interactions between the API and the components. However, this would require a pressurized AFM cell.

Adhesion for Processability

For oral formulations prepared as tablets, a common problem during processing is the sticking or picking of material to the punch resulting in tablet defects. Oftentimes, this effect isn't realized until after many tablets have been punched, and thus testing for potential sticking issues would require preparation of large batches of material and consecutive press runs of many tablets as the formulation

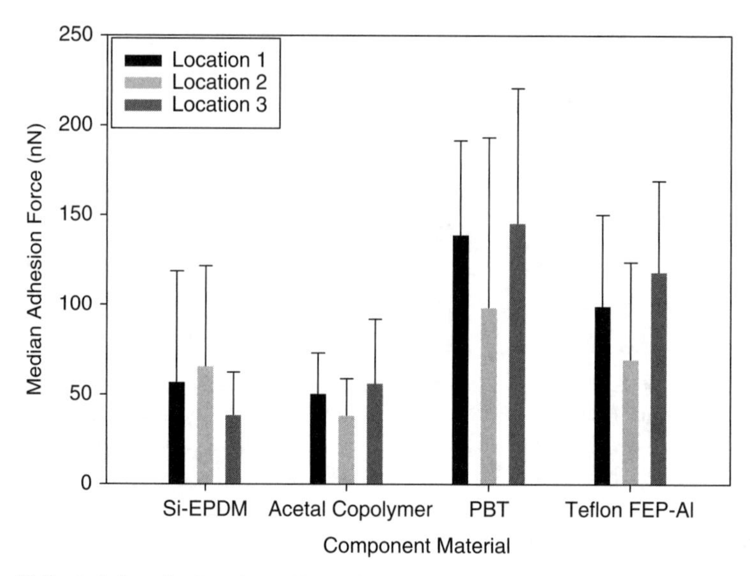

Figure 13.7 Relative adhesion of an API to various device components (measured in three locations) from a pressurized metered dose inhaler.

composition is tweaked. Strategies to use AFM as a predictive tool have been reported and generally include attaching either an API or excipient particle to the AFM cantilever and measuring adhesion versus tablet press materials or, alternatively, attaching a steel sphere or coating an AFM tip in metal and measuring adhesion against API and excipient materials [35–38]. Given that the blend to be tableted consists of many different components including the API, excipients, and lubricants, the exact mechanism for sticking is not completely understood, although the API itself is generally believed to be a major factor given that two identical blends with different APIs may have different tendencies to stick. Figure 13.8 shows the result of a study where amorphous APIs, solid dispersions, and polymer excipients were prepared as films and their adhesion to an approximately 60 μm diameter, 316 stainless steel microsphere mounted on an AFM cantilever was measured. Amorphous APIs were selected because they have a higher surface energy and thus greater adhesion than their crystalline forms. Furthermore, it is hypothesized that surface amorphization induced during tableting by shear stresses or by frictional heating causing melting of the crystalline surfaces may be a factor in sticking. Also, preparation of amorphous films allows very smooth surfaces to be prepared, thus eliminating surface roughness effects on the adhesion measurements, although the role of particle roughness itself should not be discounted as a potential factor in sticking. Clearly, some amorphous APIs have a greater adhesion to the microsphere than others, which could result in increased potential to cause sticking. Ongoing work involves developing an empirical measure of the tendency of an API to stick in a tablet press to validate the AFM adhesion method as a predictive tool.

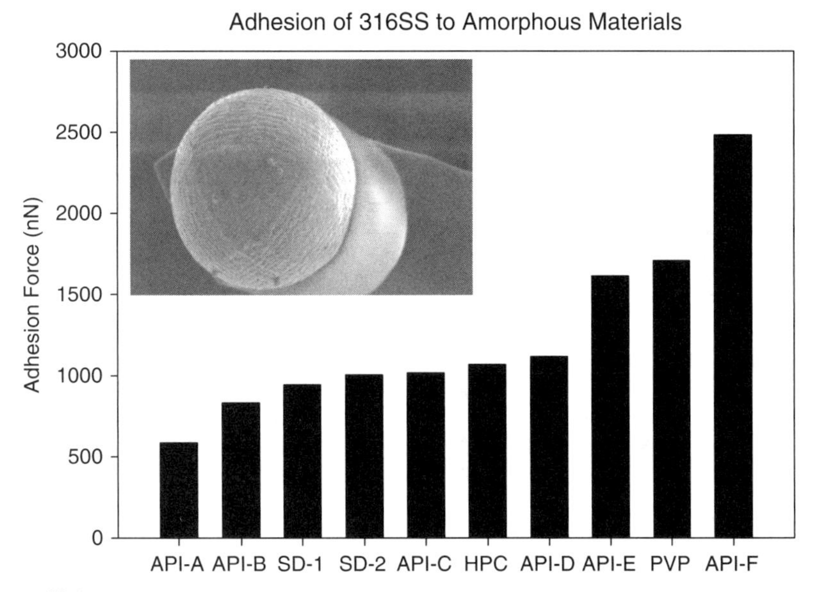

Figure 13.8 Average adhesion forces of a stainless steel microsphere to various amorphous APIs, solid dispersions (SD), and polymers polyvinylpyrrolidone (PVP) and hydroxypropylcellulose (HPC) at 30% relative humidity. (*Inset*) SEM image of 316 stainless steel microsphere attached to an AFM cantilever.

13.5 CONCLUSION

Scanning probe microscopy and nanomechanical analysis are powerful techniques with direct applications to pharmaceutical development. They provide the ability to characterize single-particle properties that can then be related to bulk-scale phenomena. A significant advantage of SPM is the small amount of material required for analysis, which is especially beneficial for screening applications where material may be limited. The techniques have broad applicability across the pharmaceutical development process from crystallization of the API and particle size reduction to formulation and preparation as the final dosage form. And, with continued advances in high-speed imaging, automation, and combination techniques including thermal analysis and spectroscopy, the utility of SPM in pharmaceutical development will be increasingly expanded.

ACKNOWLEDGMENTS

The author acknowledges the following people for their collaboration and contributions to original research published in this chapter: Bryan Benson, Amanda Simpson, Rick Nay, Rahul Gandhi, and Dan Braido. The author thanks Hanmi Xi and Mike Lowinger for their insightful comments on the manuscript.

REFERENCES

1. Chew, C. M., Ristic, R. I., Dennehy, R. D., and De Yoreo, J. J. Crystallization of Paracetamol under Oscillatory Flow Mixing Conditions, *Cryst Growth Des* **4**(5) (2004): 1045–1052.
2. Begat, P., Young, P. M., Edge, S., Kaerger, J. S., and Price, R. The Effect of Mechanical Processing on Surface Stability of Pharmaceutical Powders: Visualization by Atomic Force Microscopy, *J Pharm Sci* **92**(3) (2003): 611–620.
3. Abu Bakar, M. R., Nagy, Z. K., and Rielly, C. D. Investigation of the Effect of Temperature Cycling on Surface Features of Sulfathiazole Crystals During Seeded Batch Cooling Crystallization, *Cryst Growth Des* **10**(9) (2010): 3892–3900.
4. Ristic, R. I. Oscillatory Mixing for Crystallization of High Crystal Perfection Pharmaceuticals, *Chem Eng Res Des* **85**(A7) (2007): 937–944.
5. Perkins, M. C., Bunker, M., James, J., Rigby-Singleton, S., Ledru, J., Madden-Smith, C., Luk, S., Patel, N., and Roberts, C. J. Towards the Understanding and Prediction of Material Changes During Micronisation Using Atomic Force Microscopy, *Eur J Pharm Sci* **38**(1) (2009): 1–8.
6. Jones, M. D., Young, P. M., Traini, D., Shur, J., Edge, S., and Price, R. The Use of Atomic Force Microscopy to Study the Conditioning of Micronised Budesonide, *Int J Pharm* **357**(1–2) (2008): 314–317.
7. Li, H., Wen, H., Stowell, J. G., Morris, K. R., and Byrn, S. R. Crystal Quality and Physical Reactivity in the Case of Flufenamic Acid (FFA), *J Pharm Sci* **99**(9) (2010): 3839–3848.
8. Cassidy, A. M., Gardner, C. E., Auffret, T., Aldous, B., and Jones, W. Decoupling the Effects of Surface Chemistry and Humidity on Solid-State Hydrolysis of Aspirin in the Presence of Dicalcium Phosphate Dihydrate, *J Pharm Sci* **101**(4) (2012): 1496–1507.
9. Jones, M. D., Beezer, A. E., and Buckton, G. Determination of Outer Layer and Bulk Dehydration Kinetics of Trehalose Dihydrate Using Atomic Force Microscopy, Gravimetric Vapour Sorption and Near Infrared Spectroscopy, *J Pharm Sci* **97**(10) (2008): 4404–4415.
10. Cassidy, A. M. C., Gardner, C. E., and Jones, W. Following the Surface Response of Caffeine Cocrystals to Controlled Humidity Storage by Atomic Force Microscopy, *Int J Pharm* **379**(1) (2009): 59–66.
11. Mahlin, D., Berggren, J., Gelius, U., Engstrom, S., and Alderborn, G. The Influence of PVP Incorporation on Moisture-Induced Surface Crystallization of Amorphous Spray-Dried Lactose Particles, *Int J Pharm* **321**(1–2) (2006): 78–85.
12. Mahlin, D., Berggren, J., Alderborn, G., and Engstrom, S. Moisture-Induced Surface Crystallization of Spray-Dried Amorphous Lactose Particles Studied by Atomic Force Microscopy, *J Pharm Sci* **93**(1) (2004): 29–37.
13. Price, R., and Young, P. M. Visualization of the Crystallization of Lactose from the Amorphous State, *J Pharm Sci* **93**(1) (2004): 155–164.
14. Chen, D. B., Haugstad, G., Li, Z. J., and Suryanarayanan, R. Water Sorption Induced Transformations in Crystalline Solid Surfaces: Characterization by Atomic Force Microscopy, *J Pharm Sci* **99**(9) (2010): 4032–4041.
15. Khougaz, K., and Clas, S. Crystallization Inhibition in Solid Dispersions of MK-0591 and Poly(vinylpyrrolidone) Polymers, *J Pharm Sci* **89**(10) (2000).
16. Broman, E., Khoo, C., and Taylor, L. A Comparison of Alternative Polymer Excipients and Processing Methods for Making Solid Dispersions of a Poorly Water Soluble Drug, *Int J Pharm* **222**(1) (2001).
17. Aso, Y., Yoshioka, S., and Kojima, S. Molecular Mobility-Based Estimation of the Crystallization Rates of Amorphous Nifedipine and Phenobarbital in Poly(vinylpyrrolidone) Solid Dispersions, *J Pharm Sci* **93**(2) (2004).
18. Patterson, J. E., James, M. B., Forster, A. H., Lancaster, R. W., Butler, J. M., and Rades, T. Preparation of Glass Solutions of Three Poorly Water Soluble Drugs by Spray Drying, Melt Extrusion and Ball Milling, *Int J Pharm* **336**(1) (2007).
19. Sahin, O., Magonov, S., Su, C., Quate, C. F., and Solgaard, O. An Atomic Force Microscope tip Designed to Measure Time-Varying Nanomechanical Forces, *Nat Nano* **2**(8) (2007): 507–514.

20. Lauer, M., Grassmann, O., Siam, M., Tardio, J., Jacob, L., Page, S., Kindt, J., Engel, A., and Alsenz, J. Atomic Force Microscopy-Based Screening of Drug-Excipient Miscibility and Stability of Solid Dispersions, *Pharm Res* **28**(3) (2011): 572–584.
21. de Vegt, O., Vromans, H., den Toonder, J., and Maarschalk, K. V. Influence of Flaws and Crystal Properties on Particle Fracture in a Jet Mill, *Powder Technology* **191**(1–2) (2009): 72–77.
22. Taylor, L. J., Papadopoulos, D. G., Dunn, P. J., Bentham, A. C., Dawson, N. J., Mitchell, J. C., and Snowden, M. J. Predictive Milling of Pharmaceutical Materials Using Nanoindentation of Single Crystals, *Organic Process Research & Development* **8**(4) (2004): 674–679.
23. Zugner, S., Marquardt, K., and Zimmermann, I. Influence of Nanomechanical Crystal Properties on the Comminution Process of Particulate Solids in Spiral Jet Mills, *European Journal of Pharmaceutics and Biopharmaceutics* **62**(2) (2006): 194–201.
24. Olusanmi, D., Roberts, K. J., Ghadiri, M., and Ding, Y. The Breakage Behaviour of Aspirin under Quasi-static Indentation and Single Particle Impact Loading: Effect of Crystallographic Anisotropy, *Int J Pharm* **411**(1–2) (2011): 49–63.
25. Meier, M., John, E., Wieckhusen, D., Wirth, W., and Peukert, W. Influence of Mechanical Properties on Impact Fracture: Prediction of the Milling Behaviour of Pharmaceutical Powders by Nanoindentation, *Powder Technology* **188**(3) (2009): 301–313.
26. Kubavat, H. A., Shur, J., Ruecroft, G., Hipkiss, D., and Price, R. Investigation into the Influence of Primary Crystallization Conditions on the Mechanical Properties and Secondary Processing Behaviour of Fluticasone Propionate for Carrier Based Dry Powder Inhaler Formulations, *Pharm Res* **29**(4) (2012): 994–1006.
27. Shi, L. M., and Sun, C. C. Overcoming Poor Tabletability of Pharmaceutical Crystals by Surface Modification, *Pharm Res* **28**(12) (2011): 3248–3255.
28. Jing, Y. Y., Zhang, Y., Blendell, J., Koslowski, M., and Carvajal, M. T. Nanoindentation Method To Study Slip Planes in Molecular Crystals in a Systematic Manner, *Cryst Growth Des* **11**(12) (2011): 5260–5267.
29. Cao, X. P., Morganti, M., Hancock, B. C., and Masterson, V. M. Correlating Particle Hardness with Powder Compaction Performance, *J Pharm Sci* **99**(10) (2010): 4307–4316.
30. Kiran, M., Varughese, S., Reddy, C. M., Ramamurty, U., and Desiraju, G. R. Mechanical Anisotropy in Crystalline Saccharin: Nanoindentation Studies, *Cryst Growth Des* **10**(10) (2010): 4650–4655.
31. Picker-Freyer, K. M., Liao, X. M., Zhang, G. F., and Wiedmann, T. S. Evaluation of the Compaction of Sulfathiazole Polymorphs, *J Pharm Sci* **96**(8) (2007): 2111–2124.
32. Feng, Y. S., and Grant, D. J. W. Influence of Crystal Structure on the Compaction Properties of *n*-Alkyl 4-Hydroxybenzoate Esters (Parabens), *Pharm Res* **23**(7) (2006): 1608–1616.
33. Begat, P., Morton, D. A. V., Staniforth, J. N., and Price, R. The Cohesive-Adhesive Balances in Dry Powder Inhaler Formulations I: Direct Quantification by Atomic Force Microscopy, *Pharm Res* **21**(9) (2004): 1591–1597.
34. Hooton, J. C., Jones, M. D., and Price, R. Predicting the Behavior of Novel Sugar Carriers for Dry Powder Inhaler Formulations via the Use of a Cohesive-Adhesive Force Balance Approach, *J Pharm Sci* **95**(6) (2006): 1288–1297.
35. Bunker, M., Zhang, J. X., Blanchard, R., and Roberts, C. J. Characterising the Surface Adhesive Behavior of Tablet Tooling Components by Atomic Force Microscopy, *Drug Dev Ind Pharm* **37**(8) (2011): 875–885.
36. Lee, J. Intrinsic Adhesion Force of Lubricants to Steel Surface, *J Pharm Sci* **93**(9) (2004): 2310–2318.
37. Wang, J. J., Li, T., Bateman, S. D., Erck, R., and Morris, K. R. Modeling of Adhesion in Tablet Compression—I. *Atomic Force Microscopy and Molecular Simulation, J Pharm Sci* **92**(4) (2003): 798–814.
38. Weber, D., Pu, Y., and Cooney, C. L. Quantification of Lubricant Activity of Magnesium Stearate by Atomic Force Microscopy, *Drug Dev Ind Pharm* **34**(10) (2008): 1097–1099.

Chapter 14

Comparative Nanomechanical Study of Multiharmonic Force Microscopy and Nanoindentation on Low Dielectric Constant Materials

Katharine Walz, Robin King, Willi Volksen,
Geraud Dubois, Jane Frommer, and Kumar Virwani

IBM Almaden Research Center, San Jose, CA

14.1 INTRODUCTION

Methods to measure Young's moduli in sub-1 μm-thick films include nanoindentation [1, 2], surface acoustic wave spectroscopy [3], ellipsometric porosimetry [4], scanning probe microscopy [5–7], Brillouin light scattering [8, 9], and plane-strain bulge test [10]. In this chapter we focus on two probe-based techniques—multiple harmonics atomic force microscopy (AFM) and nanoindentation—and compare them to surface acoustic wave spectroscopy (SAWS).

In this chapter we present a comparison of experimental results from eight different low-k dielectric samples that were analyzed for their mechanical properties with ultralow load nanoindentation, SAWS, and multiharmonics AFM analysis. The sample set was chosen for its importance in evolving semiconductor industry materials. In the miniaturization of semiconductor electronics, the metal lines that carry signals to microprocessors get ever closer [11], increasing parasitic capacitive losses especially when operating at frequencies [12] in the range of hundreds of megahertz. A standard in the industry today to reduce these losses is to insert between the

Scanning Probe Microscopy in Industrial Applications: Nanomechanical Characterization,
First Edition. Edited by Dalia G. Yablon.

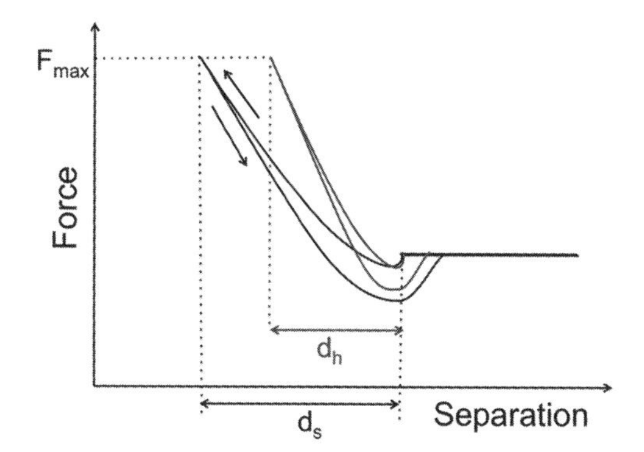

Figure 14.1 AFM tip motion into the sample for hard and soft materials. For the same maximum force the AFM tip penetrates deeper into a soft sample (d_s) than into a hard sample (d_h).

conducting elements materials of low dielectric constant [13] ("low-*k*"), lower than silicon dioxide, which has been the insulating material of choice for almost three decades. Chemical modification of the silica network, first by the introduction of fluorine and eventually by the addition of carbon, has been adopted as a path to introducing low-*k* materials in manufacturing. Using this strategy, the dielectric constant can be desirably lowered to 2.7–2.6 as a result of a decrease in network connectivity and hence a decrease in material density. These two characteristics have a strong impact on mechanical properties. For example, a reduction in dielectric constant (k) of ~ one-third is accompanied by an order of magnitude decrease in Young's modulus (E) on going from SiO_2 ($k=4.0$, $E=72$ GPa) to carbon-doped silicon oxides ($k=2.7$, $E=7$–10 GPa). By controlling porosity [13], carbon-doped oxides offer the potential for "generational extendability"—continued reduction of dielectric constant without changing elemental composition. The addition of nanometer-sized pores to existing low-*k* materials, while commonly embraced as the only manufacturing-compatible way to access the desired ultralow-*k* regime ($k<2.4$–2.5), introduces a number of new integration issues [14–16]. Most importantly, the mechanical strength of the dielectric silicon oxides, already degraded by a decrease in network connectivity in carbon-doped oxides, is further reduced on increasing porosity. This leads to films that crack on curing or during mechanically demanding processing and packaging steps [17]. For this reason, accurate techniques are essential for determining the mechanical properties of these ultralow-*k* thin films at the thicknesses used for integration into devices and on diverse substrates. Accurate Young's modulus values also strengthen the modeling input for finite-element analysis of materials in future chip design.

The field of multifrequency [18] AFM has been around since the invention of AFM in 1986 [19] and is based on capturing the multiple harmonics of the fundamental cantilever frequency. In a brief description, the AFM tip is used to indent into a sample surface; the total force experienced by the tip is plotted as a function of the distance between the tip and the sample surface in a classic static force curve (for more on force curves, see Chapter 3). Figure 14.1 shows that for the same applied maximum force the tip penetrates deeper into a softer material (d_s) than a harder material (d_h). Figure 14.1 also illustrates other features of a typical force curve such as snap-to-contact and the

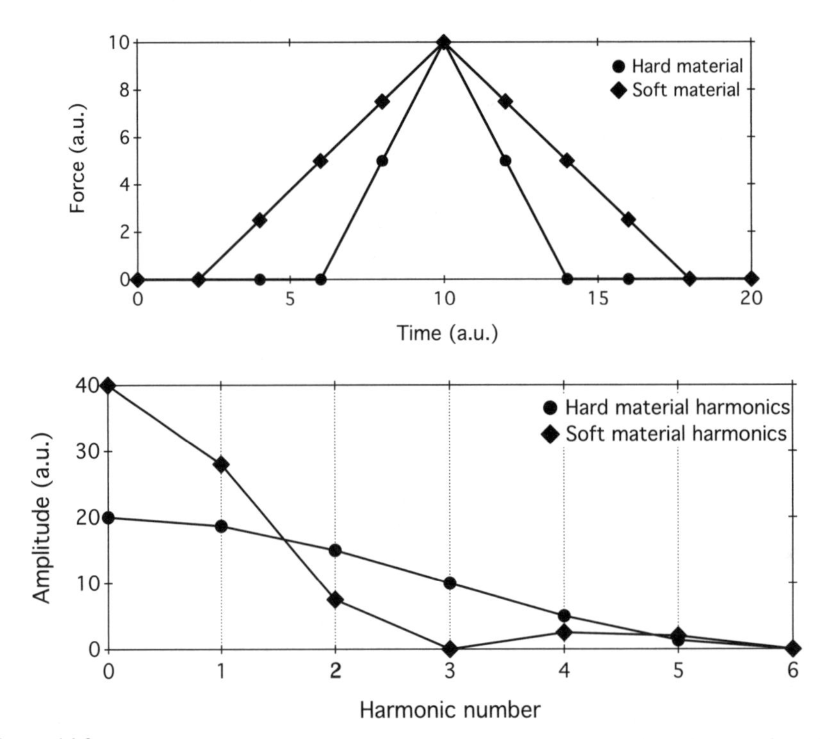

Figure 14.2 (*Top*) AFM force vs. separation plot from Figure 14.1 converted to a force vs. time plot. The tip–sample interaction occurs for a longer duration for a soft material than for a hard material where the interaction occurs for a shorter time. (*Bottom*) Fourier components of the top force vs. time plot. Harmonics 0–3 are sufficient to reconstruct interaction of the tip with a soft sample; however, harmonics 0–7 are required to reconstruct the interaction of the tip with a hard sample. Harmonic 0 refers to the DC component of the Fourier transformation.

adhesion experienced by a probe after indenting into a sample. A plot of the force as a function of time is shown in Figure 14.2 (top): Indenting into a soft material takes longer than a hard material to reach the same maximum force. The force versus time curves are then decomposed into individual Fourier components. For simplicity the first seven components are shown in Figure 14.2 (bottom). The basic features of a multiharmonic AFM experiment are illustrated by the decomposed components. Fewer harmonics are required to reconstruct a softer indent. In our example (Fig. 14.2) harmonics 0 through 3 are sufficient to reconstruct the force because the amplitude of the third harmonic is zero—harmonics greater than three contribute insignificantly to the reconstruction of the force. Harmonics 0 through 6 are required to achieve the same force reconstruction for the hard material.

This brings us to a basic challenge in multiharmonic AFM tip–sample interaction. Ideally, one would collect the amplitudes of *all* the harmonics of the tip–sample interaction, even in the intermittent or "tapping" mode. However, the AFM cantilever acts as a low-pass filter for all frequencies above the fundamental frequency of oscillation of the cantilever (Fig. 14.3). This low signal-to-noise ratio makes the measurement

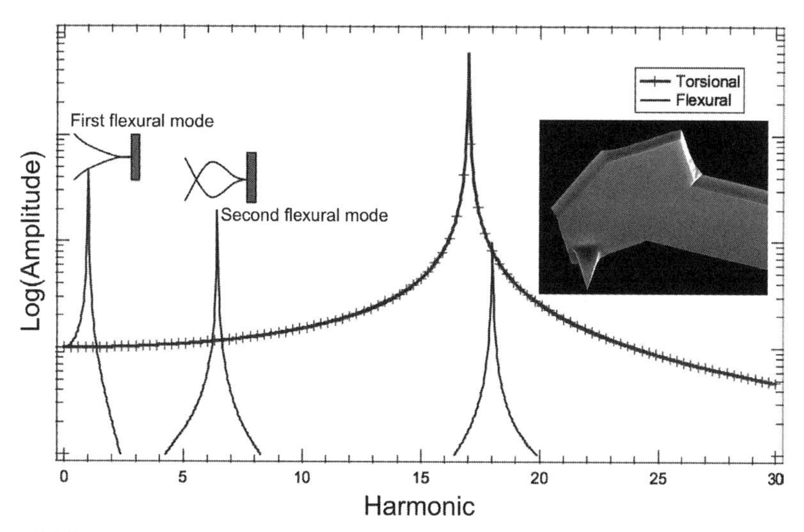

Figure 14.3 Flexural (vertical) and torsional (lateral) frequency response of multiharmonic probe with offset cantilever (SEM image of cantilever in inset). The probe maintains a good amplitude signal-to-noise ratio until torsional resonance frequency. In contrast the flexural mode reduces signals immediately above the first flexural resonance frequency with amplification being achieved only at higher modes of oscillation.

and the interpretation of the higher harmonic information from standard tapping mode measurements difficult, although multiple groups have published data from such measurements [20–25]. Ozgur Sahin [26–28] pioneered the use of a cantilever with an offset probe for multiharmonic analysis. By measuring the torsional amplitude at exact integer multiples (harmonics) of the tapping mode drive frequency, it is possible to extract the variation in force between the tip and the sample when the tip goes through a period of the tapping oscillation. A correctly chosen single harmonic can provide compositional mapping of a complex composite material while providing the same lateral resolution as a tapping image (~5 nm). Multiple harmonics can also be observed and converted back to the time domain, providing force versus distance data analogous to an averaged force curve.

It is reasonable to question whether the dynamics of an offset-probe cantilever differ from a standard rectangular cantilever that displays sinusoidal harmonic motion [29]. This is answered in Figure 14.4, which shows the variation of flexural (vertical) and torsional (lateral) deflection signals as a function of time for an offset-probe cantilever. Negative flexural deflection denotes motion of the probe toward the sample, and positive flexural deflection denotes motion of the probe away from the sample. Overall flexural motion is still sinusoidal with an additional tip–sample interaction as a spike in the torsional signal measured at the point of closest approach, nicely correlating with the minima in the flexural deflection signal. One also measures coupling of the flexural oscillations into the torsional oscillations, although this does not affect the overall analysis. Thus the use of “tapping” harmonics to describe the dynamics of an offset-probe cantilever is justified.

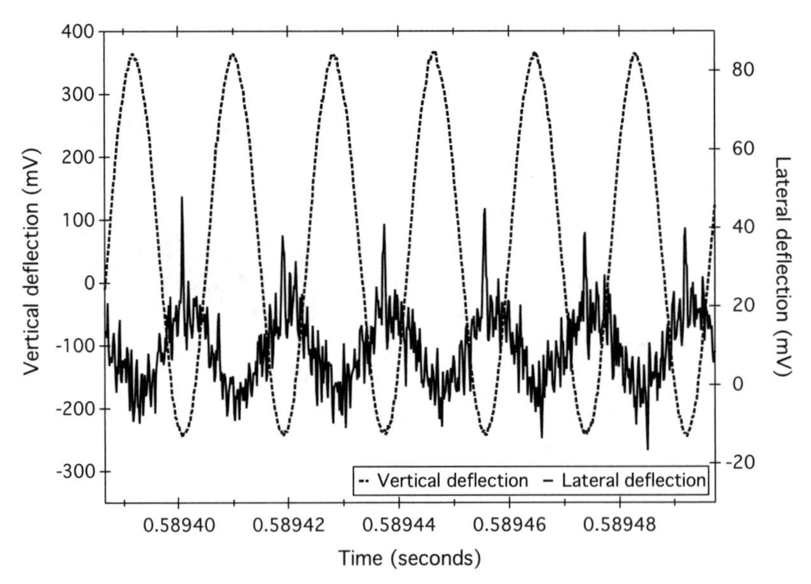

Figure 14.4 Vertical (flexural) and lateral (torsional) deflection signals of an offset tip cantilever. The flexural signal is sinusoidal in nature hence the assumptions of "tapping" dynamics can be used in the analysis of multiharmonic AFM data. The spikes on the lateral defection signal occur at the point of closest approach between the tip and the sample, and nicely correlate with the minima in the vertical deflection signal.

In AFM imaging, the contact area between the tip and sample generally determines the resolution. When compared to force volume or nanoindentation, harmonic imaging increases the resolution by decreasing the deformation depths and consequently the contact area between the tip and sample. The deformation depths in multiharmonic microscopy are similar to those experienced with hard tapping [30]. Because the deformation depths are small and lateral forces are negligible in tapping mode, there is minimal damage to the sample. The small deformation depth also decreases the substrate's effect on the measured modulus for very thin films. AFM with multiple harmonics is sensitive to the force on the tip (which includes everything observed in a force curve collected at the tapping frequency) and there is some filtering of the data due to the finite bandwidth of the torsional resonance.

The experimental results presented in this chapter were collected using special cantilevers with the tip offset to one side as shown in Figure 14.3 as described earlier. This geometry allows measurements at higher harmonics that correlate to material properties such as stiffness, adhesion, and dissipation. Peak and average tip–sample interaction forces can also be measured by reconstructing the entire tip–sample interaction force curve from amplitudes or weights collected at multiples of the fundamental vibration frequency of the cantilever. Unlike the tapping mode where the traditional phase contrast depends heavily on the amplitude setpoint, using multiple harmonics enables distinction of the origin of phase contrast in an image.

The second technique used in this chapter for mechanical property analysis is nanoindentation. Indentation-based mechanical property testing methods have been around for more than a century. Swedish mechanical engineer Johan August Brinell in 1900 developed a test based upon a 10-mm diameter steel ball indenting on a surface of interest with a force of 29 kN (3000 kGf) [31, 32]. For harder materials a tungsten carbide ball was used in place of a steel ball. The Brinell hardness test was widely used during the early part of the twentieth century with the development of various disciplines of metallurgy. In 1921, Robert L. Smith and George E. Sandland pioneered a technique—the Vickers test [33]—based on a pyramidal tip instead of a spherical ball for measuring hardness. The most attractive aspect of the Vickers hardness test was that the measured hardness was independent of the indenter size. Vickers indentation measurements could be performed on hard and soft materials but with less success on brittle materials. In 1939, Frederick Knoop and co-workers [34] of the National Bureau of Standards (now the National Institutes of Standards and Technology) developed an indentation test for brittle materials and thin sheets based upon a pyramidal diamond point. As film and coating technology progressed to increasingly thin films, the accuracy of conventional macroscale indentation tests waned. Microindentation tests were devised with applied forces in single-digit newtons and resulting deformations in tens of micrometers. Vickers and Knoop hardness tests continued to the micron-scale domain primarily because of the use of pyramidal indenters. By the mid-1970s micron-scale indentation tests proved to be inadequate for the burgeoning field of nanotechnology [35] and nanoscale indentation techniques started to be developed. In nanoindentation the loads are on the order of hundreds to thousands of micronewtons and the deformation depths hundreds of nanometers to single-digit micrometers. In our nanoindentation studies we measure Young's modulus and hardness of low-k dielectric materials using ultralow load (3–25 μN) nanoindentation.

In this study, Young's modulus measurements from AFM-based multiharmonic and nanoindentation measurements are compared with modulus measurements from surface acoustic wave spectroscopy. Acoustic waves that propagate along the surface of a material with finite elasticity are referred to as surface acoustic waves. Lord Rayleigh first described these waves in his classical paper [36] in 1885. As expected, these waves have a longitudinal as well as a shear component that can couple with the media through which the wave propagates. The higher the Young's modulus of the material, the shorter the distance the waves propagate for the same input energy. In SAWS the decay in the amplitude of surface acoustic waves is sensed with piezoelectric transducers and correlated to the mechanical properties of the thin films. Thus one can measure mechanical properties of materials with surface acoustic waves. For reliable SAWS measurements on a system consisting of two materials such as a thin film deposited on a silicon substrate, both the underlying substrate mechanical properties and the density of the overlying material must be known to a high degree of accuracy. In our studies X-ray diffraction provided the density of the overlying low dielectric constant films on polished, crystalline Si (100) wafers.

14.2 EXPERIMENTAL

14.2.1 Synthesis of Low-*k* Dielectric Films

The spin-on porous low-*k* dielectric films of this study were synthesized from a mixture of two materials—one that forms the continuous organosilicate matrix and one that forms the pores in the matrix, the porogen [37]. Porosity was systematically varied by varying the mass ratio of porogen to matrix in the starting mixture. The organosilicate matrix precursor was a copolymer of methyl trimethoxysilane and tetra-ethoxyorthosilicate with an approximate molecular weight of 2000 Dal. The porogen was a propylene-glycol-based polyol with a molecular weight of 6000 Dal. Eight different mass ratios (Table 14.1) of the two components were prepared as solutions in propylene-glycol mono-*n*-propyl ether solvent and spun onto silicon wafers. A heating cycle (ramp from 25 to 425°C over 1 hour) accomplished the next steps of (1) phase separating the two components and cross-linking the organosilicate in the 150–200°C range and (2) eliminating the porogen as the temperature further increased to 425°C. The eight resulting porous organosilicate films (600–800 nm thick) were then characterized for thickness and density using X-ray reflectivity [38, 39] (see Section 14.2.4).

14.2.2 Multiharmonic Microscopy with AFM

Multiharmonic microscopy with AFM was performed with the HarmoniX routine on a Bruker Dimension Icon. Probes used for simultaneous collection of topographic and mechanical property information from low-*k* dielectric samples had fundamental

Table 14.1 Properties of Eight Different Low-*k* Thin–Film Samples (A–H) and Young's Moduli (*E*) Measured by Surface Acoustic Wave Spectroscopy (SAWS) and Nanoindentation (NI) and Multiharmonic AFM

Sample	Porogen[a] %	Density[b] (gm/cm^3)	Thickness (nm)	E (SAWS) (GPa)	E (NI) (GPa)
A	0	1.286	586.1	4.84	5.32
B	5	1.197	599.8	3.74	3.95
C	10	1.106	618.3	2.90	3.15
D	15	1.017	625.3	2.19	2.56
E	20	0.940	641.8	1.71	1.88
F	30	0.789	686.7	0.86	1.07
G	40	0.642	715.4	0.44	0.54
H	50	0.551	736.6	0.14	0.29

[a]Porogen % refers to relative mass of pore-forming agent in the precursor mixtures of matrix-forming and pore-forming components (see text). The films were all spun under the same conditions; that the resulting thicknesses trend with porogen content could reflect an expanded film volume that results from the volatilization of the porogen component.

[b]Density and thickness determined by X-ray reflectivity. Though no correlation was expected between film thickness and density or modulus, the observed trend is under investigation.

tapping frequencies in the range of 50–80 kHz and torsional resonance frequencies in the range of 850–975 kHz. These probes were chosen because they allowed 18–20 harmonics of the fundamental tapping frequency to be collected and used for mechanical property evaluation. For topography and modulus measurements the probes were operated at an ~15% offset below the fundamental flexural resonance of the cantilever. The peak-to-peak amplitude of the probe was about 80 nm and the setpoint amplitude range was between 30 and 50 nm.

Three parameters were needed for calibration in mechanical property determination—probe end radius, flexural force, and torsional force. An etched titania standard [40–42] was scanned in tapping mode to measure the profile of the tip, hence to determine the probe end radius. Two consecutive 1-μm × 1-μm scans of the titania surface were captured under light tapping conditions to minimize tip wear. The instrument's analysis software provided an arithmetic average of tip end radii in the range of 7–20 nm at 5 and 10 nm heights from the tip apex. After acquiring the probe end radius, the amplitude sensitivity and deflection sensitivity were determined for input to the flexural and torsional force constants, described in greater detail below. The amplitude and deflection sensitivity measurements were performed on mica, freshly cleaved to minimize the influence of ambient contaminants.

For the calibration of the flexural force constant of the offset cantilever, the deflection sensitivity [43] was assumed equal to the tapping mode amplitude sensitivity. For rectangular cantilevers the ratio of the two sensitivities is called the kappa factor [44, 45]. The assumption was tested on three of the dozens of probes used for multiharmonic measurements, verifying that tapping mode amplitude and contact mode deflection sensitivities were equal to each other within experimental error. The sensitivity values were input to the spring constant measurements to determine the flexural stiffness of the probe in nanonewtons/nanometers.

The force during torsional oscillations of the probe is calibrated using the literature method [46] briefly described as follows. In both conventional tapping mode and multiharmonic measurements, the cantilever operates in a combination of attractive and repulsive forces. To calibrate the torsional force constant, we exploit the fact that the same multiharmonic cantilever is capable of operating under two tapping conditions: attractive and a combination of attractive and repulsive forces. On driving the cantilever from a purely attractive mode of oscillation to a combined attractive and repulsive mode (by operating the cantilever above its fundamental flexural resonance frequency), an abrupt transition between the two modes causes the cantilever to deflect. The flexural and torsional deflections measured in this transition are in the ratio of the respective force constants. Since the flexural force constant is known, the torsional force constant can then be deduced.

14.2.3 Nanoindentation

Quasi-static nanoindentation measurements were performed on a commercial nanoindenter, Hysitron Model TI-950, with a diamond cube corner probe of 40-nm end radius. To eliminate adsorbed moisture that could complicate measurements of mechanical properties, nanoindentation was carried out under a positive pressure of dry nitrogen in the closed indenter enclosure. Freshly loaded samples were allowed to equilibrate with dry nitrogen for at least 12 hours, also achieving thermal equilibrium. Before each of the eight silicate samples was indented, the cube corner probe was calibrated on a quartz sample with a modulus of 70 GPa. Quartz was chosen because it undergoes a plastic deformation on indentation, providing true profiles of the diamond indenter tip shape for use in calculating contact areas for silicate film indentation. AFM images of the ~60-nm shallow indents into quartz revealed that the residual deformation could be fit by a parabolic surface, justifying the use of only three coefficients to fit the tip area function instead of the customary six or eight coefficients. Thus the tip area function was represented by the following equation [47]:

$$A = C_0(h_c)^2 + C_1 h_c + C_2(h_c)^{1/2} \tag{14.1}$$

where h_c is the contact depth, and C_0, C_1, and C_2 are fitting coefficients. The indents into low-*k* silicate samples were also shallow—less than 60-nm contact depth as measured by AFM—representing less than 10% of the total film thicknesses used in this study (Table 14.1). In a preload indentation step, the indenter probe approached the low-*k* material surface with a setpoint load of 700 nN chosen to balance the requirements of minimal surface perturbation with sufficient load for stable contact. Each low-*k* sample was indented with a series of eight maximum forces: 25, 20, 15, 12.5, 10, 7.5, 5, and 3 μN. For each load, seven independent indents were performed on the surface at least 10 μm apart from each other. The load function was comprised of a sequence of three steps: load, hold, and unload. The maximum force was reached in 5 seconds in the "load" step; the maximum force was held for 2 seconds in the "hold" step; and the force was ramped down to 700 nN in 5 seconds in the "unload" step, thus constituting a 5–2–5 load function profile. The upper half (98–50%) of the unload curve [48] was fitted with the above-mentioned tip area function to yield a contact stiffness, used in calculating the reduced modulus (E_r) from classic Oliver–Pharr equations [49]. The Young's modulus of the silicate sample (E_s) was then calculated using the following formula:

$$\frac{1}{E_r} = \frac{1-\nu_i^2}{E_i} + \frac{1-\nu_s^2}{E_s} \tag{14.2}$$

where E_r is the reduced modulus, E_i is the indenter modulus of 1140 GPa, ν_i is the indenter Poisson's ratio of 0.07 for diamond, and ν_s is the sample Poisson's ratio of 0.25 [50]. A graph of reduced modulus as a function of contact depth was then plotted for the series of loads for each sample (see Section 14.3.2).

14.2.4 X-Ray Reflectivity (XRR)

Specular X-ray reflectivity was used to measure film thickness and density with a Panalytical X'Pert Pro MRD diffractometer employing a ceramic X-ray tube ($\lambda = 0.154$ nm) and a high-resolution goniometer (resolution ±0.0001°). The critical angles from the reflectivity data were obtained from the peak position of Iq^4 versus q plots ($q=4\pi/\lambda \sin\Theta$), where λ is the wavelength and Θ is the grazing incident angle of the X-ray beam.

14.2.5 Surface Acoustic Wave Spectroscopy

Elastic moduli were measured by SAWS [51] on a Fraunhofer LaWave spectrometer [52]. Acoustic waves were generated by a nitrogen-pulsed laser ($\lambda=337$ nm, pulse duration = 0.5 ns) and detected with a piezoelectric polymer-based transducer. Fitting the measured surface wave velocity as a function of frequency with a theoretical dispersion curve and factoring in the sample thickness and density obtained from XRR measurements yielded a reduced elastic modulus. The reduced modulus was converted to Young's modulus, assuming the same Poisson's ratio for the sample as above, $v_s=0.25$. We do not discuss in great detail the SAWS measurements, as the technique is well established and accepted for low-*k* dielectric measurements.

14.3 RESULTS AND DISCUSSIONS

In multiharmonic AFM the force applied by the cantilever is sufficient to slightly deform the sample surface to extract mechanical property information. In this section experimental measurements of mechanical properties from multiharmonic AFM are described and compared to those from nanoindentation and surface acoustic wave spectroscopy measurements.

14.3.1 Multiharmonic Microscopy with AFM

We first compared two samples with 0 and 15% porogen content for which SAWS provided densities of 1.286 (sample A) and 1.017 g/cm^3 (sample D), respectively, and moduli of 4.8 and 2.2 GPa (Table 14.1). The SAWS modulus values put these two samples in a sweet spot for multiharmonic measurements with probes in the frequency range of 50–80 kHz. Multiple data points were collected on each sample over different areas. While measurements on the same sample showed little variability, the results from the two samples with different moduli (according to SAWS and nanoindentation) also showed little variability from each other as shown in Figure 14.5. Measures were taken to ensure that the following parameters were kept constant across the two samples: use of the same probe, free air oscillation amplitude, tapping mode setpoint, laser position on the cantilever, and number of harmonics chosen for measurement.

These initial puzzling results were joined by additional experimental data of similarly inconsistent nature, described below. Possible sources of the lack of consistency are enumerated, together with measures taken to address them:

1. *Probe Contamination* Debris on the probe could serve as a source of signals rather than the intended sample. By intentionally dulling the tip, electrostatics and high pressure under the AFM tip as sources of contamination are minimized.

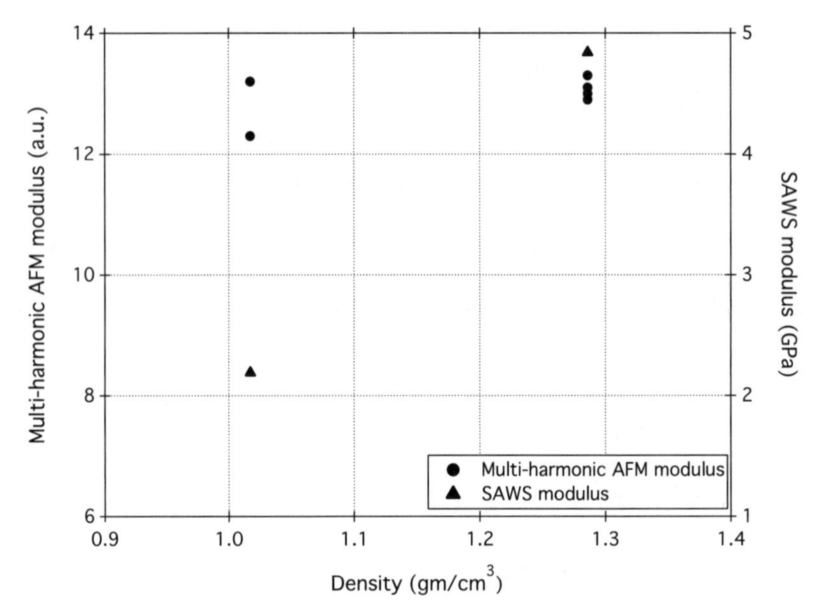

Figure 14.5 Multiharmonic AFM modulus for two samples with densities of 1.286 g/cm^3 and 1.017 g/cm^3 (circles) imaged under identical conditions with the same probe. The SAWS moduli of the exact same samples are provided for reference (triangles). The probe was unable to distinguish between the two different modulus values of the two samples.

2. *Dynamic versus Quasi-static Measurement Conditions* SAWS and nanoindentation are both performed under quasi-static conditions, whereas multiharmonic measurements are performed at tapping mode frequencies between 50 and 80kHz. Higher frequency measurements could be affected by the viscoelastic response.
3. *Sampling Depth* The sampling depth in multiharmonic AFM is on the order of single-digit nanometers, in contrast to nanoindentation's tens-to-hundreds of nanometers, and SAWS' entire film thickness of hundreds of nanometers. The AFM probe might be sampling only a sample surface skin or contamination layer common to both samples. In order to increase AFM sampling depth one would require stiffer probes.

A harmonics calibration standard composed of a film of two polymeric components—polystyrene and low-density polyethylene—of different and well-known moduli provided further evidence of inconsistency in the multiharmonic AFM technique. To check the repeatability of measurements, with a single offset tip the following sequence of modulus measurements was made: (1) the harmonics calibration standard, followed by (2) experimental low-*k* dielectric films, followed by (3) the harmonics calibration standard. For these studies we chose probes intentionally dulled on the titania tip check sample. Modulus data are plotted as a function of measurement sequence (Fig. 14.6). The polystyrene domains of the harmonics calibration standard consistently yielded a modulus of 2000MPa in measurements 1–6. Measurements 7,

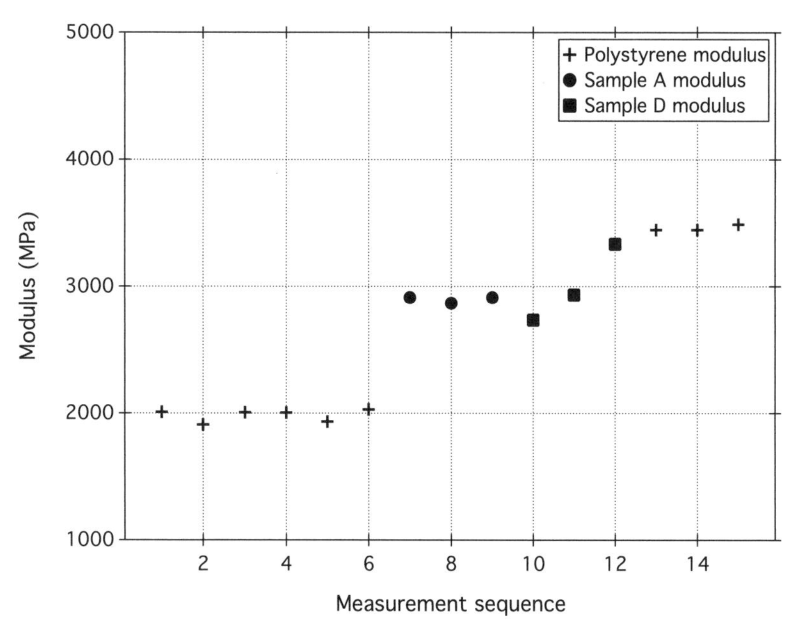

Figure 14.6 Sequence of modulus measurements by multiharmonic AFM on a polystyrene control sample, a low-*k* sample A (density 1.286 g/cm^3, SAWS modulus 4.84 GPa) low-*k* sample D (density 1.017 g/cm^3, SAWS modulus 2.19 GPa), followed by repeated measurements on the polystyrene control. The measurements were performed with the same probe. The modulus values fall into uncorrelatable clusters (see text).

8, and 9, performed on a low-*k* dielectric sample of SAWS-determined modulus of 4840 MPa (sample A), yielded a modulus of ~2900 MPa. Measurements 10, 11, and 12, performed on a low-*k* dielectric sample of SAWS-determined modulus of 2900 MPa (sample C), yielded modulus values from 2700 to 3300 MPa. The final set of measurements 13, 14, and 15 back on the polystyrene standard sample yielded a modulus of ~3450 MPa, an increase of ~70% over original measurement points 1–6. Despite attempts to closely control measurement conditions, the persistent inconsistency indicated we were not able to gain control over the experimental setup.

Systematically investigating tip contamination on the sample with a density of 1.017 g/cm^3 and a modulus of 2190 MPa (from SAWS) (sample D), we intentionally dulled two probes on a hard and abrasive titania tip check sample to a radius of about 20 nm before collecting multiharmonic measurements on the sample. The assumption is that a blunter tip is less sensitive to contamination and to electrostatic material pickup. The moduli measured with the two probes differed by a factor of 2: 1000 versus 500 MPa (Fig. 14.7). For comparison the modulus measured from SAWS was 2190 MPa. Inconsistencies continued in a check of tip cleanliness (data point 4 in Fig. 14.6). After being used to measure modulus on sample D, a probe was cleaned with ultraviolet (UV) ozone and used again on sample D, this time producing a 45% greater modulus value (750 MPa) than before tip cleaning, yet still not approaching the SAWS-determined modulus. Thus it appears that some cleaning procedures cannot

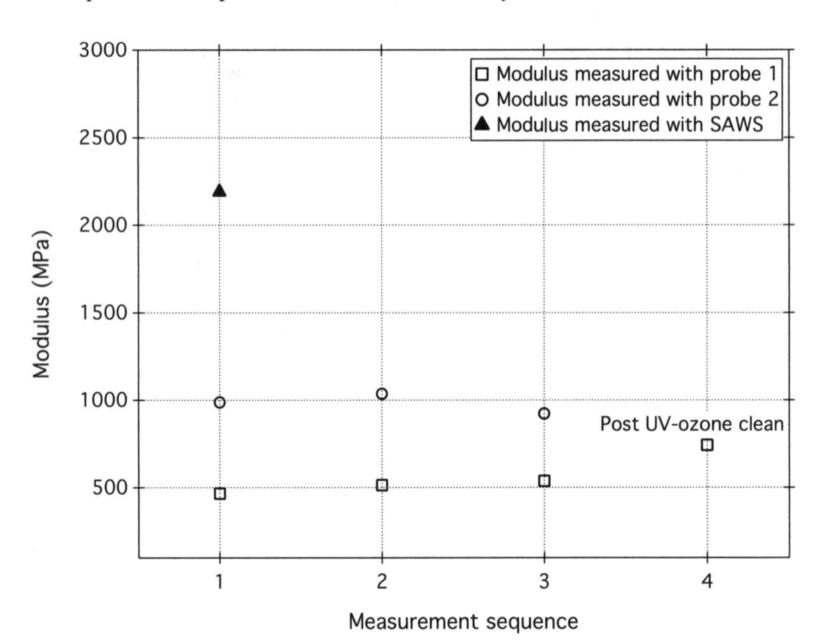

Figure 14.7 Modulus measurements with two different probes yielding two different moduli for the same sample (sample D). SAWS-determined modulus provided for reference.

overcome the modulus difference of a factor of 2 measured with multiharmonic mechanical measurements on the AFM. Electrostatic effects cannot be ruled out.

Reiterating factors that might be preventing accurate measurements of mechanical properties with multiharmonic AFM include:

1. *Sampling Depth* The AFM probes to depths of single-digit nanometers, possibly an undersampled volume for reliable mechanical response. This is in contrast to nanoindentation measurements with ~10× higher forces that were able to accurately measure low-*k* mechanical properties. Stiffer multiharmonic AFM probes would be a good starting point for addressing sample-depth dependence.
2. *Sampling Pressures* The AFM tip may be crushing the porous structure in low-*k* materials. The forces of the AFM tip are an order of magnitude smaller than of the nanoindenter tip at the lowest loads (300 nN versus 3 μN). However, the sampling pressure under an AFM tip is expected to be greater than under an indenter tip since the "area" in the force–area pressure relationship is ~2 orders of magnitude smaller for the AFM tip: 4 versus 40 nm end radii. Material modification causes us to reconsider AFM measurements as nondestructive or minimally perturbing, a commonly held belief.
3. *Machine Compliance* Our experience with nanoindentation measurements (discussed in the next section) shows that machine compliance needs to be carefully taken into account to make accurate measurements of modulus with probe-based systems. Machine compliance includes the response of the

physical setup of probe holder, piezotube, instrument frame, and the like. A macroscale instrument is being used to probe nanometer-scale properties; when trying to probe sample mechanical properties, any deformation of the instrument structure can couple directly into the measurements. This response could be highly nonlinear in nature and may also be affected by resonating components in the stack.

4. *Sampling Frequency* It is well recognized in the literature [53] that polymers exhibit different mechanical properties as a function of the measurement frequency. The exploration of low-k films' mechanical properties with non-quasi-static measurements such as the AFM's higher frequencies is still in its infancy and might follow the same behavior.
5. *Harmonics Range* The multiharmonic AFM technique relies on collection of a sufficient number of tip–sample interaction harmonics to reconstruct mechanical properties. For some sample sets, larger sets of harmonics might be required to yield sufficiently good mechanical property information. New probes that collect even greater numbers of harmonics are a step in that direction.

In sum, there are multiple ill-defined parameters directing the outcome of multiharmonic AFM measurements of modulus. The relatively young field merits further development before it can contribute significantly to the growing need for modulus measurements of thin films.

14.3.2 Nanoindentation

The same set of low-k samples studied by multiharmonic AFM were also characterized by a second probe-based technique of nanoindentation. Load versus displacement curves were recorded on the eight samples of densities ranging from 1.286 to 0.551 g/cm^3 (samples A–H in Table 14.1 and Fig. 14.8). Unusual aspects of this nanoindentation study were (1) the choice of ultralow loads (3–25 µN) to minimize the "substrate effect" (see Chapter 7) and (2) the use of sharp (40-nm end radius) cube-corner probe to cause elastic and plastic deformation of the sample in order to extract the hardness and Young's modulus information. The indenter operated in a "load control" mode where motion of the probe was constrained such that the load was controlled precisely and the sample was free to deform under the action of the applied load. While applying the load, the system continuously captured the probe's deformation into the sample thus yielding a load versus displacement graph.

The start and the endpoint of all the load versus displacement curves are a 700-nN load (discussed in detail below) to both locate the surface and also to define the end of the load versus displacement curve. This small load of 700 nN causes a displacement of almost 40 nm into the least dense sample's surface (sample H) as illustrated in Figure 14.8. A load of 5 µN results in an indentation of ~130 nm into the same sample and an indentation of ~18 nm into the highest density sample (sample A). The deformation caused by the initial 700-nN load into samples A–F was almost completely elastic, whereas less dense samples G and H exhibited some plastic deformation at such low loads.

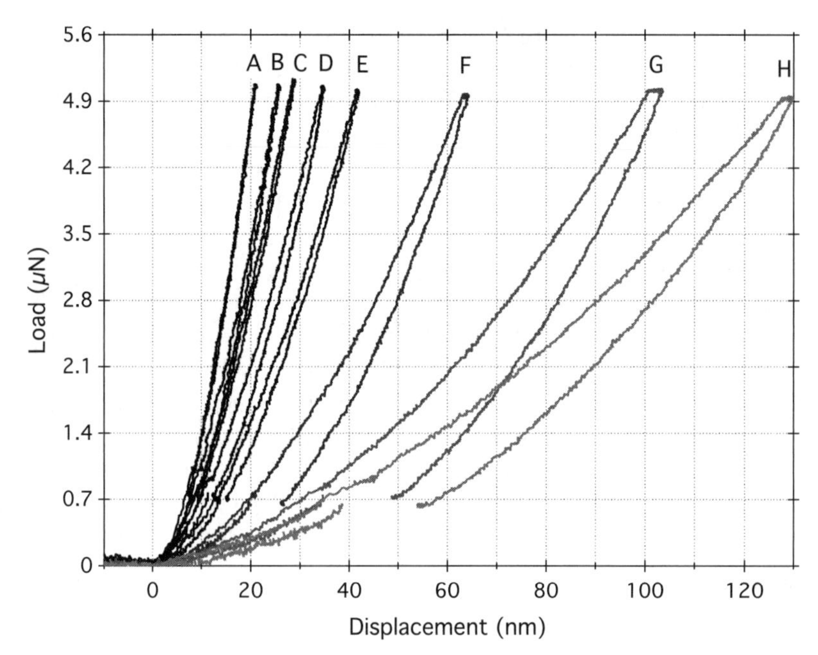

Figure 14.8 Representative force vs. displacement curves from eight different low-*k* samples. Each sample was indented in a "load-control" mode (see text). For the same 5-μN load, sample A (density 1.284 g/cm^3) had the smallest contact depth and sample H (density 0.55 g/cm^3) had the largest contact depth.

Closer examination of a representative force versus displacement curve (Fig. 14.9) is presented for a 10-μN load on an intermediate density sample (sample C). Positive displacement denotes motion of the indenter probe into the sample, whereas negative displacement denotes motion of the probe away from the sample labeled as "lift" in the figure. Once contact with the sample surface was made (marked with a diamond point at 700 nN on the curve), the probe was retracted by 50 nm from the surface and then reapproached to perform an indent into the surface shown in Figure 14.9. At the low loads of this study, this lift protocol is critical for minimizing the error in determining the indentation contact depth. Lifting the probe away from the sample serves two purposes: (1) to determine the point of zero load and zero displacement, where the approach and retract curves converge after the initial 700-nN setpoint load and (2) to reveal a documentable deformation caused by the initial setpoint load. After experiencing the 700-nN setpoint load, moderate density sample C had a nearly complete elastic recovery with little plastic deformation. In contrast, low-density sample H had a ~6-nm plastic deformation after experiencing the same setpoint load (Fig. 14.8). Once the initial approach was complete, the indentation process was carried out in the three steps of load, hold, and unload with the following observations:

- *Load* "Pop-in" events—abrupt increases in the displacement for the same applied load—were not detected during loading (or unloading), indicating the use of a cube-corner probe at ultralow loads avoided cracking and phase transformations observed at larger loads [54].

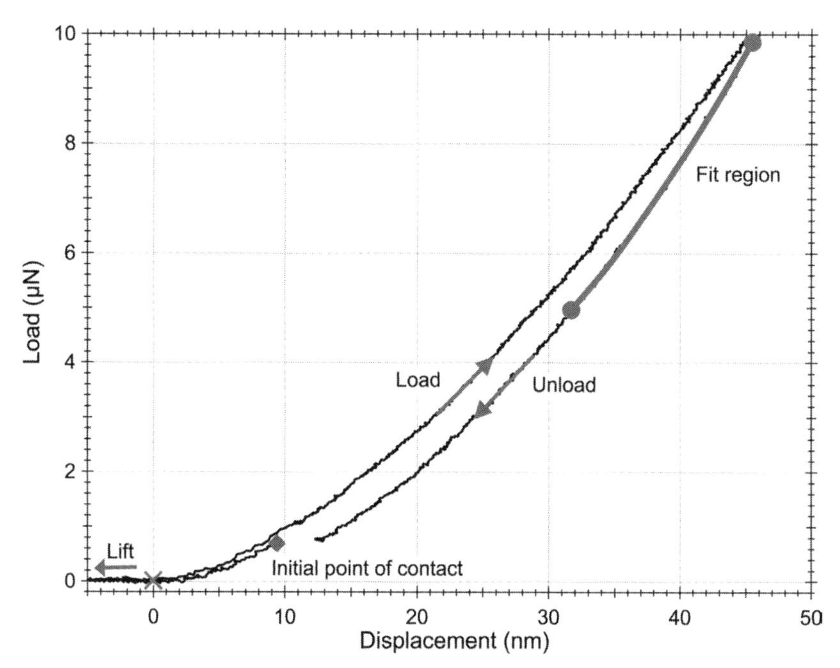

Figure 14.9 Force vs. displacement graph for a 10-μN maximum load indent on low-*k* sample C (see Table 14.1) using a Northstar cube-corner indenter with end radius of 40 nm. The diamond denotes the point of initial contact at 700 nN of load. The X denotes the offset point of zero force and zero displacement obtained using the lift protocol (see text). The fit region that is used to calculate the reduced modulus (see text) is indicated on the unload curve.

- *Hold* While holding the force at 10 μN for 2 seconds, the displacement was relatively unchanged, indicating a negligible amount of sample creep (Fig. 14.9). In contrast, creep is substantial for samples of higher porosity and density lower than 0.94 g/cm^3 (samples F, G, and H in Fig. 14.8).
- *Unload* The modulus and hardness of samples was determined from the initial 98–50% portion of the unload curve [55]. At the end of the unload curve we measured a residual indentation deformation in sample C of about 14 nm from a load of 10 μN.

The load–hold–unload process was repeated on each sample seven times at each load for statistically significant sampling. With eight test loads (25, 20, 15, 12.5, 10, 7.5, 5, and 3 μN), each sample underwent acquisition of 56 load versus displacement curves. After each sample, load versus displacement curves were recorded from a quartz calibration sample to confirm the health of the indenter probe. Reduced modulus values were calculated from the force versus displacement curves of Figure 14.9 using the Oliver–Pharr model [49]. For each load a ratio of contact depth-to-film thickness was then calculated and plotted versus reduced modulus (Fig. 14.10). A linear fit of the modulus data was used to define an "intercept modulus" on the ordinate axis to provide the sample modulus at a theoretical "zero" indentation depth [47]. For example, sample A's intercept modulus in Figure 14.10 is 5.67 GPa. Young's modulus was then calculated

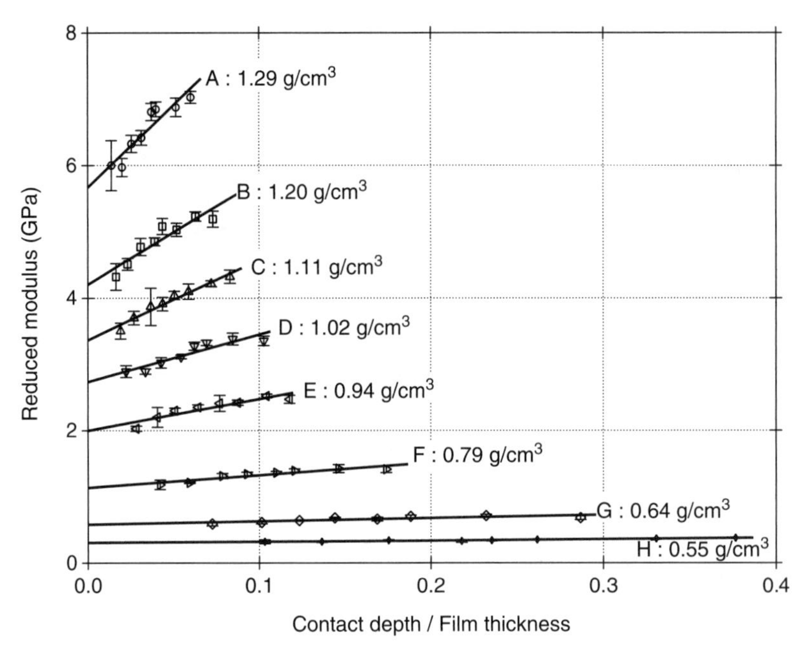

Figure 14.10 Reduced modulus from nanoindentation experiments as a function of indenter contact-depth-to-film-thickness ratio for all eight low-*k* samples (see Table 14.1). Each data point is an average of seven independent measurements performed at each predefined load. Error bars are one standard deviation plus and minus. An "intercept modulus" on the ordinate axis was defined to provide the sample modulus at a theoretical "zero" indentation depth.

from the intercept modulus using Eq. (14.2) (experimental section). From this analysis applied to all of the low-*k* films of varying density, the measured modulus was found to increase as the depth of the indentation increased. This is consistent with literature reports of a "substrate effect" [56–58]—as the probe penetrates deeper into the film, the measured modulus increases due to greater proximity to the hard silicon substrate with a modulus of 120 GPa [59]. Hence, the "intercept modulus" represents the sample modulus at the limit of minimized physical perturbation of the low-*k* film-on-silicon-substrate stack. Our data indicate that a linear fit is sufficient to obtain the intercept modulus even for films with moduli greater than 3 GPa. This differs from previously reported studies [60] where both exponential fit and linear fits were considered, and linear fits were deemed insufficient to model the data correctly. Our use of a sharp cube-corner probe and ultralow loads yielded shallower indentation depths, less impacted by the underlying hard substrate.

14.3.3 Comparison of SAWS, Nanoindentation, and Multiharmonic AFM

The intercept moduli measured from nanoindentation, the moduli measured by surface acoustic wave spectroscopy, and those from multiharmonic AFM were plotted as a function of film density (Fig. 14.11). The moduli from the nanoindentation and

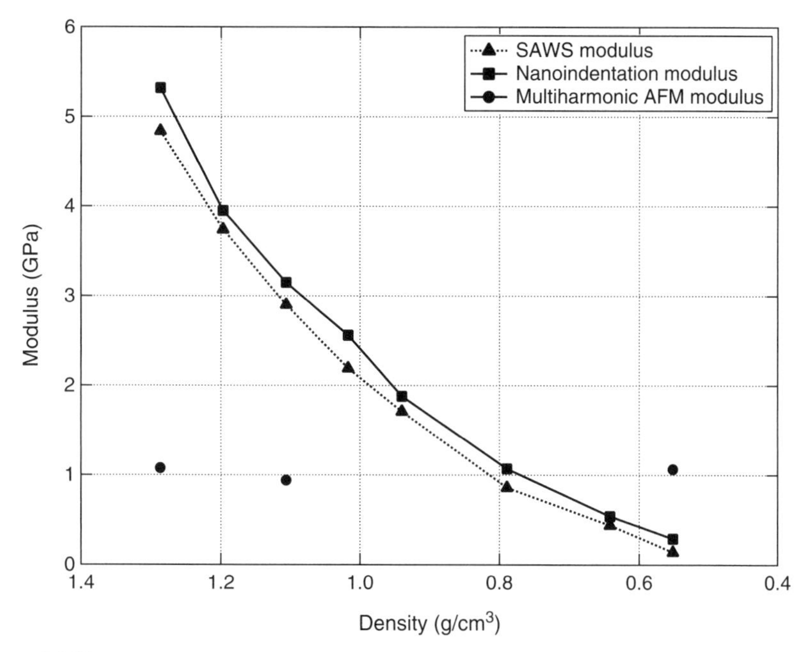

Figure 14.11 Young's modulus as a function of material density for the series of eight low-*k* samples (Table 14.1). Triangles: Young's modulus measured by surface acoustic wave spectroscopy (SAWS). Squares: Intercept Young's modulus (from Fig. 14.10) measured by nanoindentation. Circles: Modulus measured by multiharmonic AFM on three of the samples.

SAWS methods were found to be within 10% agreement. This is in contrast with previous studies [61, 62] where nanoindentation measurements consistently overestimated the modulus values of low-*k* films. The films of our study demonstrated moduli of ~5.2–0.4 GPa—a range of more than an order of magnitude—tracking well over the density range of 1.3–0.6 g/cm^3, respectively. The decrease in modulus on going from high to low density is accompanied by a decrease in dielectric constant (k) from 2.7 to 1.8.

In contrast with nanoindentation and SAWS, the moduli measured with multiharmonic AFM were independent of film density. Many rational explanations are proposed for this discrepancy, including sampling depth, differences in pressures under the AFM tip as compared to the cube corner probe used for nanoindentation, lack of accounting for machine compliance, frequency of the probe at which the measurements are performed, and choice of the number of harmonics to reconstruct the force curves used for mechanical property analysis.

14.4 CONCLUSIONS

We performed measurements on porous low-*k* materials with the three techniques of multiharmonic AFM, ultralow load nanoindentation, and SAWS. The Young's moduli measured by nanoindentation and SAWS agree over an order of magnitude change in the film density. Nanoindentation sampled tens of nanometers of low-*k* material thickness, and the SAWS method sampled the entire hundreds of nanometers thickness. In contrast, the multiharmonic AFM method sampled only the top few nanometers of the surface.

The modulus measurements with multiharmonic AFM did not correlate with modulus values from the other two techniques. We had particularly high expectations for multiharmonic AFM to provide mechanical property information for low-density/high-porosity films because the forces involved are at least a factor of 10 smaller than ultralow load nanoindentation. A multitude of reasons is proposed above for the failure of multiharmonic AFM to produce consistent results. At this point we can only say with confidence that our multiharmonic measurements need further refinement for successful application to low-*k* dielectric measurements.

Good correlation between nanoindentation and SAWS lends credence to nanoindentation and overcomes the limitations of the SAWS technique that its substrate must be single crystalline, must be of known modulus, and free of a second overlayer such as an oxide. The expectation is that nanoindentation can now be performed on various kinds of substrates. The use of ultralow load indentation takes on particular importance in ultrathin film stacks becoming more prevalent in technology. Improved determination of fundamental properties such as Young's modulus will provide more refined input to predictive modeling.

ACKNOWLEDGMENTS

The authors thank Hector Lara, Bede Pittenger, and Steve Minne at Bruker Nano Surface Division for discussions on multiharmonic measurements and Richard Nay and Lance Kuhn at Hysitron Inc. for discussions on nanoindentation measurements.

REFERENCES

1. Oliver, W. C., and Pharr, G. M. *Journal of Materials Research* **7** (1992): 1564.
2. Pharr, G. M. *Materials Science and Engineering: A* **253** (1998): 151.
3. Boissiere, C., Grosso, D., Lepoutre, S., Nicole, L., Bruneau, A. B., and Sanchez, C. *Langmuir* **21** (2005): 12362.
4. Mogilnikov, K. P., and Baklanov, M. R. *Electrochem. Solid-State Lett* **5** (2002): F29.
5. Passeri, D., Bettucci, A., Biagioni, A., Rossi, M., Alippi, A., Lucci, M., Davoli, I., and Berezina, S., *Review of Scientific Instruments* **79** (2008): 066105.
6. Hurley, D. C. In *Acoustic Scanning Probe Microscopy*, Marinello, F., Passeri, D., and Savio, E., Eds., Springer: Berlin, 2013, pp. 351–373.
7. Stan, G., King, S. W., and Cook, R.F. *Journal of Materials Research* **24** (2009): 2960.
8. Flannery, C. M., Wittkowski, T., Jung, K., Hillebrands, B., and Baklanov, M. R. *Applied Physics Letters* **80** (2002): 4594.
9. Link, A., Sooryakumar, R., Bandhu, R. S., and Antonelli, G. A. *Journal of Applied Physics* **100** (2006): 013507.
10. Xiang, Y., Chen, X., Tsui, T. Y., Jang, J.-I., and Vlassak, J. J. *Journal of Materials Research* **21** (2006): 386.

11. Ho, P. S., Leu, J., and Lee, W. W. In *Low Dielectric Constant Materials for IC Applications*, Ho, P. S., Leu, J. J., and Lee, W. W., Eds., Springer: Berlin, 2003, pp. 1–21.
12. Miller, R. D. *Science* **286** (1999): 421.
13. Volksen, W., Miller, R. D., and Dubois, G. *Chemical Review* **110** (2009): 56.
14. Mosig, K., Jacobs, T., Brennan, K., Rasco, M., Wolf, J., and Augur, R. *Microelectronic Engineering* **64** (2002): 11.
15. Havemann, R. H., Jain, M. K., List, R. S., Ralston, A. R., Shih, W.-Y., Jin, C., Chang, M. C., Zielinski, E. M., Dixit, G. A., Singh, A., Russell, S.W., Gaynor, J. F., McKerrow, A. J., and Lee, W. W. MRS Online Proceedings Library 511, null (1998).
16. Chevolleau, T., Posseme, N., David, T., Bouyssou, R., Ducote, J., Bailly, F., Darnon, M., ElKodadi, M., Besacier, M., Licitra, C., Guillermet, M., Ostrovsky, A., Verove, C., and Joubert, O. In Interconnect Technology Conference (IITC), 2010 International (2010), pp. 1–3.
17. Susko, R., Daubenspeck, T., Wassick, T., Sullivan, T., Sauter, W., and Cincotta, J. *ECS Transactions* **16** (2009): 51.
18. Giessibl, F. J. *Review of Modern Physics* **75** (2003): 949.
19. Binnig, G., Quate, C. F., and Gerber, C. *Physical Review Letters* **56** (1986): 930.
20. Crittenden, S., Raman, A., and Reifenberger, R. *Physical Review B* **72** (2005): 235422.
21. Garcia, R., and Herruzo, E. T. *Nature Nanotechnology* **7** (2012): 217.
22. Lozano, J. R., and Garcia, R. *Physical Review Letters* **100** (2008): 076102.
23. Stark, R. W., and Heckl, W. M. *Surface Science* **457** (2000): 219.
24. García, R., and Pérez, R. *Surface Science Reports* **47** (2002): 197.
25. Raman, A., Melcher, J., and Tung, R. *Nano Today* **3** (2008): 20.
26. Sahin, O. *Physical Review B* **77** (2008): 115405.
27. Sahin, O., and Erina, N. *Nanotechnology* **19** (2008): 445717.
28. Sahin, O., Magonov, S., Su, C., Quate, C. F., and Solgaard, O. *Nature Nanotechnology* **2** (2007): 507.
29. Stark, M., Stark, R.W., Heckl W.M., and Guckenberger, R. *PNAS* **99** (2002): 8473.
30. Magonov, S. N., Elings, V., and Whangbo, M.-H. *Surface Science* **375** (1997): L385.
31. Tabor, D. *The Hardness of Metals*, Oxford University Press: Oxford, UK, 1951.
32. E28 Committee, *Test Method for Brinell Hardness of Metallic Materials*, ASTM International: West Conshohocken, USA, 2001.
33. Smith, R. L., and Sandly, G. E. *Proceedings of the Institution of Mechanical Engineers* **102** (1922): 623.
34. Knoop, F., Peters, C. G., and Emerson, W. B. *A Sensitive Pyramidal-Diamond Tool for Indentation Measurements*, National Bureau of Standards, 1939.
35. Feynman, R. P., and Sykes, C. *No Ordinary Genius: The Illustrated Richard Feynman*, W. W. Norton: New York, 1995.
36. Rayleigh, L., *Proceedings of the London Mathematical Society* ***s1-17*** (1885): 4.
37. Lee, H.-J., Soles, C. L., Vogt, B. D., Liu, D.-W., Wu, W., Lin, E. K., Kim, H.-C., Lee, V. Y., Volksen, W., and Miller, R. D. *Chem Mater* **20** (2008): 7390.
38. Parratt, L. G. *Physical Review* **95** (1954): 359.
39. Holý, V., and Baumbach, T. *Physical Review B* **49** (1994): 10668.
40. Montelius, L., and Tegenfeldt, J. O. *Applied Physics Letters* **62** (1993): 2628.
41. Atamny, F., and Baiker, A. *Surface Science* **323** (1995): L314.
42. Hubner, U., Morgenroth, W., Meyer, H. G., Sulzbach, T., Brendel, B., and Mirande, W. *Applied Physics A: Materials Science & Processing* **76** (2003): 913.
43. Hutter J. L., and Bechhoefer, J. *Review of Scientific Instruments* **64** (1993): 1868.
44. Gibson, C. T., Weeks, B. L., Abell, C., Rayment, T., and Myhra, S. *Ultramicroscopy* **97** (2003): 113.
45. Sader, J. E., Chon, J. W. M., and Mulvaney, P. *Review of Scientific Instruments* **70** (1999): 3967.
46. Sahin, O. *Rev Sci Instrum* **78** (2007): 103707.
47. Fischer-Cripps, A. C. *Nanoindentation*, 2nd ed., Springer: New York, 2004.
48. Chudoba, T., Schwaller, P., Rabe, R., Breguet, J.-M., and Michler, J. *Philosophical Magazine* **86** (2006): 5265.

49. OLIVER. W. C., and PHARR, G. M. *Journal of Materials Research* **19** (2004): 3.
50. MORGEN, M., RYAN, E. T., ZHAO, J.-H., HU, C., CHO, T., and HO, P. S. *Annual Review of Materials Science* **30** (2000): 645.
51. SCHNEIDER, D., SCHWARZ, T., and SCHULTRICH, B. *Thin Solid Films* **219** (1992): 92.
52. HESS, P. *Physics Today* **55** (2002): 42.
53. YABLON, D. G., GANNEPALLI, A., PROKSCH, R., KILLGORE, J., HURLEY, D. C., GRABOWSKI, J., and TSOU, A. H. *Macromolecules* **45** (2012): 4363.
54. GERK, A. P., and TABOR, D. Published Online: 23 February 1978; Doi:10.1038/271732a0 271, 732 (1978).
55. PHARR, G. M., and BOLSHAKOV, A. *Journal of Materials Research* **17** (2002): 2660.
56. TAYEBI, N., POLYCARPOU, A. A., and CONRY, T. F. *Journal of Materials Research* **19** (2004): 1791.
57. SAHA, R., and NIX, W. D. *Acta Materialia* **50** (2002): 23.
58. TSUI, T. Y., and PHARR, G. M. *Journal of Materials Research* **14** (1999): 292.
59. WORTMAN, J. J., and EVANS, R. A. *Journal of Applied Physics* **36** (1965): 153.
60. HERRMANN, M., SCHWARZER, N., RICHTER, F., FRUHAUF, S., and SCHULZ, S. E. *Surface & Coatings Technology* **201** (2006): 4305.
61. FLANNERY, C. M., and BAKLANOV, M. R. In *Proceedings of the IEEE 2002 International Interconnect Technology Conference*, 3–5 June 2002, IEEE: Piscataway, NJ, 2002, pp. 233–235.
62. FLANNERY, C. M., WITTKOWSKI, T., JUNG, K., HILLEBRANDS, B., and BAKLANOV, M. R. *Applied Physics Letters* **80** (2002): 4594.

Chapter 15

Nanomechanical Characterization of Biomaterial Surfaces: Polymer Coatings That Elute Drugs

Klaus Wormuth[1] and Greg Haugstad[2]

[1] *Institute for Physical Chemistry, University of Cologne, Cologne, Germany*
[2] *University of Minnesota, Minneapolis, MN*

15.1 INTRODUCTION

The temporary use or long-term implantation of biomaterials into the human body for the healing, augmentation, or replacement of tissue or organs presents significant challenges. First, the material must maintain functionality in the aggressive enzymatic and mechanical stress environments within the human body. Second, the surfaces of the material must exhibit biocompatibility and not induce allergic, toxic, or carcinogenic reactions or create blood clots. Ideally, the body accepts the biomaterial, and the surface of the implanted material induces the same biochemical reactions induced by tissues in the body [1]. Most synthetic materials react nonideally with the body, inducing a cascade of biochemical processes. Tissues or blood surrounding an implant react as if an injury or infection had occurred, and inflammation cells attack the implant. In a positive scenario, biocompatibility occurs through creation of a thin layer of fibrotic (scar) tissue that contains no inflammation cells and that encapsulates the implant and prevents further biochemical reactions [1].

Often the surfaces of medical devices and implants are treated or coated with biochemically active materials to promote biocompatibility. Coatings on medical devices such as catheters, guide wires, angioplasty balloons, and stents improve

Scanning Probe Microscopy in Industrial Applications: Nanomechanical Characterization, First Edition. Edited by Dalia G. Yablon.

the biocompatibility of medical devices by favorably altering the chemical nature of the device/tissue or device/blood interface. Coatings can minimize tissue damage (reduce friction), decrease chances for blood clot formation (prevent platelet adsorption), and improve the healing response (deliver drugs).

"Biochemically active" biomaterial coatings result from binding proteins, enzymes, or biopolymers onto surfaces, which hopefully induce biochemical reactions favorable for biocompatibility. "Drug-eluting" biomaterials control the release of a drug into the tissue surrounding an implant with the goal of inducing the proper healing responses in a localized area around the device. Local controlled release of a drug offers significant advantages over systemic drug delivery (injections or tablets) by maintaining drug dosage within the therapeutic window and reducing toxic side effects. A highly successful example of a coated medical device is the "drug-eluting stent" [2].

This chapter explores the application of scanning probe microscopy (SPM) to the nanomechanical characterization of drug-eluting coatings containing high concentrations of drugs mixed with either hydrophobic polymers or hydrophobically associating hydrogels. SPM provides valuable information about the surfaces of these types of biomaterials by nanomechanically distinguishing between drug and polymer domains, and thus sensing the distribution of drug at the surface of polymer-based drug-eluting coatings.

All of the successful drug-eluting stents currently on the market consist of a metal mesh coated with a thin layer of drug mixed into a "biostable" (i.e., nondegradable) polymer. As a model system, we examine coatings of the drug dexamethasone, an anti-inflammatory steroid, mixed into poly(butyl methacrylate) with or without added poly(lauryl methacrylate). Some stent coatings use polymer blends to obtain optimal mechanical properties and drug release kinetics. Critical in the development of successful stent coatings is a fundamental understanding of the release kinetics of the drug, which depends intimately upon the degree of mixing of the drug and polymer(s). Especially important is an understanding of the morphology of the outer surface of the coating, which when implanted directly contacts tissue and blood. The physical texture and chemical heterogeneity of the coating surface directly influences how cells interact with the coating surface [1]. In our studies, SPM nanomechanical characterization reveals the mechanisms by which drug delivery occurs from the surfaces of stent coatings upon exposure to aqueous environments, which to first order mimic the environment found within the human body.

In the drug–polymer coatings examined here, the drug is initially in the amorphous (noncrystalline) state when mixed into either purely hydrophobic polymers or a copolymer containing both hydrophilic and hydrophobic blocks. As known from the literature, drugs in the amorphous state are often metastable, and environmental or mechanical stimuli often cause amorphous drugs to crystallize [3]. Indeed we find that certain temperature and humidity conditions cause crystallization of a relatively hydrophobic drug blended with hydrophobic polymer. A fascinating finding is that the interaction of the SPM tip with the surface of drug–polymer mixtures may "scan-induce" drug crystallization.

Currently in the development stage, "biodegradable" coatings for stents that contain polymers that hydrolyze and erode away are of intense interest. As a model system we examine coatings of simvastatin, a hypolipidemic drug used to treat cardiovascular disease, mixed into PolyActive degradable polymer. In these coatings of a relatively hydrophobic drug blended with a hydrophilic-hydrophobic copolymer, upon immersion in water the hydrophilic polymer domains swell and squeeze drug onto the surface. With time-lapse SPM imaging, interesting morphological changes of the surface drug domains occur as the drug elutes from the coating.

15.2 MATERIALS AND METHODS

Thin biomaterial coatings were prepared by spin-coating solutions of dexamethasone drug and polymer(s) (15 g/mL) in a volatile solvent tetrahydrofuran. The drug–polymer solution was cast onto an untreated silicon wafer spinning at 2000 rpm under ambient conditions and allowed to dry for at least 24 hours prior to examination. A variety of coatings were prepared containing 33–50 wt% of dexamethasone drug mixed with either poly(butyl methacrylate) (PBMA), a stiff polymer with a glass transition temperature (T_g) around 30°C, or a polymer blend of PBMA combined with poly(laurel methacrylate) (PLMA), a soft polymer with a T_g around –30°C [4]. Note that pure amorphous dexamethasone has a T_g of about 19°C [5].

Coatings of simvastatin drug mixed into copolymers containing hydrophobic-hydrophilic segments were prepared by treating silane-treated glass slides with a benzophenone cross-linker, dip coating solutions of drug and copolymer in chloroform (20 g/mL) onto the treated glass slides, allowing the coating to dry, and irradiating the coating with ultraviolet light to chemically bind the coating to the glass slide. Thus coatings were created containing 15 wt% simvastatin drug mixed with a copolymer containing 45 wt% of hydrophobic poly(butylene terephthalate) (PBT) segments, along with 55 wt% of fixed length (1000 Da) hydrophilic poly(ethylene oxide) (PEO) segments. These PBT–PEO polymers (marketed as PolyActive) form physically cross-linked gels upon immersion in water: the hydrophobic PBT regions associate strongly, and water swells the PEO segments [6]. These polymers are hydrolysable and slowly degrade in aqueous environments. Pure PBT polymer is highly crystalline and exhibits a T_g of 54°C, whereas pure PEO polymer is quite soft at room temperatures with a T_g of –48°C [7]. Note that pure amorphous simvastatin has a T_g of about 28°C [8].

This chapter utilizes nanomechanical response as probed with (i) fast force curves (see Chapter 6) and (ii) dynamic mode (resonant) phase measurements (see Chapter 4) to generate images that differentiate material phases in mixtures of drug and polymer(s). The particular implementation of (i) is known by the trade name assigned by the vendor: digital pulsed force mode (D-PFM, Witec). Two implementations of method (ii) were used, so-called TappingMode (Bruker) and acoustic alternating current (AC) mode (Agilent) or "intermittent contact mode" or "amplitude modulated AFM". Method (i) necessarily examines *repulsive* forces between tip and sample: the resistance of one material (sample) to penetration by another (tip). Additionally, tip–sample *attractive* forces are used to contrast materials via differences in the most negative cantilever deflection sensed during retraction, often called the pull-off force. These attractive forces are usually van der Waals in character (but may include capillary forces due to the formation of a meniscus at the tip–sample interface). Method (ii) generally can be operated in a net attractive or net repulsive regime, but in the present work the parameters were selected to stabilize the net repulsive regime, such that a nanomechanical contact is made. Phase imaging contrasts energy dissipation (tip to sample) that is dominated by viscoelastic dissipation in contact as well as the making and breaking of adhesive contact, which is a hysteretic and thus energy-dissipative phenomenon. (See Ref. [9] for details of both D-PFM and phase imaging in dynamic mode.) *Note*: When a figure contains multiple images of a given type (e.g., topography, adhesion, etc.), the contrast scales are identical within image type to facilite direct comparison between the images.

15.2.1 Digital Pulsed Force Mode (D-PFM)

D-PFM executes measurements of cantilever deflection versus *Z* scanner displacement rapidly (here 2000 cycles/second, sinusoidally in time) but well below the cantilever resonant frequency and corresponding to approximately one cycle per pixel. D-PFM grabs only select data points during each *Z* cycle to generate adhesive/nanomechanical images "on the fly" at conventional measurement site densities, usually 512×512 (with the option to collect and save every *Z* cycle of data with a second computer, typically yielding gigabyte-regime data files). These select data points enable (1) tracking of topography via variable *Z* displacement (in series with a fixed-amplitude *Z* modulation) as is driven by the usual feedback circuit so as to keep the maximum positive deflection per cycle (relative to the zero-force baseline) near a setpoint value (from one touch, i.e., image point, to the next); (2) the "adhesive force" image (from the most negative deflection during retraction); (3) either (a) the time location of the maximum positive deflection or (b) the differential deflection between two time points in contact—usually the first being shortly after a given contact is made and the second being located near the maximum positive deflection. These two image types 3a–b (only one of which is collected) are qualitatively sensitive to material stiffness and will be designated as such.

Regarding image type 3a, our observation is that soft surfaces usually remain higher in elevation following the preceding break of contact under tension, such that the subsequent time location of maximum deflection occurs sooner. Thus the effect, when present, is inherently viscoelastic. There may also be differences in the time-dependent material response as the contact force grows, contained in the same number; this would usually act in the opposite sense, with a more viscoelastic response producing a delay in the achievement of maximum force. The measured quantity is the aggregate shift in time-dependent force, containing both effects. Regarding image type 3b, a more rigid material will transition from exerting a lesser to greater resistive force more quickly in time for a given *Z* approach velocity; thus for a fixed time interval, one anticipates a larger differential deflection. What complicates this picture is the fact that velocity is variable during the designated time interval, being sinusoidal in time; and potentially the "start time" is variable relative to the time point at which contact is achieved. Thus in the case of either 3a or 3b, the image can only be qualitatively assigned to material stiffness.

It should be emphasized that method (i), D-PFM, though an intermittent contact method, is in fact quasi-static: The measured quantity, whether to drive the

Z feedback or to provide materials contrast, is a cantilever deflection (quasi-static force). This means force balance: At any point in contact the tip is not accelerating, and quasi-static contact mechanics models can be applied directly (i.e., using measured force) to interpret the images. In contradistinction, method (ii) involves intermittent *dynamic*, not quasi-static, interaction: Here a time-integrated cantilever *amplitude under resonant cantilever excitation* is actuating the *Z* feedback (for further explanation of these terms, see Chapters 4 and 6), and force balance is not achieved. That is, the tip is almost always accelerating (positive or negative) while in contact with the sample, such that force cannot be measured via any "snapshot" of deflection, and thus contact mechanics models cannot be directly applied to interpret the data; instead these models only can be employed via computer simulation.

In the following section the specified D-PFM feedback force values (the maximum during approach) by default employ the manufacturer-specified value of cantilever spring constant (3 N/m, for integrated rectangular silicon cantilever/tip of radius ~10 nm or less, AppNano), except in cases where user-calibrated values were obtained, as stated. In the latter case the operative cantilever tip was pressed against a (manufacturer-calibrated, rectangular, tipless) silicon cantilever to obtain force versus *Z* displacement data; the ratio of contact slope on this precalibrated cantilever to that on the noncompliant cantilever chip (i.e., the inverse sensitivity) was used to calculate the operative cantilever spring constant from that of the precalibrated cantilever [9].

15.3 DEXAMETHASONE IN PBMA OR PBMA–PLMA POLYMER BLENDS

We discuss the system of dexamethasone in various polymers. We start with a discussion of the polymer blend matrix (support) itself to better understand the behavior of this binary blend. We then add dexamethasone to the polymer mix and explore effects of temperature and water exposure on the material.

15.3.1 Polymer Blend in Air: PBMA–PLMA (without Drug)

We examine the ability of nanomechanically sensitive D-PFM to identify and characterize the distribution of material phases. We begin with a simple polymer coating without drug, a blend of 77/23% by weight PBMA–PLMA. Figure 15.1 contains representative results for image pairs—height (left) and adhesion (right)—acquired at three different setpoint values corresponding to maximum forces of approximately 6, 12, and 30 nN as shown. In all images and at all setpoints the phase segregation of the two polymers is obvious, with the minority phase present as circular or oval domains variable in diameter. More notable contrast is observed at higher setpoints, for both height and adhesion images (whereas stiffness images, not shown, exhibit contrast that only slightly changes with setpoint). A setpoint-dependent apparent height indicates a strongly variable depth of indentation with each touch

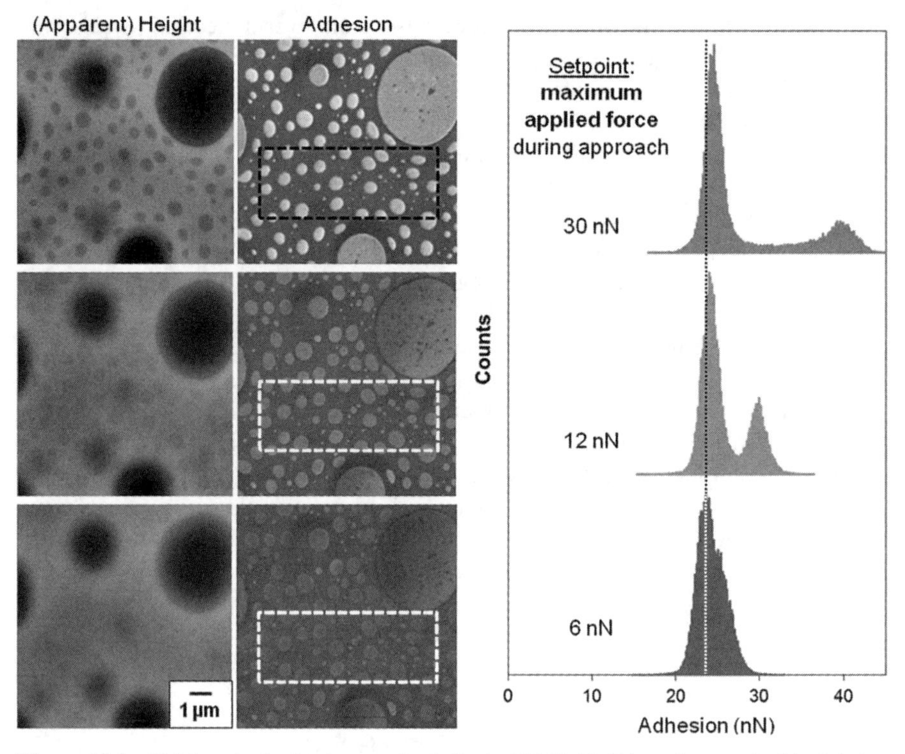

Figure 15.1 Height and adhesion image pairs (left) of a PBMA–PLMA coating, acquired in digital pulsed force mode (D-PFM) in air at three setpoint force values; corresponding adhesion histograms derived from the marked rectangular regions in the adhesion images.

into the minority phase: at higher setpoint the tip must penetrate farther into the sample to achieve the setpoint deflection value. The Z scanner achieves this higher penetration by displacing tip farther toward sample, which is then contained in the apparent height image (being a map of Z scanner displacement, regardless of the *cause* of displacement). Thus the minority phase is much softer than the majority phase, consistent with expected differences of modulus between PBMA (majority) and PLMA (minority).

The strong increase of adhesion contrast is quantified in Figure 15.1 at right, as an adhesion histogram for each of the three setpoints (of maximum force during approach), derived from data inside of the marked rectangular region. The distribution of adhesion is relatively small at the lowest setpoint but clearly spreads into two distinct peaks (i.e., becomes bimodal) for the higher two setpoints. In particular the centroid value of the distribution of adhesion on PLMA has shifted from ≈ 25 nN at the lowest setpoint to ≈ 40 nN at the highest setpoint, whereas the centroid of PBMA adhesion has only weakly shifted upward (by roughly 1 nN). The behavior of PLMA is a definitive example of *adhesion memory*: the observation that the force required to break adhesive contact depends on preceding loading history, *contrary to conventional adhesive contact mechanics models* (Ref. [9]; Chapter 2). These models do not account for irreversible (hysteretic) processes in the making and breaking of contact.

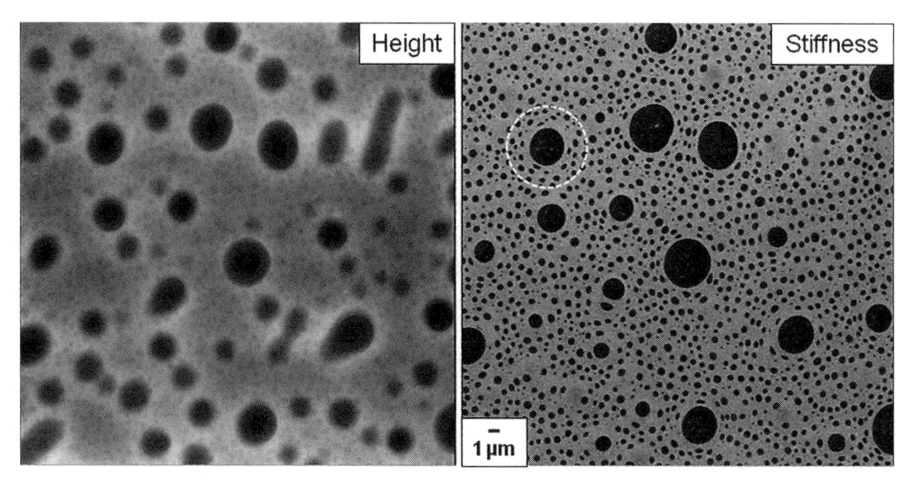

Figure 15.2 Survey height and stiffness image pair of a PBMA–PLMA coating acquired in air in D-PFM.

The idea is that larger areal contacts formed during approach (i.e., by reaching higher loading forces) require more force to break during retraction. This adhesion memory is likely to be viscoelastic in nature (i.e., rate dependent), at least in part.

It is also revealing to examine the shapes of the PLMA domains. Figure 15.2 provides a larger survey image of these morphologies in both height and qualitative stiffness (setpoint 12 nN). The stiffness (right) image starkly displays the softness of circular PLMA domains at the surface. Under close examination of the small domains (<1 µm), one sees that these shapes are mainly circular but *in some cases are more elliptical, where positioned near the large domains* (~2 µm or larger). Specifically, the long axis of each elliptical shape tends to lie parallel to the nearby rim of a large domain; this can be seen within the marked circular region, for example, as one radially nears the central PLMA domain (i.e., within 1–2 µm of its rim). Similarly, inspecting other locations near large PLMA domains, one notes similarly distorted small PLMA domains. This observation suggests the occurrence of a strain event during film formation, reasoned as follows. First, we note that although many of the large soft (PLMA) domains are centered in crater-shaped depressions as seen in the height image, *some of the large topographic depressions* (those that are not *quite* as deep) *do not exhibit softness*, meaning PLMA is not present at the surface, and in these locations the small PLMA domains are mainly circular in shape (i.e., not laterally distorted). In further experiments (not shown) we have discovered that the creation of *plastic indents* at such locations can *liberate PLMA beneath the surface*, which then oozes upward through the "perforations." These observations suggest that in fact all of the large topographic depressions are locations of *subsurface PLMA*, and that the large, soft circular domains in Figure 15.2 are sites where a PBMA skin spontaneously ruptured during film preparation to expose the underlying PLMA, perhaps during the later stages of drying. This rupture event presumably distorts the shape of nearby (initially circular) small PLMA domains, as seen adjacent to the rims of all large, soft domains in Figure 15.2.

15.3.2 Drug in Polymer in Air: Dexamethasone–PBMA

Whereas PLMA segregates from PBMA over relatively large length scales (domains hundreds of nanometers to microns in size separated by distances of similar scale), a much lower-molecular-weight ingredient blended with PBMA may produce phase segregation over much smaller scales. This is demonstrated in Figure 15.3 for the case of dexamethasone mixed with PBMA (50/50 dexamethasone–PBMA by weight). Height images (top) show relatively flat, raised, and often circular domains (plus some more complicated shapes); corresponding adhesion images show little

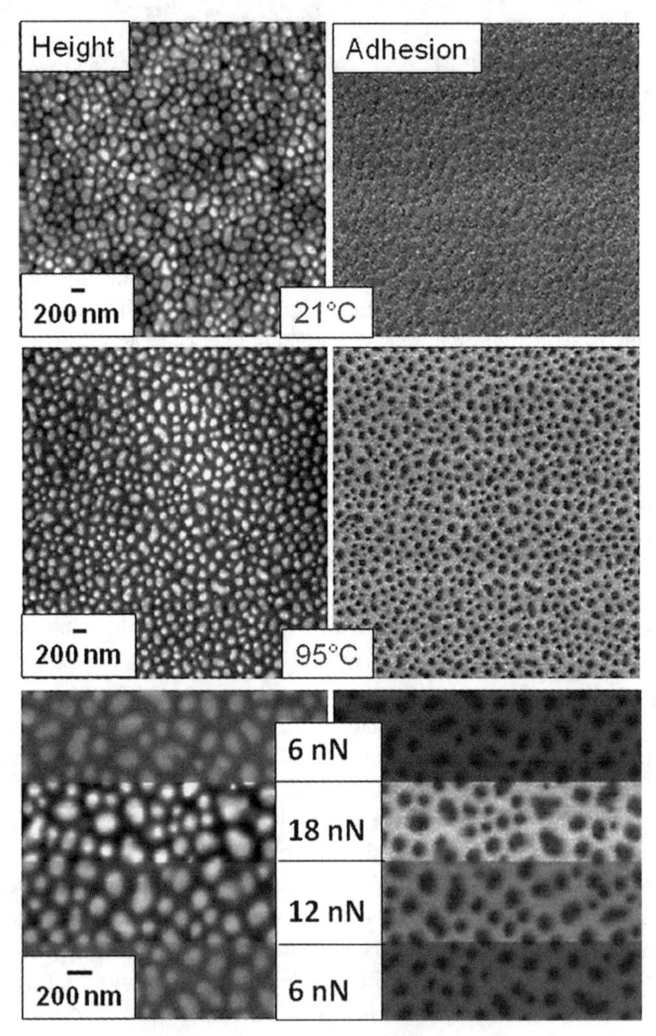

Figure 15.3 Height and adhesion image pairs of a PBMA–dexamethasone coating, acquired in D-PFM in air at two temperatures: room (top) and 95°C (middle). Bottom image pair shows strong changes of both height and adhesion images at 95°C for variable setpoint forces.

contrast initially (nor do qualitative stiffness images, not shown). In both cases negligible load dependence of contrast was observed (not shown). Recognizing that a glassy polymer of predominantly hydrophobic character is not expected to strongly contrast with small-molecule domains of similar character (T_g of PBMA = 30°C, T_g of dexamethasone = 19°C), we explored variable sample temperature to yield contrast. Raising the temperature to 95°C indeed produced fairly strong contrast in adhesion (middle right); moreover, pronounced load dependence was apparent in both "height" and adhesion as shown in Figure 15.3 bottom. The regions that soften (sink or have darker appearance in "height" images) become stickier (brighter) in adhesion. These regions also slightly grow in surface presence and shrink again with temperature decrease (not shown). Such apparent (small) changes in surface presence may in truth relate to *mechanical* presence: greater indentation on a softened material can reduce the apparent presence of the more rigid material in *boundary* regions, where a contact formed with both materials becomes net softer.

15.3.3 Dexamethasone–PBMA: Effect of Soaking in Water

Upon water immersion, the surface topography radically changes within the time required to set up and acquire the first image (10 minutes), and continues to evolve over the 24-hour period that it was monitored (Fig. 15.4). In particular, many of the roughly circular protrusions observed when dry are replaced by roughly circular holes (darks spots). At other sites taller protrusions (bright spots) are observed (relative to "baseline" heights between holes) and appear to reverse-image the tip shape, implying steep protrusions. With passing time on the scale of hours, the holes become deeper and broader, while the tall protrusions collapse. A subset of these protrusions convert to holes (i.e., over a much longer time scale than the larger ensemble of holes that appeared in the first immersion image) as exemplified in the zoomed images of Figure 15.4 (circled locations). On the 24-hour time scale, larger lateral domains become relatively high or low in height, suggesting subsurface aggregation of drug (e.g., at lower left and right). Some of these larger regions were visible using an overhead video microscope, indicating characteristic length scales above the diffraction limit of visible light.

To assess the "efficiency" of drug release from the domains observed in Figure 15.4, we found it instructive to compare the *size distribution* of protrusions to that of holes, using grain size analysis via the watershed algorithm (generated using the freeware program Gwyddion). Histograms of equivalent disk radius were compared between initial protrusions and later holes observed after 23 hours immersed. The peak in the histograms (statistical mode) was evaluated to be 80 nm for the initial protrusions when dry and 70 nm after 23 hours immersed. The difference can be attributed to broadening due to the shape of the AFM tip (radius ca. 10 nm). Thus we conclude that essentially all drug has been evacuated from the formerly drug-containing domains later imaged as "holes" to within measurement uncertainty.

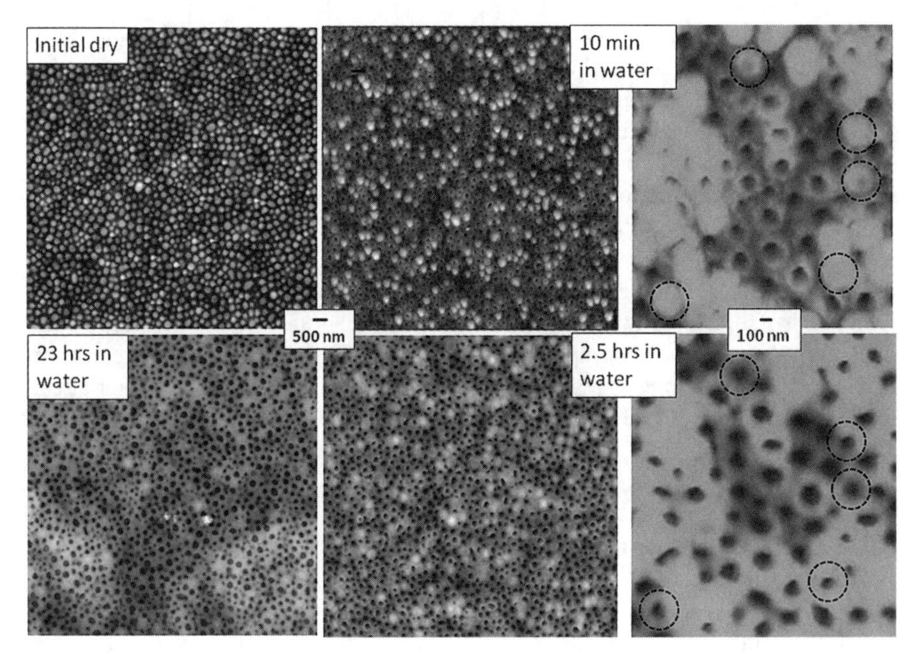

Figure 15.4 Height images of a PBMA–dexamethasone coating acquired in D-PFM in air as initially dry (upper left) and evolving while immersed in water for variable times as indicated. Two rightmost images zoom to the same region to note circled locations that initially contain protrusions but later contain holes.

15.3.4 Drug in Polymer Blend in Air: Dexamethasone–PBMA–PLMA

We now consider a sample that combines the elements of the two preceding cases: drug mixed into a blend of two polymers, dexamethasone–PBMA–PLMA (43.5/43.5/13 by weight), keeping the ratio of the two polymers nearly the same as in Figures 15.1 and 15.2, and the ratio of PBMA to drug the same as in Figure 15.3. Here we find an even richer case as seen in the representative example of Figure 15.5 (showing both height and differential height to bring out the topographic richness of the surface, as well as adhesion and stiffness to again reveal property differences). Similar to Figure 15.2 we find both large and relatively small circular domains that are soft at room temperature, implying PLMA; we additionally find even smaller (mainly circular) *protruding* domains similar in stiffness to the intervening narrow valleys, akin to Figure 15.3 (top). These smallest domains exhibit markedly lesser adhesion than the large circular domains. We thus tentatively identify the small protruding domains as arising from dexamethasone. (Further results will support this assignment.)

Given the softness of PLMA at room temperature, it is difficult to unambiguously characterize the "true" surface topography, divorced from mechanical compliance effects (i.e., variable indentation). To aid in obtaining true topography, we cooled the sample down to approximately –30°C (at low humidity via air circulated through a dessicator column) to make all surface components rigid. In Figure 15.6 we compare this sample at

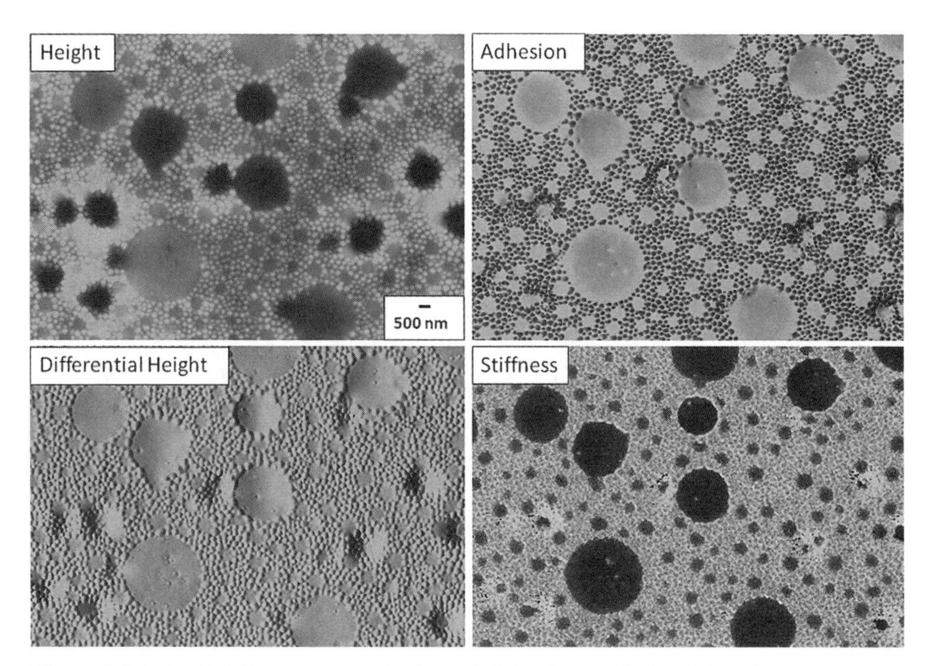

Figure 15.5 Multiple image types acquired in D-PFM in air on a PBMA–PLMA–dexamethasone coating. For color details, please see color plate section.

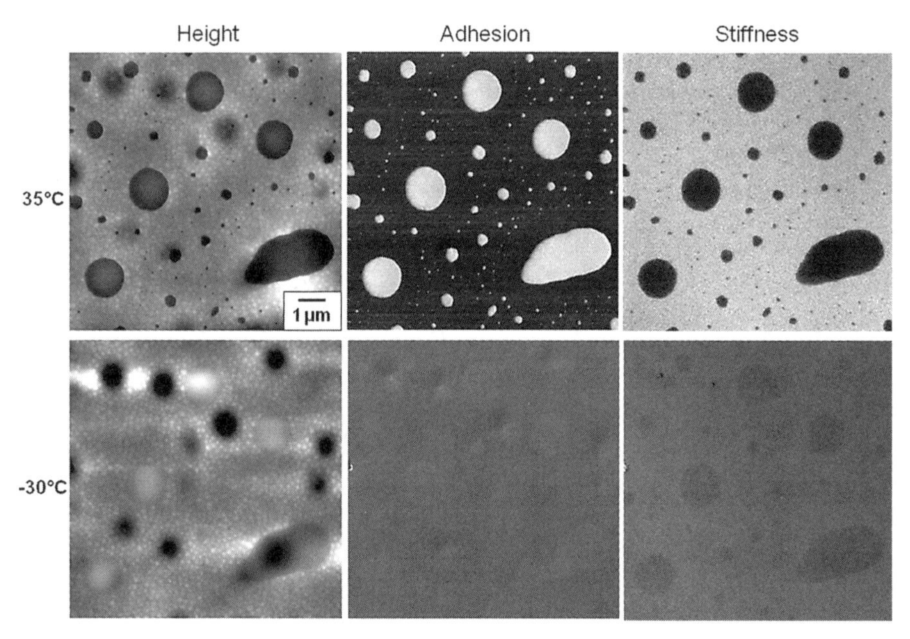

Figure 15.6 Multiple image types acquired in D-PFM in air on a PBMA–PLMA–dexamethasone coating at two temperatures, 35 and –30°C.

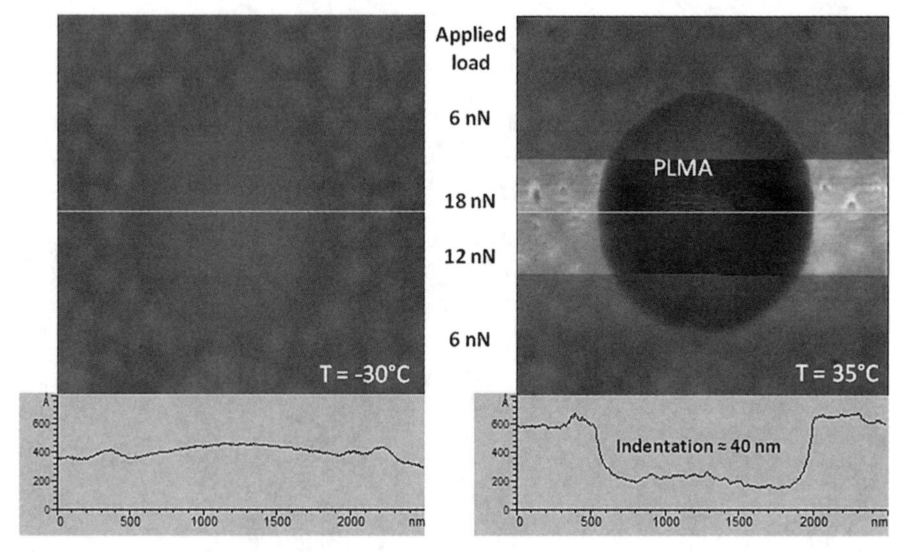

Figure 15.7 Height images acquired in D-PFM in air on a PBMA–PLMA–dexamethasone coating at two temperatures, 35 and –30°C. (Bottom) Cross-sectional height data from the locations marked with horizontal lines in the images, at 18 nN setpoint force, revealing a 40-nm elastic indentation during imaging.

low temperature and higher temperature of 35°C. At low temperature, nearly all mechanical and adhesion contrast has vanished and the height image looks distinctly different from that at 35°C: Many apparent "holes" at the higher temperature vanish or even become slight "domes" at low temperature. Figure 15.7 zooms in on one of these domes at each temperature and examines *apparent* height images in cross section, for three maximum loading force values: 6, 12, and 18 nN. At the higher temperature one observes, as expected, the increasing "sinking" of the PLMA circular domain with increasing loading force, whereas at the lower temperature no such virtual sinking is measured. Assuming negligible indentation in the latter case, we can take the measured "depth" of the PLMA region at different forces as a measure of indentation. Assuming simple Hertzian contact mechanics as a first approximation for mechanical indentation by an approximately 10-nm radius tip, we estimate a modulus of order ~10 MPa. See Chapter 2 for detailed information on Hertz contact mechanics. (Finer quantitation would require an adhesive contact mechanics model and an appropriate tip shape such as spherically capped conical, as well as possible accounting of viscoelastic creep.)

15.3.5 Dexamethasone–PBMA–PLMA in Air: Crystallization at High Temperature

The three-component case was also found to be unstable to tip interactions at high temperature. Figure 15.8 compares height and adhesion images before (top) and after (bottom) raster scanning a 20-μm region (designated by a dashed square in each image) at a temperature of 80°C. The region scanned at high temperature, and some locations slightly outside of this region, contains large crystal-shaped objects. These

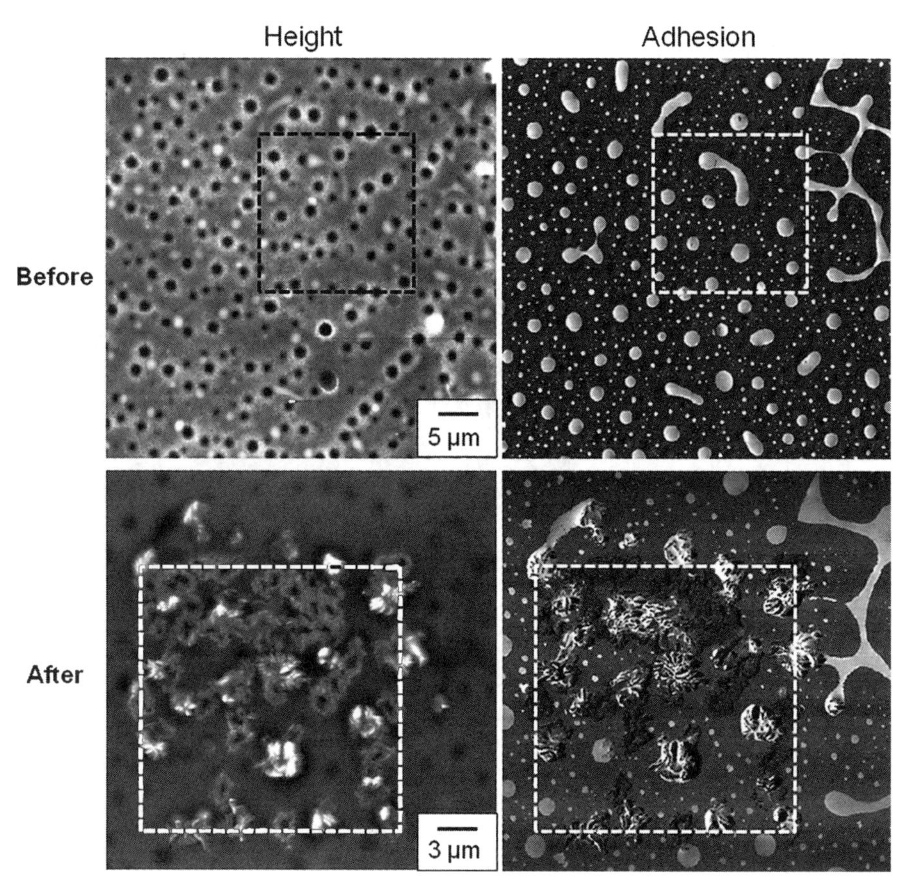

Figure 15.8 Height and adhesion image pairs of a PBMA–PLMA–dexamethasone coating acquired in D-PFM in air before (top) and after (bottom) a subregion was raster scanned at an elevated temperature of 80°C (as designated with dashed squares). For color details, please see color plate section.

tend to form on top or around the periphery of PLMA domains, suggesting that the presence of this polymer provides an environment in which the drug attraction to itself overcomes drug intermixing with the polymers. From other studies using confocal Raman microscopy [10], we found that in blends of PBMA and PLMA, dexamethasone preferentially partitions into the PBMA-rich phase and appears to have limited miscibility (solubility) in the more hydrophobic PLMA. Thus, the presence of PLMA interfaces could provide a surface upon which dexamethasone crystallizes.

15.3.6 Dexamethasone–PBMA–PLMA: Effect of Soaking in Water

Upon exposure to aqueous environment, as shown in Figure 15.9, the three-ingredient case exhibits rampant transformations as seen with PBMA. Holes or pits replace small circular protrusions, and these holes grow in size and somewhat in number,

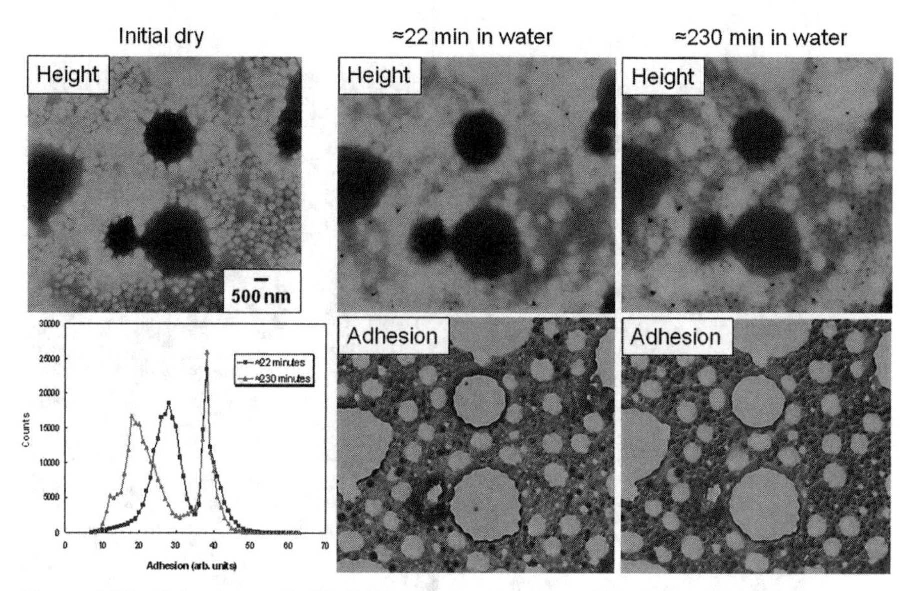

Figure 15.9 Height image of a PBMA–PLMA–dexamethasone coating acquired in D-PFM in air (top left) and height/adhesion image pairs of the same region while immersed in water for 22 minutes (middle) and 230 minutes (right). Histogram (bottom left) of the two adhesion images. For color details, please see color plate section.

much akin to the case of PBMA–drug. Another similarity is the outward bulging of a small subpopulation of the original small bumps. The intermediate-sized, PLMA-rich domains more significantly bulge outward, by ≈ 200 nm, suggesting local penetration of water and/or subsurface aggregation of drug; whereas the large PLMA domains seemingly remain unchanged in relative surface height. With passing time the adhesion signatures also evolve, generally toward lower adhesion, except for the large PLMA domains, which remain remarkably constant in adhesion. These observations are quantified as histograms of adhesion images, also shown in Figure 15.9. The tall, sharp histogram peak is the signal derived from the large PLMA domains. Its constancy indicates a lack of surface changes within these domains.

The reduction of adhesion for all other surface constituents (the "matrix" around the bright yellow PLMA domains) with passing time is intriguing. The adhesion measurement is fundamentally a probe of contact hysteresis, in turn a gauge of molecular mobility during the formation and break of tip–sample contact. Molecular rearrangements during contact allow the interfacial system to achieve a lower free-energy state that requires a pulling force to disrupt [9]. Our observations suggest a decrease in surface molecular mobility as drug is eluting from the coating. Certainly the dexamethasone molecules are mobile, and as they evacuate from the near-surface region one might expect a decrease in overall molecular mobility in the matrix; this is especially likely because the presence of dexamethasone in PBMA should plasticize the polymer. The evolution in time of what appear to be at least two surface components—a low adhesion, broad peak (including a slight shoulder) and a high-adhesion shoulder on the narrow (and stable) high-adhesion peak—seems to suggest that *the removal of dexamethasone reduces PBMA conformational mobility* and perhaps some mobility within a PLMA-interfacial region. Importantly, *immersion control experiments on drugless coatings show no changes in adhesion* on PBMA or PLMA.

15.4 SIMVASTATIN IN PEO–PBT COPOLYMERS

15.4.1 Hydrophilic-Hydrophobic Copolymer in Air: PEO–PBT (without Drug)

In this section we examine the nanomechanical differentiation between polymer domains in the hydrophilic-hydrophobic copolymer PEO–PBT, a biodegradable material for drug delivery applications [6]. Because it is a copolymer rather than a blend, one anticipates much smaller scale phase segregation compared to the PBMA–PLMA blend. At room temperature PBT is expected to be quite rigid since it is predominately crystalline and below T_g (54°C) while the short chains of PEO are expected to be soft, amorphous, and well above T_g (−48°C). We examine changes in relative mechanical behavior with increasing humidity; in PEO one anticipates absorption of water into the matrix and resulting softening via plasticization. To lessen possible capillary effects we operate in (repulsive regime) dynamic mode, where contact times are of order 1 μs or less and utilize phase imaging for property contrast.

Figure 15.10 shows the measured values of (a) ramping relative humidity per scan line during imaging, along with (b) the mean value of phase per each corresponding scan line of phase data. The arrow indicates the elapsing time direction

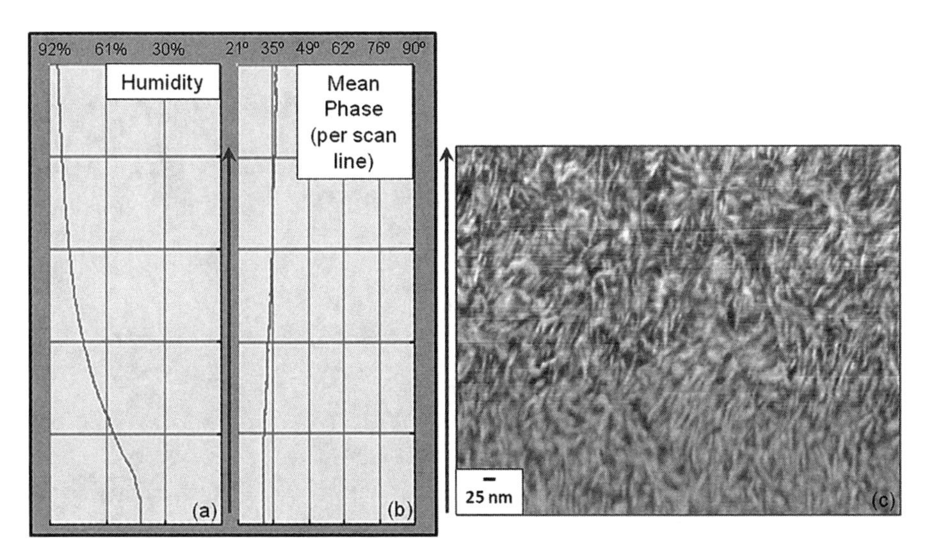

Figure 15.10 Phase image acquired in dynamic AFM in air on a PEO–PBT block copolymer coating during a humidity ramp. Graphed values of humidity and phase correspond to individual horizontal lines of the image.

during which humidity is ramped and the image region is correspondingly scanned. The trend of the change of mean phase with increasing humidity is consistent with an increasingly *dissipative* response. Figure 15.10(c) contains the phase image from which the mean phase values were extracted, prior to performing a line-by-line offset to keep the mean constant from image bottom to top for improving viewing purposes (sometimes known as "flattened" to zero order). Here we see (i) spatial variations due to mechanical differences between PEO matrix and PBT fibrils and (ii) an increase of these differences with rising humidity. At low humidity at the bottom of the image, contrast is barely perceivable, whereas with increasing humidity the PBT fibrils become starkly apparent at relatively bright phase within a darker surrounding matrix of PEO. This is consistent with a softening and perhaps increasingly viscous and sticky PEO matrix, together with PBT fibrils that remain relatively rigid (and probably less viscous and sticky).

In summary, the hydrophilic PEO domains are plasticized and swell with water, while hydrophobic PBT domains remain highly crystalline and strongly associated and thus provide "physical cross-links," which prevent the polymer from diffusing and dissolving.

15.4.2 Drug in Copolymer in Air: Simvastatin–PEO–PBT

Without drug present and under ambient conditions, traditional dynamic phase imaging in the repulsive regime [Fig. 15.11(a)] shows wormlike copolymer nanostructures. Upon mixing 15 wt% simvastatin into the PEO–PBT polymer [Fig. 15.11(b)], the phase (dissipation) contrast becomes somewhat stronger, and qualitatively different features are observed compared to the no-drug case. In particular, "dots" of bright material also appear. This suggests that upon mixing simvastatin with PEO–PBT, the

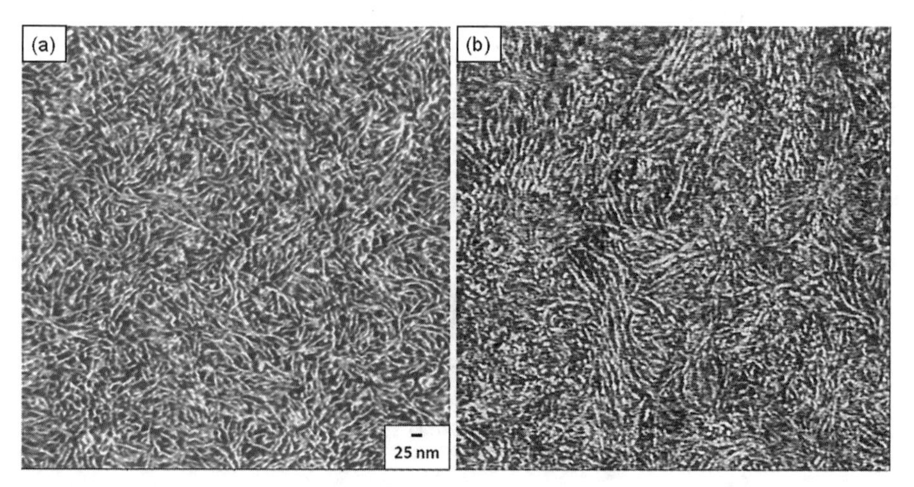

Figure 15.11 Phase images acquired in dynamic AFM in air on coatings containing (a) PEO–PBT block copolymer and (b) a mixture of PEO–PBT block copolymer and simvastatin.

relatively hydrophobic simvastatin associates with the PBT domains to perhaps further stiffen compared to PEO, and some excess simvastatin resides in the dot-shaped regions. However, these results point out the limitations in using phase imaging alone to determine the nature of drug–polymer mixtures: the results are qualitative and sometimes difficult to interpret.

15.4.3 Simvastatin–PEO–PBT: Effect of Soaking in Water

The same coatings as examined above, 15 wt% simvastatin in PEO–PBT, were later submerged in water and imaged as a function of time. Here D-PFM was used for its simplicity of operation in liquid compared to acoustically excited dynamic mode (where additional system resonances can complicate the cantilever response). Within the first scan period (time required to create a completely scanned image, about 10 minutes), the sample appearance changed dramatically: large and somewhat circular bumps appeared over the surface of the coating as seen topographically in Figure 15.12.

Upon closer examination with delicate imaging conditions and a very sharp tip, Figure 15.13, the water-swollen nanostructure of the copolymer is resolved as a faint background, with the larger structures (typically ~50 nm in diameter) being higher in surface elevation ("bumps"). As soaking time progressed, the shape and number of the bumps evolved. As shown in Figure 15.13 with passing time on the scale of many tens of minutes, some of the bumps shrink and even disappear; whereas other bumps

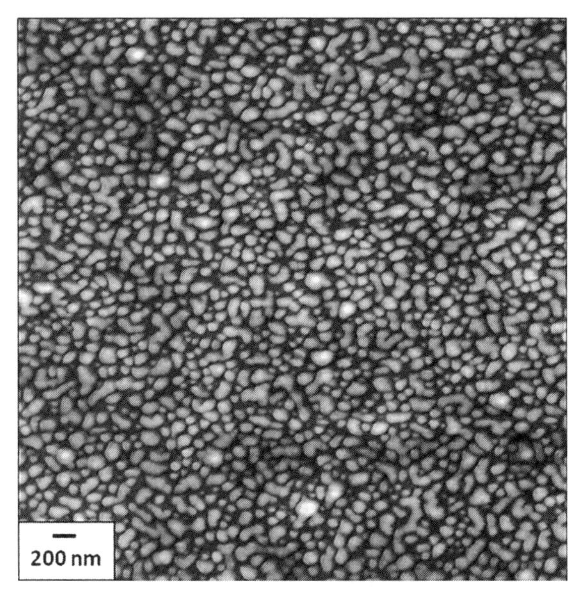

Figure 15.12 Low-resolution height image acquired in dynamic AFM shortly after immersion in water, on a coating containing a mixture of PEO–PBT block copolymer and simvastatin.

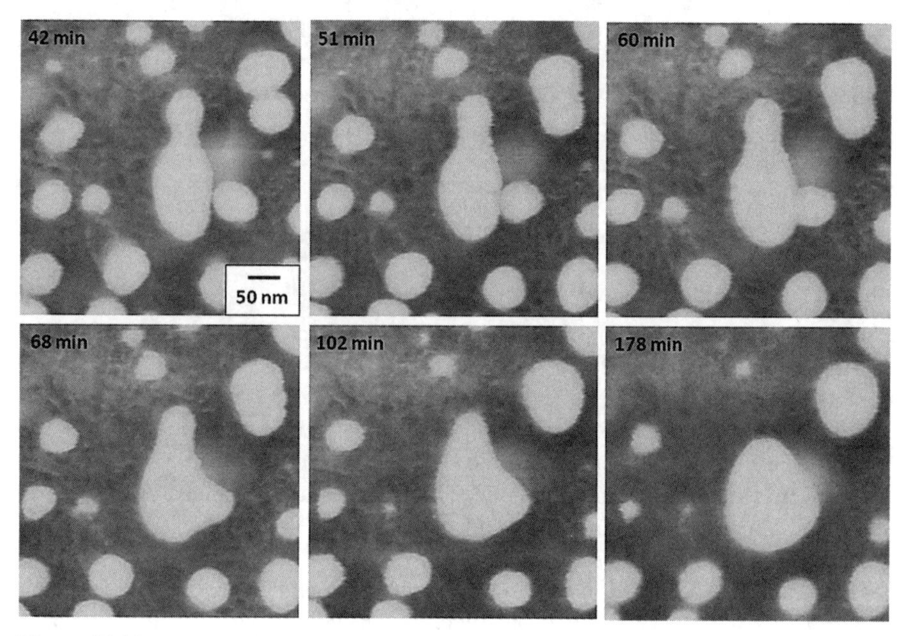

Figure 15.13 High-resolution height images acquired in dynamic AFM at different time points in water, on a coating containing a mixture of PEO–PBT block copolymer and simvastatin.

remain relatively unchanged in size and shape, and still other bumps grow in size and evolve in shape. In some cases one clearly observes the merging of bumps to form larger ones. This coalescence, together with the disappearance or shrinking of some smaller bumps, implies an Ostwald ripening mechanism of drug particle growth.

Because the polymer is cross-linked and thus only water swellable (not water soluble), the polymer itself is presumably not free to diffuse. But water likely *plasticizes* the copolymer, perhaps enhancing diffusion of simvastatin though the copolymer. Amorphous simvastatin is also likely plasticized by water: Though only slightly miscible in simvastatin, water often exhibits a dramatic plasticizing and thus mobilizing effect on amorphous drugs [3]. Molecules of simvastatin may dissolve in the surrounding water, but once the solubility limit is reached, one anticipates the simvastatin recondensing onto the coating surface. This concept is consistent with the results shown in Figure 15.13: the bumps must be domains of simvastatin squeezed out of the copolymer upon soaking in water. The free diffusion of simvastatin in the copolymer and surrounding water facilitates a coalescence and Ostwald ripening phenomenon, driven by the minimization of interfacial area between simvastatin and water.

15.5 CONCLUDING COMMENTS

The results shown here reveal the fascinating variety of behavior found upon exposing biomaterials, specifically polymer coatings that elute drug, to high temperatures or humidity, or exposing the coatings to aqueous environments. Coatings of amorphous drug in polymer exhibit morphologies that depend upon the specifics of the preparation process: The mixtures are not completely mixed, the drug is in the amorphous state, and thus the systems are inherently metastable. Environmental changes result in nanostructural changes as drug diffuses within and out of the polymer coating. In some cases, the drug changes form and crystallizes, dependent in part on its interaction with different polymer chemistries.

Because SPM nanomechanical characterization indirectly measures changes in the chemical composition of regions on the surface of the drug–polymer mixtures, some challenges arise in determining the chemical composition in the regions of interest. However, with knowledge of the glass transition temperature of the amorphous drug, and the glass transition temperature(s) of the polymer(s) applied, along with the viscoelastic nature of the materials, much can be learned from nanomechanical measurements. For example, in mixtures of dexamethasone in PBMA, after exposure to water, small protrusions of segregated dexamethasone elute drug and leave pits as elution proceeds. In mixtures of simvastatin in PBT–PEO polymer, after exposure to water, amorphous simvastatin is forced to the surface of the coatings, and with time these surface domains coalesce and exhibit Ostwald ripening.

Thus, nanomechanical characterization of the surfaces of biomaterial coatings that elute drugs by using scanning probe microscopy methods allows qualitative and quantitative characterization of the physical nature of the surface morphologies found in drug–polymer coatings. Such information is of fundamental relevance to understanding how drug–polymer biomaterials release drug and hopefully, as in the case of drug-eluting stents, accelerate the healing processes in atherosclerotic heart vessels.

ACKNOWLEDGMENTS

Support from SurModics Inc., the Industrial Partnership for Research in Interfacial and Materials Engineering (I-PRIME), and the Characterization Facility of the University of Minnesota (funded in part by the NSF MRSEC program), is gratefully acknowledged.

REFERENCES

1. Ratner, B., Hoffman, A., Schoen, F., and Lemons, J. *Biomaterials Science: An Introduction to Materials in Medicine*, 2nd ed., Academic: San Diego, 2004.
2. Serruys P., and Gershlick, A. *Handbook of Drug-Eluting Stents*, Taylor & Frances: London, 2005.
3. Hilfiker, R. *Polymorphism in the Pharmaceutical Industry*, Wiley-VCH: Weinheim, 2006.
4. Beiner, M., Schröter, K., Hempel, E., Reissig, S., and Donth, E. *Macromolecules* **32** (1999): 6278.
5. Panyam, J., Williams, D., Dash, A., Leslie-Pelecky, D., and Labhasetwar, V. *J Pharm Sci* **93** (2004): 1804.
6. Bezemer, J., Grijpma, D., Dijkstra, P., van Blitterswijk, C., and Feijen, J. *J Control Rel* **62** (1999): 393.
7. Deschamps, A. *Polymer* **42** (2001): 9335.
8. Graeser, K., Strachan, C., Patterson, J., Gordon, K., and Rades, T. *Cryst Growth Des* **8** (2008): 128.
9. Haugstad, G. *Atomic Force Microscopy*, Wiley: Hoboken, NJ, 2012.
10. Haugstad, G., and Wormuth, K. *Mat Res Soc Symp Proc* **1318** (2011): 2–11.

Index

Scanning Probe Microscopy in Industrial Applications: Nanomechanical Characterization, First Edition. Edited by Dalia G. Yablon.